Composite Materials (Second Edition)

Mathematical theory and exact relations

Online at: https://doi.org/10.1088/978-0-7503-6249-8

Composite Materials (Second Edition)

Mathematical theory and exact relations

Yury Grabovsky
Department of Mathematics, Temple University, Philadelphia, PA, USA

IOP Publishing, Bristol, UK

ISBN 978-0-7503-6249-8 (ebook)
ISBN 978-0-7503-6247-4 (print)
ISBN 978-0-7503-6250-4 (myPrint)
ISBN 978-0-7503-6248-1 (mobi)

DOI 10.1088/978-0-7503-6249-8

Version: 20250601

IOP ebooks

British Library Cataloguing-in-Publication Data: A catalogue record for this book is available from the British Library.

Published by IOP Publishing, wholly owned by The Institute of Physics, London

IOP Publishing, No.2 The Distillery, Glassfields, Avon Street, Bristol, BS2 0GR, UK

US Office: IOP Publishing, Inc., 190 North Independence Mall West, Suite 601, Philadelphia, PA 19106, USA

To my family: Irina, David and Phillip

Contents

Preface

Preface to the 2nd edition

After the publication of the first edition of the book, Guerra and Raiţă ('Quasiconvexity, null Lagrangians, and Hardy space integrability under constant rank constraints.' *Arch. Ration. Mech. Anal.*, **245**:279–320, 2022) developed the tools needed to prove the compensated compactness theorem for all constant rank homogeneous differential operators of arbitrary degrees. As a result, the homogenization theory could now be developed with full mathematical rigor in all contexts of physical interest.

Also, in the intervening time between editions, all exact relations and links for 2D thermoelectricity have been computed. This latest addition brings to a completion the project of computing all exact relations and links for two- and three-dimensional conductivity, elasticity, thermoelectricity, thermoelasticity and piezoelectricity.

Finally, the preparation of the second edition gave me the opportunity to revise the exposition of the entire text. As a result, part I has been substantially rewritten in light of Guerra and Raiţă's results. Part II has been edited for clarity and slightly expanded with the inclusion of a technical part of the theory that did not make it into the first edition (see theorem 6.1 and chapter 14). Its importance was realized when it proved to be very useful for the computation of all exact relations and links for 2D thermoelectricity—the new addition to part III of the book. One more appendix chapter characterizing all Jordan subalgebras of real symmetric matrices has been added. It sheds light on how the example of a polycrystalline L-relation (exact relation for laminate materials only) that is not exact has been discovered. This chapter could also be of independent interest on its own.

Having been afforded the luxury of time and hindsight, the second edition improves on mathematics, covered material, and exposition in substantial ways.

Preface to the 1st edition

Since the appearance of the general theory of exact relations in 1998, *all* microstructure-independent formulas and *all* links between pairs of composites with the same microstructure could theoretically be obtained in a systematic and largely mechanical way. However, actually doing so involves quite a bit of effort, and in some cases, a massive amount of effort. Over the intervening 18 years the author, sometimes alone, and often with the help of a group of bright undergraduates or a graduate student, was steadily computing complete lists of exact relations and links for conducting, elastic, piezoelectric, thermoelectric and thermoelastic composites. This book contains a compilation of all the results obtained through all these years of work. At the same time, the book includes mathematically rigorous and self-contained development of the general theory of exact relations and links, which is based on the theory of homogenization. Traditionally, homogenization theorems are proved separately in each physical context. The novelty of the approach taken in this book is that the development occurs in the general L^2 framework, common to all types of composites. This level of generality has made the theory cleaner, but also raised several interesting questions that were previously obscured by the particulars of each physical context. Therefore, this book can be read on three different levels: as a single source for all exact relations and links for effective properties of composite materials (part III); as the definitive exposition of the general theory of exact relations (part II); and as a new streamlined development of homogenization-based mathematical theory of composite materials (part I).

Notwithstanding the order in which the book is written, part III, containing complete lists of exact relations and links, can be consulted independently, without the need to refer to parts I and II. Part III is aimed at the broadest possible audience, requiring from the reader only the understanding of tensorial description of properties of anisotropic materials in a given physical context. Complete calculations leading to the results listed in part III are often very lengthy and are not presented in the book. Instead, we refer the reader to published journal articles detailing such calculations, unless the results are new and unpublished.

Part I of the book develops a mathematically rigorous general theory of composite materials that encompasses in a single abstract framework every material property of interest, be it conductivity, elasticity, piezoelectricity or thermoelectricity. While the abstract mathematical framework used in the book is well-known among experts, it was never used as a sole basis for developing the theory. The author hopes that the more mathematically inclined reader will appreciate the main technical feature of part I: its complete divorce from partial differential equations of classical physics, leading to elegant proofs of fundamental results. The story unfolds in scaling and translation-invariant subspaces of L^2 described by the Fourier multiplier operators encoding all relevant physics. The author also hopes that these readers will take an interest in some of the unsolved problems that arose in the process. Even so, all serious technicalities are placed into chapters marked with the † sign, which can be skipped, or sought, depending on reader's proclivities.

The author wishes to acknowledge continued financial support of the National Science Foundation, through 6 grants over the course of 18 years of theory development. One of the grants—the REU site grant made it possible to assemble groups of talented undergraduate students from all over the country in the course of three summers from 2002 to 2004. Specifically, the author thanks Thilagavathi Murugesan and Saule Zhoshina–two graduate students at the University of Utah who were helping to compute exact relations for three-dimensional elastic composites during the summer of 1998; three REU teams of undergraduate students from all over the country, who were computing all exact relations and links for fiber-reinforced conducting composites with Hall effect: Erin R. Blew, David Carchedi, Edward Corcoran, Ryan Fuoss, Joseph Galante, Jerome Hodges IV, Russell Howes, Matthew Jacobs, Matthew Macauley, John Quah, Austin Roberts, Elianna Ruppin, Steven Stewart, and Peter Tom-Wolverton; Hansen To—author's graduate student at Temple University, who used Maple to compute all exact relations for three-dimensional conducting composites with Hall effect in 2004; Meredith Hegg—author's graduate student at Temple University, who computed an amazing link for fiber-reinforced elastic composites in 2012, Tatyana Nuzhnaya, a graduate student at Temple University who during the summer 2013 computed with Matlab all algebras and ideals for fiber-reinforced elastic composites; Mark Mikida and Andrew Schneider—undergraduate researchers who during 2013–14 academic year computed all exact relations for fiber-reinforced elastic composites; Adam Michael Jacoby, a graduate student at Temple University who during the summer 2014 computed with Maple all factor-algebras, their isomorphisms, and wrote the elimination program that made it feasible to compute all links for fiber-reinforced elastic composites. The links were computed by three Temple undergraduate researchers Patrick Wynne, Hansen Pei and Abraham Lyle, who worked both alone and as a team during 2014–16 academic years; Daniel Lapsley, an undergraduate researcher, who during summer 2016, wrote Maple program that multiplies irreducible representations of SO(3) in the space of piezoelectric tensors. This has lead a quick calculation of all links for three-dimensional piezoelectric composites. Last but not least, the author wishes to extend special thanks to Graeme Milton who suggested that I write this book in the first place, and whose continued encouragement sustained author's efforts to apply the theory in more and more complicated physical problems.

Acknowledgements

This material is based upon work supported by the National Science Foundation under Grants No. DMS-2005538 and DMS-2305832.

Author biography

Yury Grabovsky

Yury Grabovsky is a Professor at the Department of Mathematics at Temple University, where his research interests include Calculus of Variations on the mathematics side and Continuum Mechanics on the physics side. His latest work was on mathematical theory of composite materials, buckling of slender bodies, such as plates, rods, and shells, and understanding stability of equilibrium configurations with phase boundaries in nonlinear elasticity. After graduating with a PhD from Courant Institute of Mathematical Sciences, New York University, Yury spent a year as a postdoc at the Center for Nonlinear Analysis at Carnegie Mellon University and four years as a Wylie Instructor at the University of Utah. In 1999 he was hired as a tenure-track Assistant Professor by Temple University. Yury enjoys teaching and doing research with students. His web page on using continued fractions in the design of calendar systems routinely draws attention of media every leap year. It has been translated into Portuguese and published in a journal for mathematics school teachers 'Educacao e Matematica'.

IOP Publishing

Composite Materials (Second Edition)
Mathematical theory and exact relations
Yury Grabovsky

Chapter 1

Introduction

Man-made composite materials play an increasingly important role in our everyday lives. They can be found everywhere, from skies and golf clubs to airplanes and spaceships. The utility of manufactured composites comes from their combination of properties of raw constituents that have some, but not all, of the desired properties of the composite. The quest to understand mathematically the relation between material properties of the components, the microstructure, and the observed macroscopic behavior of composites began as early as the 19th century. For example, Poisson [28], Faraday [3] and Maxwell [19] considered arrays of conducting spheres embedded in an insulating or conducting matrix. In this setting the mathematical problem is to find the limit at infinity of the electric or magnetic field values as the sizes and distances between spheres go to zero, thereby identifying the effective macroscopic properties of microscopically heterogeneous medium. In order to describe macroscopic properties of materials with complicated, possibly disordered microstructures, a concept of representative volume element (RVE) was introduced [12, 13]. However, from the mathematical standpoint the notion of RVE is ambiguous and ultimately mathematically unsatisfactory, since the power of its wide applicability in mechanics lies in its very ambiguity. For the purposes of mathematical analysis we return to the original investigations in physics, where the heterogeneous media is modeled by an *infinite sequence* of structures whose properties vary on a smaller and smaller length scale. If early models of Poisson, Faraday and Maxwell involved simple periodic geometry, modern analytical tools permit us to prove general homogenization theorems for structures of arbitrary complexity. Macroscopic properties of such media can be defined mathematically by means of the G-limit, a concept introduced by Spagnolo [31], and generalized by Murat and Tartar [27], who called their more general notion the H-limit. This method of defining composites and their effective properties is adopted in this book from the very beginning. In the mathematical theory of homogenization it is as versatile as the concept of RVE in mechanics, but without the ambiguity of the RVE.

doi:10.1088/978-0-7503-6249-8ch1

Among the many important questions one can ask about properties of composites we single out the *G-closure problem*, introduced by Lurie and Cherkaev [16]. This is a very natural question in situations where the geometric details of the microstructure are unavailable, and only the properties of constituents and possibly their volume fractions are known. The G-closure problem asks to describe the set of all effective tensors of composites that can be made from a given set of component materials. Even though several G-closure sets have been computed exactly (e.g., [4, 5, 9, 16–18]), the current consensus is that the problem of an explicit analytic description of such sets is largely intractable.

In parallel with this development, researchers kept discovering other, somewhat counter-intuitive, facts about effective behavior of composites. In certain special situations it could be shown that the effective tensor of a composite satisfies one or more equations that hold *regardless of the microstructure*. A typical, and probably the earliest example of this is the formula discovered by Hill [13, 14], whereby any composite made of isotropic elastic materials with the same shear modulus, must necessarily be isotropic with that same shear modulus, while its bulk modulus can be computed by an explicit formula involving the bulk and shear moduli of the constituents and their volume fractions. In the language of G-closures these results meant that there are special sets of materials, whose G-closures have *empty interiors* in the space of all material tensors. Viewed geometrically, such results can be represented as G-closed submanifolds of positive codimension in the space of tensors of material properties. We call such submanifolds *exact relations*.

Exact relations indicate which combinations of material moduli would be insensitive to the changes in the design of the composite structure. They can be used to guide the design of new composites, validate testing procedures, and probe the underlying modeling assumptions by serving as benchmarks. In a series of papers [6–8, 10] the author with his collaborators developed a general theory of exact relations that provides effective methods for explicit description of *all* exact relations in every physical context. Such broad scope of the theory was achieved by means of the application of a unified functional analytic framework developed for random and periodic composites in [2, 15, 21, 22, 25]. In this book this unified framework is expanded further to serve as a basis for the theory of effective behavior of composites based on H-convergence, while building on the methods of Murat and Tartar, developed in [26, 27, 34, 35].

The abstraction of the theory of exact relations from any specific physical context permitted us to go beyond the original concept of an exact relation and obtain deeper and more general properties of composites, which we call *links*. A link specifies relations between effective tensors of a pair of composite materials with the same microstructure, that differ only in the values of material properties of their constituents. Mathematically, links are transformation rules that are stable under homogenization. The simplest, and probably earliest example of a non-obvious link is for two-dimensional conducting composites. It says that if the conductivities of every constituent in an arbitrary composite are transformed according to the rule

$\sigma \mapsto \sigma/\det\sigma$, then the effective conductivity of the 'transformed' composite will be $\sigma_*/\det\sigma_*$, where σ_* is the effective conductivity tensor of the original composite[1]. This result is due to Mendelson [20].

Part I of the book describes the general mathematical framework of homogenization, based on the concept of H-convergence. The main mathematical tool for proving key homogenization theorems is compensated compactness, developed by Murat and Tartar in the context of first order linear differential constraints [26, 32, 33]. Recent work of Guerra and Raiţă [11, 29] permitted us to generalize compensated compactness to all physical contexts, where material response is described in terms of systems of linear homogeneous partial differential equations. Historically, however, the development of the theory of exact relations has been largely based on an even more general Hilbert space framework expounded in the first Milton's book [23] and significantly reinforced in his second [24]. In the course of writing this book it became clear that in a framework of such sweeping generality proving the compensated compactness property, which is the cornerstone of the entire theory, does not seem to be possible. Instead of narrowing the scope to the one in Guerra and Raiţă's work, which covers each and every example of physical interest, the author has chosen to retain the full generality of the original Hilbert space framework and include compensated compactness as an additional postulate. The benefit of this approach is that it clearly shows that it is the compensated compactness property alone, that makes the theory work. It shows which statements do and do not depend on compensated compactness. It also poses a challenge to seek an effective understanding of which features of the general Hilbert space framework guarantee compensated compactness. Part I of the book also contains a clean and transparent proof of the locality of H-closure, first formulated by Kohn and Dal Maso and subsequently proved for both conductivity [30, 35] and elasticity [1].

As the theory of exact relations and links was developing, it produced new tools capable of harvesting *all* exact relations and links in virtually any physical context, including coupled properties, such as piezoelectricity or thermoelasticity, and even fibrous, or fiber-reinforced, composites. The general theory of exact relations together with the tools for their effective computation, described in part II of the book, made it possible to produce complete lists of all exact relations and links for conducting, elastic, piezoelectric, thermoelastic and thermoelectric composites. These lists are collected in part III and represent years of massive effort by the author and many students both graduate and undergraduate. In view of this, part III contains only the summary of the results. The details of analysis have been documented either in published or unpublished papers. In the latter case a reference to arXiv preprints are given.

[1] The effective conductivity tensor σ_* does not have to be constant throughout the composite material, since we do not assume that the microstructure in the vicinity of different macroscopic points is statistically the same.

References

[1] Babadjian J-F and Barchiesi M 2009 A variational approach to the local character of g-closure: the convex case *Ann. Inst. Henri Poincare (C) Non Linear Anal.* **26** 351–73

[2] Dell'Antonio G F, Figari R and Orlandi E 1986 An approach through orthogonal projections to the study of inhomogeneous random media with linear response *Ann. Inst. Henri Poincaré* **44** 1–28

[3] Faraday M 1838 Experimental researches in electricity-fourteenth series *Phil. Trans. R. Soc.* **128** 265–82

[4] Francfort G A and Milton G W 1987 Optimal bounds for conduction in two-dimensional, multiphase, polycrystalline media *J. Stat. Phys.* **46** 161–77

[5] Grabovsky Y 1993 The G-closure of two well-ordered anisotropic conductors *Proc. R. Soc. Edinburgh* **123A** 423–32

[6] Grabovsky Y 1998 Exact relations for effective tensors of polycrystals I: necessary conditions *Arch. Ration. Mech. Anal.* **143** 309–30

[7] Grabovsky Y 2004 Algebra, geometry and computations of exact relations for effective moduli of composites *Advances in Multifield Theories of Continua with Substructure* Modeling and Simulation in Science, Engineering and Technology G Capriz and P M Mariano (Boston, MA: Birkhäuser) 167–97

[8] Grabovsky Y 2009 Exact relations for effective conductivity of fiber-reinforced conducting composites with the Hall effect via a general theory *SIAM J. Math Anal.* **41** 973–1024

[9] Grabovsky Y and Milton G W 1998 Rank one plus a null-Lagrangian is an inherited property of two-dimensional compliance tensors under homogenization *Proc. R. Soc. Edinburgh* A **128** 283–99

[10] Grabovsky Y, Milton G W and Sage D S 2000 Exact relations for effective tensors of polycrystals: necessary conditions and sufficient conditions *Commun. Pure Appl. Math.* **53** 300–53

[11] Guerra A and Raiţă B 2022 Quasiconvexity, null Lagrangians, and Hardy space integrability under constant rank constraints *Arch. Ration. Mech. Anal.* **245** 279–320

[12] Hashin Z 1964 Theory of mechanical behavior of heterogeneous media *Appl. Mech. Rev.* **17** 1–9

[13] Hill R 1963 Elastic properties of reinforced solids: some theoretical principles *J. Mech. Phys. Solids* **11** 357–72

[14] Hill R 1964 Theory of mechanical properties of fibre-strengthened materials: I. Elastic behaviour *J. Mech. Phys. Solids* **12** 199–212

[15] Kohler W and Papanicolaou G C 1982 Bounds for effective conductivity of random media *Macroscopic Properties of Disordered Media* ed R Burridge, S Childress and G Papanicolaou (Berlin: Springer) 111–30

[16] Lurie K A and Cherkaev A V 1981 *G*-closure of a set of anisotropic conducting media in the case of two dimensions *Dokl. Akad. Nauk SSSR* **259** 328–31 In Russian https://www.mathnet.ru/eng/dan44597

[17] Lurie K A and Cherkaev A V 1984 Exact estimates of conductivity of composites formed by two isotropically conducting media taken in prescribed proportion *Proc. R. Soc. Edinburgh* **99A** 71–87

[18] Lurie K A and Cherkaev A V 1986 Exact estimates of a binary mixture of isotropic components *Proc. R. Soc. Edinburgh* **104A** 21–38

[19] Maxwell J C 1881 *A Treatise on Electricity and Magnetism* **vol 1** (Oxford: Clarendon)

[20] Mendelson K S 1975 A theorem on the conductivity of two-dimensional heterogeneous medium *J. Appl. Phys.* **46** 4740–1
[21] Milton G W 1987 Multicomponent composites, electrical networks and new types of continued fraction. I *Commun. Math. Phys.* **111** 281–327
[22] Milton G W 1987 Multicomponent composites, electrical networks and new types of continued fraction. II *Commun. Math. Phys.* **111** 329–72
[23] Milton G W 2002 *The Theory of Composites Cambridge Monographs on Applied and Computational Mathematics* (Cambridge: Cambridge University Press)
[24] Milton G W 2016 *Extending the Theory of Composites to Other Areas of Science* (Salt Lake City, UT: Milton-Patton Publishers)
[25] Milton G W and Kohn R V 1988 Variational bounds on the effective moduli of anisotropic composites *J. Mech. Phys. Solids* **36** 597–629
[26] Murat F 1981 Compacité par compensation: condition nécessaire et suffisante de continuité faible sous une hypothèse de rang constant *Ann. Scuola Norm. Sup. Pisa Cl. Sci. (4)* **8** 69–102
[27] Murat F and Tartar L 1997 H-convergence *Topics in the Mathematical Modelling of Composite Materials* **vol 31** ed A Cherkaev and R V Kohn (Boston, MA: Birkhäuser Boston) 21–43
[28] Poisson S D 1826 Mémoire sur la théorie du magnétisme *Mémoires de l'académie royale des sciences de l'Institut de France* **5** 247–338
[29] Raiţă B 2019 Potentials for $\mathscr{A}$-quasiconvexity *Calc. Var. Partial Differ. Equ.* **58** 1–16
[30] Raitums U 2001 On the local representation of G-closure *Arch. Ration. Mech. Anal.* **158** 213–34
[31] Spagnolo S 1968 Sulla convergenza di soluzioni di equazioni paraboliche ed ellittiche *Ann. Scuola Norm. Sup. Pisa (3) 22 (1968), 571–597; errata, ibid. (3)* **22** 673
[32] Tartar L 1979 Estimation de coefficients homogénéisés *Computing Methods in Applied Sciences and Engineering (Proc. 3rd Int. Sympos., Versailles, 1977), I* Lecture Notes in Mathematics **vol 704** (Berlin: Springer) 364–73
[33] Tartar L 1983 Compacité par compensation: résultats et perspectives *Nonlinear Partial Differential equations and Their Applications Collège de France Seminar, vol IV (Paris, 1981/1982)* (London: Pitman Publishing) 350–69
[34] Tartar L 1990 H-measures, a new approach for studying homogenisation, oscillations and concentration effects in partial differential equations *Proc. R. Soc. Edinburgh* A **115** 193–230
[35] Tartar L 2000 An introduction to the homogenization method in optimal design *Optimal Shape Design (Tróia, 1998) vol 1740 Lecture Notes in Math.* (Berlin: Springer) pp 47 156

Part I

Mathematical theory of composite materials

Chapter 2

Material properties and governing equations

2.1 Introduction

Fundamental laws of classical physics are often insufficient when the outcome depends on the properties of matter involved in the observed phenomena. A typical example is the original system of Maxwell's equations requiring additional relations between a magnetic field and a magnetic induction and the electric field and the dielectric displacement. Postulating these relations to be linear and local in space covers a vast spectrum of physical phenomena addressed in this book. A closer look at ordinary materials reveals a complicated picture at the microscopic level, from fibrillar structure of wood to polycrystalline structure of metals. Man-made composites, widely used in industry, represent another class of materials with microstructure. Mathematical tools used to analyze the relation between the microstructure, the properties of constituent materials, and macroscopic properties of composites are referred to collectively as *homogenization*.

When we talk about composites we may refer to conducting, elastic, or piezoelectric composites. It is well understood that mathematical equations governing the underlying physical phenomena, like Maxwell's equations, or equations of continuum mechanics, or both, share a common mathematical structure. For this reason many technical results in the existing literature are often established in the simpler context of conductivity with the understanding that the same approach would also apply in other contexts. This understanding has lead to the development of an abstract formalism in the works of Kohler and Papanicolaou [6], Dell-Antonio, Figari and Orlandi [3], Milton and Kohn [11] and Milton [7, 8], where the governing systems of partial differential equations are replaced with scale and translation (and often, rotation) invariant subspaces of $L^2(\mathbb{R}^d)$. Our exposition here follows closely the development in [9] (see also [10]). While this abstract approach is natural for analysis of periodic cell problems, it has never been adapted to homogenization problems in the general context of H-convergence, as is done in this book. As we will see, this last task requires the compensated compactness property of the

doi:10.1088/978-0-7503-6249-8ch2

corresponding subspaces of $L^2(\mathbb{R}^d)$, which can be guaranteed to hold, provided the constraints on the fields come from differential equations. This formal narrowing of scope is still sufficient to capture all examples coming from physics.

The main theme of any theory of composite materials is the interplay between the micro-scale and the macro-scale properties, called 'homogenization'. Its study is governed by the properties of boundary value problems for partial differential equations or systems of partial differential equations defined on the set $\Omega \subset \mathbb{R}^d$ occupied by the material. The boundary value problem for conductivity is usually formulated in terms of the scalar potential field ϕ, giving rise to the electric field $\boldsymbol{E} = -\nabla\phi$. Similarly, the boundary value problem for elasticity is usually formulated in terms of the displacement vector field $\boldsymbol{u}$, giving rise to the strain field $\varepsilon = (\nabla\boldsymbol{u} + (\nabla\boldsymbol{u})^T)/2$. In the abstract Hilbert space approach we operate on the level of physical fields, such as the current and electric fields for conductivity and the stress and strain fields for elasticity. The price we have to pay for this generality is the loss of access to naturally defined potentials. This difficulty has already been encountered in [4]. In our case, the same approach works along the same lines as in [4], once the methods from [5] are brought in to handle differential constraints of arbitrary degree.

To motivate the formal axioms of the abstract theory we examine equations of conductivity and elasticity and exhibit the common features that form the basis of the Hilbert space formalism described in this chapter.

2.2 Conductivity and elasticity

In conductivity, the electric field $\boldsymbol{e}$ is a gradient of the scalar potential ϕ, while the current density $\boldsymbol{j}$ is divergence-free:

$$\boldsymbol{e} = -\nabla\phi, \qquad \nabla\cdot\boldsymbol{j} = 0. \tag{2.1}$$

If we think of $\boldsymbol{e}$ and $\boldsymbol{j}$ as elements of $L^2(\mathbb{R}^3; \mathbb{R}^3)$, then we can write equation (2.1) in Fourier space as follows

$$\hat{\boldsymbol{e}}(\boldsymbol{\xi}) = -2\pi i\boldsymbol{\xi}\hat{\phi}(\boldsymbol{\xi}), \qquad \boldsymbol{\xi}\cdot\hat{\boldsymbol{j}}(\boldsymbol{\xi}) = 0, \tag{2.2}$$

where

$$\hat{f}(\boldsymbol{\xi}) = \int_{\mathbb{R}^3} f(\boldsymbol{x})e^{-2\pi i\boldsymbol{x}\cdot\boldsymbol{\xi}}d\boldsymbol{x}$$

is the Fourier transform. We observe that for every $\boldsymbol{\xi} \in \mathbb{R}^3\setminus\{0\}$ equation (2.2) restrict the vectors $\hat{\boldsymbol{e}}(\boldsymbol{\xi})$ and $\hat{\boldsymbol{j}}(\boldsymbol{\xi})$ to lie in specific subspaces of $\mathbb{C}^3$, namely

$$\hat{\boldsymbol{e}}(\boldsymbol{\xi}) \in \mathbb{C}\boldsymbol{\xi} = \{c\boldsymbol{\xi}: c \in \mathbb{C}\} = \mathscr{E}_{\xi} \otimes \mathbb{C}, \qquad \hat{\boldsymbol{j}}(\boldsymbol{\xi}) \in \{\boldsymbol{a} \in \mathbb{C}^3: \boldsymbol{a}\cdot\boldsymbol{\xi} = 0\} = \mathscr{J}_{\xi} \otimes \mathbb{C},$$

where $\mathscr{E}_{\xi}$ and $\mathscr{J}_{\xi}$ are the real versions of these subspaces

$$\mathscr{E}_{\xi} = \{c\boldsymbol{\xi}: c \in \mathbb{R}\}, \qquad \mathscr{J}_{\xi} = \{\boldsymbol{a} \in \mathbb{R}^3: \boldsymbol{a}\cdot\boldsymbol{\xi} = 0\}.$$

We also observe that $\mathbb{R}^3 = \mathcal{E}_\xi \oplus \mathcal{J}_\xi$, where the sum is orthogonal. This orthogonality in Fourier space defines the orthogonal decomposition $L^2(\mathbb{R}^3; \mathbb{R}^3) = \mathcal{E} \oplus \mathcal{J}$, where

$$\mathcal{E} = \{\boldsymbol{E} \in L^2(\mathbb{R}^3; \mathbb{R}^3): \hat{\boldsymbol{E}}(\boldsymbol{\xi}) \in \mathcal{E}_\xi \otimes \mathbb{C}\},$$

$$\mathcal{J} = \{\boldsymbol{J} \in L^2(\mathbb{R}^3; \mathbb{R}^3): \hat{\boldsymbol{J}}(\boldsymbol{\xi}) \in \mathcal{J}_\xi \otimes \mathbb{C}\}.$$

The operator Γ of orthogonal projection onto the subspace $\mathcal{E}$ is a Fourier multiplier:

$$(\widehat{\Gamma \boldsymbol{f}})(\boldsymbol{\xi}) = \frac{(\hat{\boldsymbol{f}} \cdot \boldsymbol{\xi})\boldsymbol{\xi}}{|\boldsymbol{\xi}|^2} = \boldsymbol{\Gamma}\left(\frac{\boldsymbol{\xi}}{|\boldsymbol{\xi}|}\right)\hat{\boldsymbol{f}}(\boldsymbol{\xi}),$$

where

$$\boldsymbol{\Gamma}(\boldsymbol{\eta}) = \boldsymbol{\eta} \otimes \boldsymbol{\eta}. \tag{2.3}$$

The Fourier multiplier operators commute with translations, expressing the translation-invariant nature of physical laws. The homogeneity, i.e., the dependence of the Fourier multiplier only on $\boldsymbol{\xi}/|\boldsymbol{\xi}|$ expresses the scaling-invariant nature of physical laws. The rotational invariance of physical laws governing conductivity can be expressed mathematically in terms of the Fourier multiplier symbol $\boldsymbol{\Gamma}(\boldsymbol{\eta})$:

$$\boldsymbol{R}\boldsymbol{\Gamma}(\boldsymbol{\eta})\boldsymbol{R}^T = \boldsymbol{\Gamma}(\boldsymbol{R}\boldsymbol{\eta}), \qquad \boldsymbol{R} \in SO(3).$$

Next we examine the equations of linear elasticity in all-space. The strain field ε is a symmetrized gradient, while the stress field $\boldsymbol{\sigma}$ is symmetric and divergence-free:

$$\varepsilon = e(\boldsymbol{u}) = \frac{1}{2}(\nabla \boldsymbol{u} + (\nabla \boldsymbol{u})^T), \qquad \nabla \cdot \boldsymbol{\sigma} = 0, \quad \boldsymbol{\sigma}^T = \boldsymbol{\sigma}. \tag{2.4}$$

In Fourier space $\hat{\varepsilon}(\boldsymbol{\xi}) \in \mathcal{E}_\xi \otimes \mathbb{C}$ and $\hat{\boldsymbol{\sigma}}(\boldsymbol{\xi}) \in \mathcal{J}_\xi \otimes \mathbb{C}$

$$\mathcal{E}_\xi - \{\boldsymbol{a} \otimes \boldsymbol{\xi} + \boldsymbol{\xi} \otimes \boldsymbol{a}: \boldsymbol{a} \in \mathbb{R}^3\}, \qquad \mathcal{J}_\xi = \{\boldsymbol{S} \in \mathrm{Sym}(\mathbb{R}^3): \boldsymbol{S}\boldsymbol{\xi} = 0\},$$

where $\mathrm{Sym}(\mathbb{R}^3)$ denotes a vector space of all real symmetric 3×3 matrices. We can easily verify the orthogonal decomposition

$$\mathrm{Sym}(\mathbb{R}^3) = \mathcal{E}_\xi \oplus \mathcal{J}_\xi, \qquad \boldsymbol{\xi} \in \mathbb{R}^3 \setminus \{0\},$$

with respect to the Frobenius inner product on $\mathrm{Sym}(\mathbb{R}^3)$:

$$(\boldsymbol{A}, \boldsymbol{B}) = \mathrm{Tr}\,(\boldsymbol{A}\boldsymbol{B}). \tag{2.5}$$

This leads to the orthogonal decomposition $L^2(\mathbb{R}^3; \mathrm{Sym}(\mathbb{R}^3)) = \mathcal{E} \oplus \mathcal{J}$,

$$\mathcal{E} = \{\boldsymbol{E} \in L^2(\mathbb{R}^3; \mathrm{Sym}(\mathbb{R}^3)): \hat{\boldsymbol{E}}(\boldsymbol{\xi}) \in \mathcal{E}_\xi \otimes \mathbb{C}\},$$

$$\mathcal{J} = \{\boldsymbol{J} \in L^2(\mathbb{R}^3; \mathrm{Sym}(\mathbb{R}^3)): \hat{\boldsymbol{J}}(\boldsymbol{\xi}) \in \mathcal{J}_\xi \otimes \mathbb{C}\}.$$

The orthogonal projection Γ onto the space $\mathscr{E}$ is a homogeneous of degree zero Fourier multiplier: $(\widehat{\Gamma f})(\boldsymbol{\xi}) = \boldsymbol{\Gamma}(\boldsymbol{\xi}/|\boldsymbol{\xi}|)\hat{\boldsymbol{f}}(\boldsymbol{\xi})$, where

$$\boldsymbol{\Gamma}(\boldsymbol{\eta})\boldsymbol{S} = \boldsymbol{S\eta} \otimes \boldsymbol{\eta} + \boldsymbol{\eta} \otimes \boldsymbol{S\eta} - (\boldsymbol{S\eta} \cdot \boldsymbol{\eta})\boldsymbol{\eta} \otimes \boldsymbol{\eta}. \tag{2.6}$$

The rotational invariance of elasticity equations can be expressed by the corresponding property of the projections $\boldsymbol{\Gamma}(\boldsymbol{\eta})$:

$$\boldsymbol{R}(\boldsymbol{\Gamma}(\boldsymbol{\eta})\varepsilon)\boldsymbol{R}^T = \boldsymbol{\Gamma}(\boldsymbol{R\eta})(\boldsymbol{R}\varepsilon\boldsymbol{R}^T).$$

2.3 Abstract Hilbert space framework

We now part with explicitly stated differential equations like (2.1) or (2.4) and work instead with the abstract fields $\boldsymbol{E}(\boldsymbol{x})$ and $\boldsymbol{J}(\boldsymbol{x})$ taking their values in an abstract finite dimensional inner product space $\mathscr{T}$. (For conductivity $\mathscr{T} = \mathbb{R}^3$ with the standard dot product and, for elasticity $\mathscr{T} = \mathrm{Sym}(\mathbb{R}^3)$, with the Frobenius inner product (2.5).) The inner product of $\{\boldsymbol{u}, \boldsymbol{v}\} \subset \mathscr{T}$ will always be denoted as $(\boldsymbol{u}, \boldsymbol{v})$. The same notation will be used for the Hermitian inner product on $\mathscr{T} \otimes \mathbb{C}$. The field $\boldsymbol{E}$ will be called the *intensity field* and $\boldsymbol{J}$—the *flux field*. Instead of the differential constraints on $\boldsymbol{E}$ and $\boldsymbol{J}$ we postulate a geometric structure governing these fields, best formulated in the Fourier space (hence the need for the complexification $\mathscr{T} \otimes \mathbb{C}$ of $\mathscr{T}$).

Postulate 2.1. (Orthogonality). *Assume that $\boldsymbol{\Gamma}(\boldsymbol{\eta})$ is an orthogonal projection operator onto a proper subspace $\mathscr{E}_{\boldsymbol{\eta}} \subset \mathscr{T}$, for every unit vector $\boldsymbol{\eta} \in \mathbb{S}^{d-1}$. Furthermore, $\boldsymbol{\Gamma}(\boldsymbol{\eta})$ is assumed to be a continuous even function on the unit sphere $\mathbb{S}^{d-1} \subset \mathbb{R}^d$, in other words,*

$$\mathscr{T} = \mathscr{E}_{\boldsymbol{\eta}} \oplus \mathscr{J}_{\boldsymbol{\eta}}, \qquad |\boldsymbol{\eta}| = 1, \quad \mathscr{E}_{-\boldsymbol{\eta}} = \mathscr{E}_{\boldsymbol{\eta}}, \quad \mathscr{J}_{-\boldsymbol{\eta}} = \mathscr{J}_{\boldsymbol{\eta}}, \tag{2.7}$$

where $\mathscr{J}_{\boldsymbol{\eta}} = \mathscr{E}_{\boldsymbol{\eta}}^{\perp}$ is the orthogonal complement of $\mathscr{E}_{\boldsymbol{\eta}}$ in $\mathscr{T}$. We will refer to the function $\boldsymbol{\eta} \mapsto \boldsymbol{\Gamma}(\boldsymbol{\eta})$ as the symbol of the orthogonal decomposition (2.7).

The continuity of $\boldsymbol{\Gamma}(\boldsymbol{\eta})$ implies, in particular, that all subspaces $\mathscr{E}_{\boldsymbol{\eta}}$ are of the same dimension (equal to $\mathrm{Tr}\, \boldsymbol{\Gamma}(\boldsymbol{\eta})$). The subspaces $\mathscr{E} \subset L^2(\mathbb{R}^d; \mathscr{T})$ and $\mathscr{J} = \mathscr{E}^{\perp}$ are then *defined by*

$$\begin{cases} \mathscr{E} = \{\boldsymbol{E} \in L^2(\mathbb{R}^d; \mathscr{T}) \colon \hat{\boldsymbol{E}}(\boldsymbol{\xi}) \in \mathscr{E}_{\frac{\boldsymbol{\xi}}{|\boldsymbol{\xi}|}} \otimes \mathbb{C} \text{ for almost all } \boldsymbol{\xi} \in \mathbb{R}^d\}, \\ \mathscr{J} = \{\boldsymbol{J} \in L^2(\mathbb{R}^d; \mathscr{T}) \colon \hat{\boldsymbol{J}}(\boldsymbol{\xi}) \in \mathscr{J}_{\frac{\boldsymbol{\xi}}{|\boldsymbol{\xi}|}} \otimes \mathbb{C} \text{ for almost all } \boldsymbol{\xi} \in \mathbb{R}^d\}. \end{cases} \tag{2.8}$$

The operator Γ of the orthogonal projection onto $\mathscr{E}$ is a Fourier multiplier

$$L^2(\mathbb{R}^d;\ \mathscr{T}) \ni \boldsymbol{f} \mapsto \widehat{\Gamma \boldsymbol{f}}(\boldsymbol{\xi}) = \boldsymbol{\Gamma}\left(\frac{\boldsymbol{\xi}}{|\boldsymbol{\xi}|}\right)\hat{\boldsymbol{f}}(\boldsymbol{\xi}). \tag{2.9}$$

In order to treat the homogenization of boundary value problems on specific domains Ω, we need *local versions* of the spaces $\mathscr{E}$ and $\mathscr{J}$. We describe them using the extension and restriction operators. The extension operator $X_0(\Omega)$ extends a field from Ω to $\mathbb{R}^d$ by zero:

$$X_0(\Omega)\colon L^2(\Omega;\ \mathscr{T}) \to L^2(\mathbb{R}^3;\ \mathscr{T}), \quad (X_0(\Omega)\boldsymbol{f})(\boldsymbol{x}) = \begin{cases} \boldsymbol{f}(\boldsymbol{x}), & \boldsymbol{x} \in \Omega, \\ 0, & \boldsymbol{x} \in \Omega^c, \end{cases} \tag{2.10}$$

where $\Omega^c = \mathbb{R}^d \backslash \Omega$ denotes the complement of Ω. The restriction operator maps $L^2(\mathbb{R}^d;\ \mathscr{T})$ onto $L^2(\Omega;\ \mathscr{T})$ by restriction to Ω. Its action is denoted by $\boldsymbol{f}|\Omega$ for $\boldsymbol{f} \in L^2(\mathbb{R}^d;\ \mathscr{T})$. We now define the local versions of the spaces $\mathscr{E}$ and $\mathscr{J}$.

$$\mathscr{E}(\Omega) = \{\boldsymbol{E}|\Omega\colon \boldsymbol{E} \in \mathscr{E}\}, \qquad \mathscr{E}_0(\Omega) = \{\boldsymbol{f} \in L^2(\Omega;\ \mathscr{T})\colon X_0(\Omega)\boldsymbol{f} \in \mathscr{E}\}, \tag{2.11}$$

and

$$\mathscr{J}(\Omega) = \{\boldsymbol{J}|\Omega\colon \boldsymbol{J} \in \mathscr{J}\}, \qquad \mathscr{J}_0(\Omega) = \{\boldsymbol{f} \in L^2(\Omega;\ \mathscr{T})\colon X_0(\Omega)\boldsymbol{f} \in \mathscr{J}\}. \tag{2.12}$$

The orthogonality of the global spaces $\mathscr{E}$ and $\mathscr{J}$ has a local version.

Lemma 2.2. *The Hilbert space $L^2(\Omega;\ \mathscr{T})$ can be decomposed into the orthogonal sums as follows*

$$L^2(\Omega;\ \mathscr{T}) = \mathscr{J}_0(\Omega) \oplus \overline{\mathscr{E}(\Omega)} = \mathscr{E}_0(\Omega) \oplus \overline{\mathscr{J}(\Omega)}, \tag{2.13}$$

where $\overline{V}$ denotes the closure in $L^2(\Omega;\ \mathscr{T})$ of the subspace $V \subset L^2(\Omega;\ \mathscr{T})$.

Proof. Let $\boldsymbol{f} \in \mathscr{E}(\Omega)^{\perp}$. Then for any $\boldsymbol{E} \in \mathscr{E}$ we have

$$\int_{\mathbb{R}^d} (\boldsymbol{E},\ X_0(\Omega)\boldsymbol{f})d\boldsymbol{x} = \int_{\Omega} ((\boldsymbol{E}|\Omega)(\boldsymbol{x}), \boldsymbol{f}(\boldsymbol{x}))d\boldsymbol{x} = 0,$$

since $\boldsymbol{E}|\Omega \in \mathscr{E}(\Omega)$. Thus, $X_0(\Omega)\boldsymbol{f} \subset \mathscr{E}^{\perp} = \mathscr{J}$. It follows that $\boldsymbol{f} \in \mathscr{J}_0(\Omega)$ by definition of $\mathscr{J}_0(\Omega)$. It is obvious that if $\boldsymbol{f} \in \mathscr{J}_0(\Omega)$ then $\boldsymbol{f} \in \mathscr{E}(\Omega)^{\perp}$. Hence, $\mathscr{J}_0(\Omega) = \mathscr{E}(\Omega)^{\perp}$. The decomposition (2.13) follows from the fact that $\mathscr{J}_0(\Omega)^{\perp} = (\mathscr{E}(\Omega)^{\perp})^{\perp} = \overline{\mathscr{E}(\Omega)}$. The second decomposition in (2.13) is proved similarly. □

It is easy to see that the spaces $\mathscr{E}_0(\Omega)$ and $\mathscr{J}_0(\Omega)$ are closed for any measurable subset Ω of $\mathbb{R}^d$. It would certainly be sufficient for our needs if subspaces $\mathscr{E}(\Omega)$ and $\mathscr{J}(\Omega)$ were closed. However, the precise conditions on the symbol $\boldsymbol{\Gamma}(\boldsymbol{\eta})$ that guarantee that are unknown. Fortunately, we don't need this for our theory to proceed. Instead, we require a seemingly weaker compensated compactness property.

Postulate 2.3. *Let B_l denote the open unit ball in $\mathbb{R}^d$. We will say that the symbol* $\boldsymbol{\Gamma}(\boldsymbol{\eta})$ *(defined in postulate 2.1) has the compensated compactness property, if* $(\boldsymbol{E}_n, \boldsymbol{J}_n) \overset{*}{\rightharpoonup} 0$ *in the sense of measures on B_l, for all sequences* $\boldsymbol{E}_n \in \mathscr{E}(B_1)$, $\boldsymbol{J}_n \in \mathscr{J}(B_1)$ *such that* $\boldsymbol{E}_n \rightharpoonup 0$, $\boldsymbol{J}_n \rightharpoonup 0$ *in* $L^2(B_1; \mathscr{T})$.

The somewhat technical postulates 2.1 and 2.3 represent the minimal set of assumptions on the symbol $\boldsymbol{\Gamma}(\boldsymbol{\eta})$ that are needed for the general homogenization theory to work. In practice, every single example that comes from physics is formulated in terms of systems of partial differential equations that fields $\boldsymbol{E}$ and $\boldsymbol{J}$ must satisfy, as in the examples of conductivity and elasticity.

Postulate 2.4. *There exists a differential operator*

$$\mathscr{A}\boldsymbol{u} = \sum_{|\alpha|=m} \boldsymbol{A}_\alpha D^\alpha \boldsymbol{u},$$

so that

$$\mathscr{J} = \{\boldsymbol{J} \in L^2(\mathbb{R}^d; \mathscr{T}): \mathscr{A}\boldsymbol{J} = 0\} \tag{2.14}$$

in the sense of distributions. Here $\boldsymbol{A}_\alpha$ are constant linear maps between two finite dimensional inner product spaces $\boldsymbol{A}_\alpha$: $\mathscr{T} \rightarrow \mathscr{V}$, such that

$$\operatorname{rank}(\boldsymbol{A}(\boldsymbol{\xi})) = \text{const}, \quad \forall \boldsymbol{\xi} \in \mathbb{R}^d \setminus \{0\}, \qquad \boldsymbol{A}(\boldsymbol{\xi}) = \sum_{|\alpha|=m} \boldsymbol{A}_\alpha \boldsymbol{\xi}^\alpha. \tag{2.15}$$

When the subspace $\mathscr{J}$ is given by (2.14), the subspace $\mathscr{E} = \mathscr{J}^\perp$ could also be described by (2.14) but with another linear homogeneous differential operator $\mathscr{B}$ with constant rank property (2.15), [5, 12]. The differential constraints $\mathscr{B}\boldsymbol{E} = 0$, $\mathscr{A}\boldsymbol{J} = 0$ directly generalize the differential constraints placed on the electric and current fields in conductivity or the strain and stress fields in elasticity. Moreover, as we will show in section 2.5, postulate 2.4 implies postulates 2.1 and 2.3. While postulate 2.4 alone could be taken as the foundation of the theory, we opt to assume only postulate 2.1 always and without any further mention. In this way it will be clear which statements require the compensated compactness postulate 2.3 and which ones do not. This choice of exposition style emphasizes that it is the compensated compactness property, rather than the differential nature of the constraints that enables us to develop the general homogenization theory. In this regard, it is important to note that the relation between postulates 2.1 and 2.3 is not understood at all. Currently, it is unknown if there are any continuous, even, projection-valued functions $\boldsymbol{\Gamma}(\boldsymbol{\eta})$ that don't possess compensated compactness property, much less any effective characterization of symbols $\boldsymbol{\Gamma}(\boldsymbol{\eta})$ that do. We can only say that the set of symbols $\boldsymbol{\Gamma}(\boldsymbol{\eta})$ with the compensated compactness property is closed in $C(\mathbb{S}^{d-1})$ (see section 2.5) and contains all symbols of systems obeying postulate 2.4.

The laws of classical physics governing our examples transform in a special way under rigid body motion. This behavior can be expressed as the rotational equivariance property of the projection-valued function $\boldsymbol{\Gamma}(\boldsymbol{\eta})$.

Postulate 2.5. (Rotational equivariance). *We assume that rotations $SO(d)$ act by orthogonal transformations on the inner product space $\mathscr{T}$. Moreover,*

$$\boldsymbol{R} \cdot (\boldsymbol{\Gamma}(\boldsymbol{\eta})\boldsymbol{e}) = \boldsymbol{\Gamma}(\boldsymbol{R\eta})(\boldsymbol{R} \cdot \boldsymbol{e}), \qquad \boldsymbol{R} \in \mathrm{SO}(d),\ \boldsymbol{\eta} \in \mathbb{S}^{d-1},\ \boldsymbol{e} \in \mathscr{T}, \tag{2.16}$$

where $\boldsymbol{R} \cdot \boldsymbol{e}$ denotes the action[1] *of the rotation $\boldsymbol{R} \in \mathrm{SO}(d)$ on the vector $\boldsymbol{e} \in \mathscr{T}$.*

The term 'rotational equivariance' refers to operators, whose action commutes with rotations. In this postulate we view $\boldsymbol{\Gamma}$ as operating on pairs of vectors $(\boldsymbol{\eta}, \boldsymbol{e})$. The rotational equivariance postulate says that applying $\boldsymbol{\Gamma}$ to the pair $(\boldsymbol{\eta}, \boldsymbol{e})$ and then rotating the result is the same as rotating $\boldsymbol{\eta}$ and $\boldsymbol{e}$ first, and then applying $\boldsymbol{\Gamma}$ to the rotated pair $(\boldsymbol{R\eta}, \boldsymbol{R} \cdot \boldsymbol{e})$. While rotational equivariance reflects the fundamental symmetries of space, in some cases these symmetries could be broken, for example, by the presence of a magnetic field. For this reason this postulate will not be assumed outright in parts I and II of the book, as the entire theory does not require it. Nonetheless we will focus on the consequences of assuming it throughout the exposition. In particular, in all case studies of part III, with one exception, the rotational equivariance postulate will be satisfied.

We also remark that if the rotational equivariance postulate holds, then the function $\boldsymbol{\Gamma}(\boldsymbol{\eta})$ is determined by a *single* finite dimensional orthogonal projection operator $\boldsymbol{\Gamma}(\boldsymbol{\eta}_0)$ with the property $\boldsymbol{R} \cdot \boldsymbol{\Gamma}(\boldsymbol{\eta}_0) = \boldsymbol{\Gamma}(\boldsymbol{\eta}_0)$ for all rotations $\boldsymbol{R}$ for which $\boldsymbol{R\eta}_0 = \pm\boldsymbol{\eta}_0$. Since $\mathrm{SO}(d)$ is an algebraic group, the function $\boldsymbol{\Gamma}(\boldsymbol{\eta})$ will necessarily agree with an even homogeneous polynomial on the unit sphere, which means, in particular, that postulate 2.5 implies postulate 2.4 that, in turn, implies postulates 2.1 and 2.3.

Example 2.6. (Maxwell equations).

Consider the nondimensionalized Maxwell system

$$\nabla \times \boldsymbol{e} = -\boldsymbol{b}_t, \quad \nabla \cdot \boldsymbol{b} = 0, \quad \nabla \times \boldsymbol{h} = \boldsymbol{d}_t, \quad \nabla \cdot \boldsymbol{d} = 0.$$

We define $\boldsymbol{E} = (\boldsymbol{e}, \boldsymbol{b})$ and $\boldsymbol{J} = (-\boldsymbol{d}, \boldsymbol{h})$, so that $\mathscr{T} = \mathbb{R}^6$ with a standard dot product. We therefore compute

$$\mathscr{E}_\eta = \{(\boldsymbol{e}_1, \boldsymbol{e}_2) \in \mathbb{R}^6\colon \boldsymbol{\xi} \times \boldsymbol{e}_1 + \xi_4\boldsymbol{e}_2 = 0\}, \quad \mathscr{J}_\eta = \{(\boldsymbol{j}_1, \boldsymbol{j}_2) \in \mathbb{R}^6\colon \boldsymbol{\xi} \times \boldsymbol{j}_2 + \xi_4\boldsymbol{j}_1 = 0\},$$

where $\boldsymbol{\eta} = (\boldsymbol{\xi}, \xi_4) \in \mathbb{S}^3 \subset \mathbb{R}^4$. If $\xi_4 \neq 0$, then

[1] Rotations may act in a different way depending on the nature of the space $\mathscr{T}$. For example, $\boldsymbol{R} \cdot \boldsymbol{e} = \boldsymbol{Re}$, when $\mathscr{T} = \mathbb{R}^d$ in the case of conductivity and $\boldsymbol{R} \cdot \boldsymbol{\varepsilon} = \boldsymbol{R\varepsilon R}^T$, when $\mathscr{T} = \mathrm{Sym}(\mathbb{R}^d)$ in the case of elasticity.

$$\mathscr{E}_\eta = \left\{ \left(e_1, -\frac{1}{\xi_4}\boldsymbol{\xi} \times e_1 \right) : e_1 \in \mathbb{R}^3 \right\}, \quad \mathscr{J}_\eta = \left\{ \left(-\frac{1}{\xi_4}\boldsymbol{\xi} \times j_2, j_2 \right) : j_2 \in \mathbb{R}^3 \right\}$$

If $\xi_4 = 0$, then

$$\mathscr{E}_\eta = \mathbb{R}\boldsymbol{\xi} \oplus (\mathbb{R}\boldsymbol{\xi})^\perp, \quad \mathscr{J}_\eta = (\mathbb{R}\boldsymbol{\xi})^\perp \oplus \mathbb{R}\boldsymbol{\xi}.$$

From these formulas it is evident that $\dim \mathscr{E}_\eta = \dim \mathscr{J}_\eta = 3$ for all $\boldsymbol{\eta} \in \mathbb{S}^3$ and that $\mathscr{J}_\eta = (\mathscr{E}_\eta)^\perp$.

It is well-known that we can arrange the 6 components of vectors in $\mathscr{T} = \mathbb{R}^6$ into a 4×4 skew-symmetric matrix:

$$\mathscr{T} \ni \boldsymbol{t} = (\boldsymbol{e}, \boldsymbol{b}) \mapsto \boldsymbol{F}(\boldsymbol{e}, \boldsymbol{b}) = \begin{bmatrix} 0 & -e_3 & e_2 & -b_1 \\ e_3 & 0 & -e_1 & -b_2 \\ -e_2 & e_1 & 0 & -b_3 \\ b_1 & b_2 & b_3 & 0 \end{bmatrix},$$

This permits us to define the action of SO(4) on $\mathscr{T}$:

$$\boldsymbol{R} \cdot (\boldsymbol{e}, \boldsymbol{b}) = (\boldsymbol{e}^{\boldsymbol{R}}, \boldsymbol{b}^{\boldsymbol{R}}), \quad \boldsymbol{R}\boldsymbol{F}(\boldsymbol{e}, \boldsymbol{b})\boldsymbol{R}^T = \boldsymbol{F}(\boldsymbol{e}^{\boldsymbol{R}}, \boldsymbol{b}^{\boldsymbol{R}}).$$

It is clear that the inner product on $\mathscr{T}$ is SO(4)-invariant, since it is half of the Frobenius inner product of corresponding skew-symmetric matrices $\boldsymbol{F}$. One can also check (using a computer algebra system) that for any $\boldsymbol{R} \in$ SO(4) we have

$$\boldsymbol{R} \cdot \mathscr{E}_\eta = \mathscr{E}_{\boldsymbol{R}\eta} \quad \forall \boldsymbol{\eta} \in \mathbb{S}^3.$$

This can be done using the following parameterization of SO(4):

$$\mathrm{SO}(4) = \left\{ \begin{bmatrix} \rho \boldsymbol{I}_3 + \boldsymbol{r}\times & -\boldsymbol{r} \\ \boldsymbol{r} & \rho \end{bmatrix} \begin{bmatrix} \boldsymbol{R}_0 & 0 \\ 0 & 1 \end{bmatrix} : \boldsymbol{R}_0 \in SO(3), (\rho, \boldsymbol{r}) \in \mathbb{S}^3 \right\},$$

where $\boldsymbol{r}\times$, suggested by the notation in [13] denotes the skew-symmetric 3×3 matrix defined by

$$(\boldsymbol{r} \times)\boldsymbol{u} = \boldsymbol{r} \times \boldsymbol{u}, \quad \forall \boldsymbol{u} \in \mathbb{R}^3. \tag{2.17}$$

2.4 Boundary value problems

The fields $\boldsymbol{E}(\boldsymbol{x})$ and $\boldsymbol{J}(\boldsymbol{x})$ in the physical material can be found from the solution of a boundary value problem, if one postulates a linear constitutive relation $\boldsymbol{J} = \mathsf{L}\boldsymbol{E}$, where $\mathsf{L} \in \mathrm{End}^+(\mathscr{T})$ is a positive definite linear operator on $\mathscr{T}$. While in almost all examples the tensor of material properties L is symmetric, we don't make this assumption in our general theory. For example, the presence of the magnetic field can break the symmetry leading to the Hall effect, encoded in the antisymmetric part of the conductivity tensor. If the tensor of material properties L is a constant operator, then we will refer to a *homogeneous* material, and if L depends on $\boldsymbol{x} \in \Omega$, then we will refer to an *inhomogeneous* material. In this book we do not treat high contrast composites, and hence we always assume that $\mathsf{L}(\boldsymbol{x})$ takes its values in a compact subset of $\mathrm{End}^{+}(\mathscr{T})$—the set of all positive definite linear operators on $\mathscr{T}$. Specifically, for $0 < \alpha < \beta$ we define

$$\mathscr{M}(\alpha, \beta) = \{\mathsf{L} \in \mathrm{End}^{+}(\mathscr{T}): (\mathsf{L}\boldsymbol{E}, \boldsymbol{E}) \geqslant \alpha|\boldsymbol{E}|^2, \ (\mathsf{L}^{-1}\boldsymbol{E}, \boldsymbol{E}) \geqslant \beta^{-1}|\boldsymbol{E}|^2\}. \tag{2.18}$$

We observe that $\mathsf{L} \in \mathscr{M}(\alpha, \beta)$ is equivalent to $\mathsf{L}^{-1} \in \mathscr{M}(\beta^{-1}, \alpha^{-1})$.

Remark 2.7. *It is important to point out that the present theory is also applicable to complex valued operators and fields. A complex Hermitian vector space* $\mathscr{T}$ *would simply be regarded as a real vector space* $\mathscr{T}_{\mathbb{R}}$ *of twice the dimension with the inner product* $(\boldsymbol{e}, \boldsymbol{e}')_{\mathscr{T}_{\mathbb{R}}} = \Re(\boldsymbol{e}, \boldsymbol{e}')$, *where* $(\boldsymbol{e}, \boldsymbol{e}')$ *is the original complex-Hermitian inner product on* $\mathscr{T}$. *In this interpretation* $\mathsf{L} \in \mathrm{End}_{\mathbb{C}}(\mathscr{T})$ *would belong to* $\mathrm{End}^{+}_{\mathbb{R}}(\mathscr{T}_{\mathbb{R}})$ *if and only if its Hermitian part, often denoted by* $\Re\mathsf{L} = (\mathsf{L} + \mathsf{L}^*)/2$, *is positive definite on* $\mathscr{T}$, *where* $\mathsf{L}^* = \overline{\mathsf{L}}^T$ *is the Hermitian conjugate of* L *on the complex Hermitian vector space* $\mathscr{T}$.

Our goal is not a solution of specific boundary value problems, but rather an understanding of how the macroscopic properties of composite materials arise from their microstructure. For that purpose it is sufficient to 'test' the response of a composite to prescribed Dirichlet boundary conditions. However, since we do not have a direct access to potentials, the boundary value problem needs to be stated in a somewhat circuitous way. Let Γ_Ω be an orthogonal projection onto $\mathscr{E}_0(\Omega)$. The governing equations can therefore be written in the following form

$$\boldsymbol{E} \in \mathscr{E}_0(\Omega), \qquad \boldsymbol{J} = \mathsf{L}(\boldsymbol{x})\boldsymbol{E}, \qquad \Gamma_\Omega \boldsymbol{J} = \boldsymbol{f}, \qquad \boldsymbol{f} \in \mathscr{E}_0(\Omega). \tag{2.19}$$

Theorem 2.8. *For every* $\boldsymbol{f} \in \mathscr{E}_0(\Omega)$ *the boundary value problem (2.19) has a unique solution* $\boldsymbol{E} \in \mathscr{E}_0(\Omega)$.

Proof. Indeed, the linear operator

$$\mathsf{T}_{\mathsf{L}}: \mathscr{E}_0(\Omega) \to \mathscr{E}_0(\Omega), \qquad \mathsf{T}_{\mathsf{L}}\boldsymbol{E} = \Gamma_\Omega(\mathsf{L}(\boldsymbol{x})\boldsymbol{E}) \tag{2.20}$$

is bounded and coercive:

$$\|\mathsf{T}_{\mathsf{L}}\| \leqslant \beta, \qquad (\mathsf{T}_{\mathsf{L}}\boldsymbol{E}, \boldsymbol{E})_{L^2(\Omega)} = (\mathsf{L}\boldsymbol{E}, \boldsymbol{E})_{L^2(\Omega)} \geqslant \alpha\|\boldsymbol{E}\|^2_{L^2(\Omega)}. \tag{2.21}$$

Therefore, it is invertible by the Lax-Milgram lemma. In fact,

$$\|T_{\mathsf{L}}^{-1}\| \leqslant \frac{1}{\alpha}. \tag{2.22}$$

□

Equation (2.19) does not mean that we are restricted to the zero boundary conditions on $\partial\Omega$. The case of arbitrary Dirichlet boundary conditions can be reduced to (2.19). Indeed, we may choose any field $\boldsymbol{g}_0 \in \mathscr{E}(\Omega)$ satisfying the desired Dirichlet boundary conditions, and then solve $\Gamma_\Omega(\mathsf{L}(\boldsymbol{x})(\boldsymbol{E} + \boldsymbol{g}_0)) = \boldsymbol{f}$, which is equivalent to $\mathsf{T}_\mathsf{L}\boldsymbol{E} = \boldsymbol{f} - \Gamma_\Omega(\mathsf{L}(\boldsymbol{x})\boldsymbol{g}_0)$.

In the context of conductivity, for example,

$$\mathscr{E}_0(\Omega) = \{\nabla\phi\colon \phi \in H_0^1(\Omega)\},$$

if the domain Ω is diffeomorphic to a ball[2]. In this case the operator Γ_Ω is the Helmholtz projection onto the space of gradients:

$$\Gamma_\Omega \boldsymbol{f} = \nabla\psi, \qquad \Delta\psi = \nabla\cdot\boldsymbol{f}, \qquad \psi \in H_0^1(\Omega).$$

There is an isomorphism between $\mathscr{E}_0(\Omega)$ and $H^{-1}(\Omega)$, whereby every $\boldsymbol{f} \in \mathscr{E}_0(\Omega)$, given by $\boldsymbol{f} = \nabla g$, $g \in H_0^1(\Omega)$ corresponds to $F = \Delta g \in H^{-1}(\Omega)$. Then, the system of equations (2.19) can be rewritten as an elliptic Dirichlet boundary value problem:

$$\nabla\cdot(\mathsf{L}(\boldsymbol{x})\nabla\phi) = F, \qquad \phi \in H_0^1(\Omega).$$

From the theory of elliptic partial differential equations it is well-known that this boundary value problem has a unique solution.

2.5 The compensated compactness property

We first show that the compensated compactness postulate 2.3 can be formulated on any open subset of $\mathbb{R}^d$ and in a more general form. In the theorem below Π_Ω denotes the orthogonal projection onto $\mathscr{J}_0(\Omega)$ in $L^2(\Omega; \mathscr{T})$.

Theorem 2.9. *Suppose that postulate 2.3 holds. Let $\Omega \subset \mathbb{R}^d$ be open. Let $\boldsymbol{f}_n \rightharpoonup \boldsymbol{f}_0$ and $\boldsymbol{g}_n \rightharpoonup \boldsymbol{g}_0$ as $n \to \infty$ weakly in $L^2(\Omega; \mathscr{T})$. Let us assume that $\Gamma_\Omega \boldsymbol{f}_n \to \Gamma_\Omega \boldsymbol{f}_0$ and $\Pi_\Omega \boldsymbol{g}_n \to \Pi_\Omega \boldsymbol{g}_0$ as $n \to \infty$ in $L^2(\Omega; \mathscr{T})$ norm. Then $(\boldsymbol{f}_n, \boldsymbol{g}_n) \stackrel{*}{\rightharpoonup} (\boldsymbol{f}_0, \boldsymbol{g}_0)$ in the sense of measures on Ω.*

Proof. Let $\widetilde{\boldsymbol{f}}_n = \boldsymbol{f}_n - \boldsymbol{f}_0$ and $\widetilde{\boldsymbol{g}}_n = \boldsymbol{g}_n - \boldsymbol{f}_0$. Then $\widetilde{\boldsymbol{f}}_n \rightharpoonup 0$, $\widetilde{\boldsymbol{g}}_n \rightharpoonup 0$, $\Gamma_\Omega\widetilde{\boldsymbol{f}}_n \to 0$ and $\Pi_\Omega\widetilde{\boldsymbol{g}}_n \to 0$. The calculation

$$(\boldsymbol{f}_n, \boldsymbol{g}_n) = (\widetilde{\boldsymbol{f}}_n, \widetilde{\boldsymbol{g}}_n) + (\widetilde{\boldsymbol{f}}_n, \boldsymbol{g}_0) + (\widetilde{\boldsymbol{g}}_n, \boldsymbol{f}_0) + (\boldsymbol{f}_0, \boldsymbol{g}_0)$$

shows that we may assume, without loss of generality, that $\boldsymbol{f}_0 = \boldsymbol{g}_0 = 0$.

Let $\widetilde{\boldsymbol{E}}_n = \widetilde{\boldsymbol{g}}_n - \Pi_\Omega\widetilde{\boldsymbol{g}}_n \in \overline{\mathscr{E}(\Omega)}$ and $\widetilde{\boldsymbol{J}}_n = \widetilde{\boldsymbol{f}}_n - \Gamma_\Omega\widetilde{\boldsymbol{f}}_n \in \overline{\mathscr{J}(\Omega)}$. Then

$$(\widetilde{\boldsymbol{f}}_n, \widetilde{\boldsymbol{g}}_n) = (\widetilde{\boldsymbol{E}}_n, \widetilde{\boldsymbol{J}}_n) + (\widetilde{\boldsymbol{E}}_n, \Gamma_\Omega\boldsymbol{f}_n) + (\widetilde{\boldsymbol{J}}_n, \Pi_\Omega\boldsymbol{g}_n) + (\Pi_\Omega\boldsymbol{g}_n, \Gamma_\Omega\boldsymbol{f}_n).$$

The above calculation shows that the conclusion of the theorem follows if the theorem is proved for all sequences $\widetilde{\boldsymbol{E}}_n \in \overline{\mathscr{E}(\Omega)}$, $\widetilde{\boldsymbol{J}}_n \in \overline{\mathscr{J}(\Omega)}$ that converge weakly to zero in $L^2(\Omega; \mathscr{T})$.

[2] If $\partial\Omega$ has several connected components, like a spherical shell, then the space $\mathscr{E}_0(\Omega)$ can be larger than $\{\nabla\phi\colon \phi \in H_0^1(\Omega)\}$.

Since $\widetilde{E}_n \in \overline{\mathscr{E}(\Omega)}$ and $\widetilde{J}_n \in \overline{\mathscr{J}(\Omega)}$ there exist $E_n \in \mathscr{E}(\Omega)$ and $J_n \in \mathscr{J}(\Omega)$, such that $\|\widetilde{E}_n - E_n\|_{L^2(\Omega)} \leqslant 1/n$, and $\|\widetilde{J}_n - J_n\|_{L^2(\Omega)} \leqslant 1/n$. It follows that $R_n = \widetilde{E}_n - E_n \to 0$ and $S_n = \widetilde{J}_n - J_n \to 0$ in $L^2(\Omega; \mathscr{T})$, and E_n and J_n converge weakly to zero. Now the calculation

$$(\widetilde{E}_n, \widetilde{J}_n) = (E_n, J_n) + (E_n, S_n) + (J_n, R_n) + (R_n, S_n)$$

shows that it is sufficient to prove the theorem for all sequences $E_n \in \mathscr{E}(\Omega)$ and $J_n \in \mathscr{J}(\Omega)$ that converge weakly to zero.

The sequence (E_n, J_n) is bounded in $L^1(\Omega)$ and therefore has a weak-* convergent subsequence in $C_0^*(\Omega)$. Let $\mu \in C_0^*(\Omega)$ denote this weak-* limit. To see that measure μ has to be zero it is enough to show that any $x_0 \in \Omega$ is not in the support of the measure μ. Indeed, since Ω is open for any $x_0 \in \Omega$ there exists $r(x_0) > 0$, such that $B(x_0, r(x_0)) \subset \Omega$. But then for any $\phi \in C_0(B(x_0, r(x_0)))$ we have

$$\begin{aligned}\int_{\Omega} \phi(x)d\mu(x) &= \lim_{n\to\infty} \int_{\Omega} \phi(x)(E_n(x), J_n(x))dx \\ &= r(x_0)^d \lim_{n\to\infty} \int_{B_1} \phi(r(x_0)z + x_0)(E_n(r(x_0)z + x_0), J_n(r(x_0)z + x_0))dz.\end{aligned}$$

If we denote

$$\psi(z) = \phi(r(x_0)z + x_0), \quad e_n(z) = E_n(r(x_0)z + x_0), \quad j_n(z) = J_n(r(x_0)z + x_0),$$

then we observe that $\psi \in C_0(B_1)$, $e_n \in \mathscr{E}(B_1)$, $j_n(z) \in \mathscr{J}(B_1)$, and $e_n \rightharpoonup 0$, $j_n \rightharpoonup 0$, as $n \to \infty$ weakly in $L^2(B_1; \mathscr{T})$. By the compensated compactness postulate 2.3 we conclude that $(e_n, j_n) \overset{*}{\rightharpoonup} 0$, as $n \to \infty$ and therefore $\int_{\Omega} \phi(x)d\mu(x) = 0$ for any $\phi \in C_0(B(x_0, r(x_0)))$. This implies that every $x_0 \in \Omega$ is not in the support of μ. It follows that $\mu = 0$. Since we have proved that the weak-* limit of any weak-* convergent subsequence of (E_n, J_n) has to be zero, we conclude that $(E_n, J_n) \overset{*}{\rightharpoonup} 0$, as $n \to \infty$ in the weak-* topology of $C_0^*(\Omega)$. The theorem is now proved. □

Our next lemma shows that compactness of certain linear operators is behind the compensated compactness theorem, partially explaining the name of the property.

Lemma 2.10. *For $\phi \in C_0^\infty(B_1)$ we define the operators*

$$K_\phi: \mathscr{E}(B_1) \to \mathscr{J}, \qquad K_\phi e = (I - \Gamma)X_0(B_1)[\phi e]$$

and

$$T_\phi: \mathscr{J}(B_1) \to \mathscr{E}, \qquad T_\phi j = \Gamma X_0(B_1)[\phi j].$$

Then the compensated compactness postulate 2.3 holds if and only if the operators K_ϕ and T_ϕ are compact for every $\phi \in C_0^\infty(B_1)$.

Proof. Suppose first that the compensated compactness property holds and that $\boldsymbol{e}_n \rightharpoonup 0$ weakly in $\mathscr{E}(B_1)$. Then

$$\|K_\phi \boldsymbol{e}_n\|^2_{L^2(\mathbb{R}^d)} = \int_{B_1} \phi(\boldsymbol{j}_n, \boldsymbol{e}_n) d\boldsymbol{x},$$

where $\boldsymbol{j}_n = (K_\phi \boldsymbol{e}_n)|_{B_1} \in \mathscr{J}(B_1)$. By compensated compactness we conclude that $K_\phi \boldsymbol{e}_n \to 0$ in $L^2(\mathbb{R}^d; \mathscr{T})$. This proves that K_ϕ is compact (for every $\phi \in C_0(B_1)$). The proof of compactness of T_ϕ is completely analogous.

Now, let us assume that the operators K_ϕ and T_ϕ are compact for every $\phi \in C_0^\infty(B_1)$. Let $\mathscr{E}(B_1) \ni \boldsymbol{e}_n \rightharpoonup 0$, and $\mathscr{J}(B_1) \ni \boldsymbol{j}_n \rightharpoonup 0$, as $n \to \infty$. We first observe that since the sequence $f_n = (\boldsymbol{e}_n, \boldsymbol{j}_n)$ is bounded in $L^1(B_1)$ it is sufficient to prove that for any $\{\phi, \psi\} \subset C_0^\infty(B_1)$ we must have $(\phi \boldsymbol{e}_n, \psi \boldsymbol{j}_n)_{L^2(B_1)} \to 0$, as $n \to \infty$. Clearly,

$$(\phi \boldsymbol{e}_n, \psi \boldsymbol{j}_n)_{L^2(B_1)} = (X_0(B_1)[\phi \boldsymbol{e}_n], X_0(B_1)[\psi \boldsymbol{j}_n])_{L^2(\mathbb{R}^d)}.$$

Writing

$$X_0(B_1)[\phi \boldsymbol{e}_n] = (I - \Gamma) X_0(B_1)[\phi \boldsymbol{e}_n] + \Gamma X_0(B_1)[\phi \boldsymbol{e}_n]$$

we obtain

$$(\phi \boldsymbol{e}_n, \psi \boldsymbol{j}_n)_{L^2(B_1)} = (K_\phi \boldsymbol{e}_n, \psi \boldsymbol{j}_n)_{L^2(B_1)} + (\phi \boldsymbol{e}_n, T_\psi \boldsymbol{j}_n)_{L^2(B_1)}$$

Since compact operators convert weakly convergent sequences into strongly convergent ones we conclude that $(\boldsymbol{e}_n, \boldsymbol{j}_n) \rightharpoonup 0$ in $\mathscr{D}'$. The boundedness of the sequence $(\boldsymbol{e}_n, \boldsymbol{j}_n)$ in $L^1(B_1)$ implies that $(\boldsymbol{e}_n, \boldsymbol{j}_n) \overset{*}{\rightharpoonup} 0$ in $C_0^*(B_1)$, as required. □

Before we describe a large class of symbols $\boldsymbol{\Gamma}(\boldsymbol{\eta})$ that have the compensated compactness property we note that this class must be closed in $C(\mathbb{S}^{d-1}; \mathrm{Sym}(\mathscr{T}))$. Indeed, suppose $\boldsymbol{\Gamma}_n(\boldsymbol{\eta})$ is a sequence of symbols satisfying postulate 2.1 that converge to $\boldsymbol{\Gamma}(\boldsymbol{\eta})$ uniformly on the sphere $\mathbb{S}^{d-1}$. Our claim follows from the fact that for any $\phi \in C_0^\infty(B_1)$ the sequence of operators $K_\phi^{(n)}$ and $T_\phi^{(n)}$ converge in operator norm to K_ϕ and T_ϕ corresponding to the limit symbol $\boldsymbol{\Gamma}(\boldsymbol{\eta})$. Indeed, for any $\boldsymbol{j} \in \mathscr{J}(B_1)$

$$\|T_\phi^{(n)} \boldsymbol{j} - T_\phi \boldsymbol{j}\|^2_{L^2(\mathbb{R}^d)} = \int_{\mathbb{R}^d} \left| \left(\boldsymbol{\Gamma}_n\left(\frac{\boldsymbol{\xi}}{|\boldsymbol{\xi}|}\right) - \boldsymbol{\Gamma}\left(\frac{\boldsymbol{\xi}}{|\boldsymbol{\xi}|}\right)\right) \widehat{\phi \boldsymbol{j}}(\boldsymbol{\xi}) \right|^2 d\boldsymbol{\xi}.$$

This shows that

$$\|T_\phi^{(n)} \boldsymbol{j} - T_\phi \boldsymbol{j}\|_{L^2(\mathbb{R}^d)} \leqslant \|\boldsymbol{\Gamma}_n - \boldsymbol{\Gamma}\|_{C(\mathbb{S}^{d-1})} \|\widehat{\phi \boldsymbol{j}}\|_{L^2(\mathbb{R}^d)} =$$
$$\|\boldsymbol{\Gamma}_n - \boldsymbol{\Gamma}\|_{C(\mathbb{S}^{d-1})} \|\phi \boldsymbol{j}\|_{L^2(B_1)} \leqslant \|\boldsymbol{\Gamma}_n - \boldsymbol{\Gamma}\|_{C(\mathbb{S}^{d-1})} \|\phi\|_{C_0(B_1)} \|\boldsymbol{j}\|_{L^2(B_1)}.$$

Our next theorem shows that in all cases describable by differential equations (which covers each and every example from physics, as well as all symbols satisfying postulate 2.5) compensated compactness property holds.

Theorem 2.11. (Compensated compactness). *Suppose that $\mathscr{A}$ is a linear homogeneous differential operator satisfying the constant rank condition (2.15). Then the symbol* $\mathbf{\Gamma}$ *corresponding to*

$$\mathscr{J} = \{\boldsymbol{f} \in L^2(\mathbb{R}^d; \mathscr{T}): \mathscr{A}\boldsymbol{f} = 0 \text{ in } \mathscr{D}'\} \tag{2.23}$$

has the compensated compactness property.

Proof. To prove the theorem we appeal to lemma 2.10 and prove that the operators K_ϕ and T_ψ are compact. This is done using the technique due to Guerra and Raiţă [5].

Let us derive the formula for Γ in terms of $\boldsymbol{A}(\boldsymbol{\xi})$, defined in (2.15). We recall the definition of the Moore–Penrose inverse, or pseudo-inverse $\boldsymbol{A}^\dagger(\boldsymbol{\xi})$ of $\boldsymbol{A}(\boldsymbol{\xi})$.

$$\boldsymbol{A}^\dagger = \begin{cases} (\boldsymbol{A} \mid_{(\ker \boldsymbol{A})^\perp})^{-1}, & \text{on } \mathscr{R}(\boldsymbol{A}), \\ 0, & \text{on } \mathscr{R}(\boldsymbol{A})^\perp. \end{cases}$$

It is also uniquely defined by the properties

$$\boldsymbol{A}\boldsymbol{A}^\dagger = \mathscr{P}_{\mathscr{R}(\boldsymbol{A})}, \qquad \boldsymbol{A}^\dagger\boldsymbol{A} = \mathscr{P}_{\mathscr{R}(\boldsymbol{A}*)},$$

where $\boldsymbol{A}$ is a linear map between two Euclidean finite dimensional spaces.

According to (2.23) the subspace $\mathscr{J}$ consists of $L^2(\mathbb{R}^d; \mathscr{T})$ vector fields whose Fourier transforms are in the kernel of $\boldsymbol{A}(\boldsymbol{\xi})$. Then $\mathscr{E} = \mathscr{J}^\perp$ must be the subspace of $L^2(\mathbb{R}^d; \mathscr{T})$ vector fields whose Fourier transforms are in $(\ker \boldsymbol{A}(\boldsymbol{\xi}))^\perp = \mathscr{R}(\boldsymbol{A}^*(\boldsymbol{\xi}))$. Thus,

$$\widehat{\Gamma \boldsymbol{f}} = \mathbf{\Gamma}(\boldsymbol{\xi})\hat{\boldsymbol{f}}(\boldsymbol{\xi}) = \boldsymbol{A}^\dagger(\boldsymbol{\xi})\boldsymbol{A}(\boldsymbol{\xi})\hat{\boldsymbol{f}}(\boldsymbol{\xi}).$$

The constant rank property (2.15) implies that $\mathbf{\Gamma}(\boldsymbol{\xi}) = \boldsymbol{A}^\dagger(\boldsymbol{\xi})\boldsymbol{A}(\boldsymbol{\xi})$ is a continuous function on $\mathbb{R}^d \setminus \{0\}$, which is homogeneous of degree zero, [2, 12]. We first observe that if $\boldsymbol{f} \in \mathscr{J}(B_1)$, then $\mathscr{A}\boldsymbol{f} = 0$ in $\mathscr{D}'(B_1)$. Indeed, for any $\boldsymbol{f} \in \mathscr{J}$, and any $\phi \in C_0^\infty(B_1; \mathscr{V})$ we have

$$(\boldsymbol{f}, \mathscr{A}^*\phi)_{L^2(B_1)} = (\boldsymbol{f}, \mathscr{A}^*\phi)_{L^2(\mathbb{R}^d)} = \int_{\mathbb{R}^d} (\hat{\boldsymbol{f}}(\boldsymbol{\xi}), \boldsymbol{A}^*(\boldsymbol{\xi})\hat{\phi}(\boldsymbol{\xi}))d\boldsymbol{\xi} = 0.$$

Since the subspace

$$Z = \{\boldsymbol{f} \in L^2(B_1; \mathscr{T}): \mathscr{A}\boldsymbol{f} = 0 \text{ in } \mathscr{D}'(B_1)\}$$

is closed in $L^2(B_1; \mathscr{T})$ we conclude that $\overline{\mathscr{J}(B_1)} \subset Z$. Now for any $\phi \in C_0^\infty(B_1)$ and for any $\boldsymbol{f} \in Z$ we have

$$\mathscr{A}(\phi\boldsymbol{f}) = \mathscr{M}\boldsymbol{f} \in H^{-(m-1)}(B_1),$$

since $\mathscr{M}$ is a differential operator of order $m - 1$ with $C_0^\infty(B_1)$ coefficients. Let $\boldsymbol{j}_n \in \mathscr{J}(B_1)$ be such that $\boldsymbol{j}_n \rightharpoonup 0$ in $L^2(B_1; \mathscr{T})$, as $n \to \infty$. Since $H^{-(m-1)}(B_1)$ is compactly embedded into $H^{-m}(B_1)$ the weak convergence of $\boldsymbol{j}_n \in \mathscr{J}(B_1)$ in $L^2(B_1)$ implies the weak convergence of $\mathscr{A}(\phi\boldsymbol{j}_n) = \mathscr{M}\boldsymbol{j}_n$ in $H^{-(m-1)}(B_1)$, which, in turn, implies strong convergence in $H^{-m}(B_1)$. Next we observe that if we extend $\phi \boldsymbol{j}_n$ by 0 outside B_1, then $\mathscr{A}(X_0(B_1)\phi \boldsymbol{j}_n) = X_0(B_1)\mathscr{A}(\phi \boldsymbol{j}_n)$. Thus, $\mathscr{A}(X_0(B_1)\phi \boldsymbol{j}_n) \to 0$ in the norm of $H^{-m}(\mathbb{R}^d)$. But this means that

$$\frac{\boldsymbol{A}(\boldsymbol{\xi})\hat{\boldsymbol{v}}_n(\boldsymbol{\xi})}{1 + |\boldsymbol{\xi}|^m} \to 0 \text{ in } L^2(\mathbb{R}^d),$$

where $\boldsymbol{v}_n = X_0(B_1)\phi \boldsymbol{j}_n$. We also observe that $\boldsymbol{v}_n \rightharpoonup 0$ in $L^2(B_1)$ and hence for any $\boldsymbol{\xi}$ we have

$$\hat{\boldsymbol{v}}_n(\boldsymbol{\xi}) = \int_{B_1} \boldsymbol{v}_n(\boldsymbol{x})e^{-i\boldsymbol{\xi}\cdot\boldsymbol{x}}d\boldsymbol{x} \to 0, \quad \text{as } n \to \infty.$$

The same formula also shows that $|\hat{\boldsymbol{v}}_n(\boldsymbol{\xi})| \leqslant C\|\boldsymbol{j}_n\|_{L^2(B_1)}$. It follows that $\boldsymbol{\Gamma}(\boldsymbol{\xi})\hat{\boldsymbol{v}}_n(\boldsymbol{\xi}) \to 0$ in $L^2(\hat{B}_1)$, where $\hat{B}_1$ is the unit ball in the Fourier space. When $|\boldsymbol{\xi}| \geqslant 1$ we obtain by homogeneity of $\boldsymbol{A}^\dagger(\boldsymbol{\xi})$

$$(1 + |\boldsymbol{\xi}|^m)\boldsymbol{A}^\dagger(\boldsymbol{\xi}) = \frac{(1 + |\boldsymbol{\xi}|^m)}{|\boldsymbol{\xi}|^m}\boldsymbol{A}^\dagger\left(\frac{\boldsymbol{\xi}}{|\boldsymbol{\xi}|}\right),$$

which shows that $(1 + |\boldsymbol{\xi}|^m)\boldsymbol{A}^\dagger(\boldsymbol{\xi}) \in L^\infty(\hat{B}_1^c)$. Thus,

$$\boldsymbol{\Gamma}(\boldsymbol{\xi})\hat{\boldsymbol{v}}_n(\boldsymbol{\xi}) = (1 + |\boldsymbol{\xi}|^m)\boldsymbol{A}^\dagger(\boldsymbol{\xi})\frac{\boldsymbol{A}(\boldsymbol{\xi})\hat{\boldsymbol{v}}_n(\boldsymbol{\xi})}{1 + |\boldsymbol{\xi}|^m} \to 0 \text{ in } L^2(\hat{B}_1^c)$$

We conclude that $\boldsymbol{\Gamma}(\boldsymbol{\xi})\hat{\boldsymbol{v}}_n(\boldsymbol{\xi}) \to 0$ in $L^2(\mathbb{R}^d)$. In particular, $T_\phi \boldsymbol{j}_n \to 0$ in $L^2(B_1)$, and compactness of T_ϕ is established. To prove compactness of K_ϕ we appeal to a result in [12] that the constant rank condition (2.15) implies the existence of a constant rank homogeneous differential operator $\mathscr{B}$ with constant coefficients, such that $\mathscr{R}(\boldsymbol{B}(\boldsymbol{\xi})) = \ker(\boldsymbol{A}(\boldsymbol{\xi}))$ for all $\boldsymbol{\xi} \neq 0$. But then

$$\begin{aligned}\mathscr{E} &= \{\boldsymbol{f} \in L^2(\mathbb{R}^d; \mathscr{T}): \hat{\boldsymbol{f}}(\boldsymbol{\xi}) \in (\ker(\boldsymbol{A}(\boldsymbol{\xi})))^\perp\} \\ &= \{\boldsymbol{f} \in L^2(\mathbb{R}^d; \mathscr{T}): \hat{\boldsymbol{f}}(\boldsymbol{\xi}) \in \ker(\boldsymbol{B}^*(\boldsymbol{\xi}))\} = \{\boldsymbol{f} \in L^2(\mathbb{R}^d; \mathscr{T}): \mathscr{B}^*\boldsymbol{f} = 0 \text{ in } \mathscr{D}'\}.\end{aligned}$$

Thus, the same argument applies to K_ϕ. □

We conclude this section by presenting several properties of the local spaces $\mathscr{E}(\Omega)$ and $\mathscr{J}(\Omega)$ that are sufficient for compensated compactness. Our first result shows that if the subspaces $\mathscr{E}(B_1)$ and $\mathscr{J}(B_1)$ are closed, then the compensated compactness property holds.

Theorem 2.12. *Assume that the subspaces $\mathscr{E}(B_1)$ and $\mathscr{J}(B_1)$ are closed. Then the compensated compactness property holds.*

We begin the proof by establishing a continuous extension property of the closed subspaces $\mathscr{E}(B_1)$ and $\mathscr{J}(B_1)$.

Lemma 2.13. *Let* Ω *be a measurable subset of* $\mathbb{R}^d$. *Then,* $\mathscr{E}(\Omega)$ *is a closed subspace of* $L^2(\Omega; \mathscr{T})$ *if and only if there exists a bounded linear extension operator* $X_{\mathscr{E}}(\Omega)$: $\mathscr{E}(\Omega) \to \mathscr{E}$.

Proof. To prove sufficiency we assume that there exists a bounded linear extension operator $X_{\mathscr{E}}(\Omega)$: $\mathscr{E}(\Omega) \to \mathscr{E}$. For any $\boldsymbol{E} \in \overline{\mathscr{E}(\Omega)}$ there exists $\boldsymbol{E}_n \in \mathscr{E}(\Omega)$, $\boldsymbol{E}_n \to \boldsymbol{E}$, as $n \to \infty$ in $L^2(\Omega; \mathscr{T})$. Then, by the boundedness of the extension operator the sequence $\widetilde{\boldsymbol{E}}_n = X_{\mathscr{E}}(\Omega)\boldsymbol{E}_n \in \mathscr{E}$ is Cauchy in $L^2(\mathbb{R}^d; \mathscr{T})$. Therefore, there exists $\boldsymbol{E}_0 \in L^2(\mathbb{R}^d; \mathscr{T})$, such that $\widetilde{\boldsymbol{E}}_n \to \boldsymbol{E}_0$, as $n \to \infty$ in $L^2(\mathbb{R}^d; \mathscr{T})$. Thus, $\boldsymbol{E}_0 \in \mathscr{E}$, since the subspace $\mathscr{E}$ is closed in $L^2(\mathbb{R}^d; \mathscr{T})$. But then $\boldsymbol{E}_n = \widetilde{\boldsymbol{E}}_n|\Omega \to \boldsymbol{E}_0|\Omega \in \mathscr{E}(\Omega)$, as $n \to \infty$. Hence, $\boldsymbol{E} = \boldsymbol{E}_0|\Omega \in \mathscr{E}(\Omega)$ and, thus, $\mathscr{E}(\Omega)$ is closed.

To prove necessity we assume that $\mathscr{E}(\Omega)$ is a closed subspace of $L^2(\Omega; \mathscr{T})$. Let $D \subset \mathbb{R}^d$ be measurable, and we define

$$\mathscr{E}_0^0(D) = X_0(D)\mathscr{E}_0(D) = \{X_0(D)\boldsymbol{E}: \boldsymbol{E} \in \mathscr{E}_0(D)\}. \tag{2.24}$$

In other words, $\mathscr{E}_0^0(D)$ consists of functions in $\mathscr{E}_0(D)$ extended by 0 to all of $\mathbb{R}^d$. It is obvious that $\mathscr{E}_0^0(D)$ is a closed subspace of $\mathscr{E}$. Our goal is to construct the bounded inverse of the restriction operator $\mathscr{E} \ni \boldsymbol{E} \mapsto \boldsymbol{E}|\Omega$. The kernel of that operator is the subspace $\mathscr{E}_0^0(\Omega^c) \subset \mathscr{E}$. Hence, we need to replace $\mathscr{E}$ with the orthogonal complement V of $\mathscr{E}_0^0(\Omega^c)$ in $\mathscr{E}$, and show that the restriction map $V \ni \boldsymbol{v} \mapsto R\boldsymbol{v} = \boldsymbol{v}|\Omega$ has a bounded inverse. It is obvious that R: $V \to \mathscr{E}(\Omega)$ is a bounded linear map. Let us show that it is also a bijection. If $R\boldsymbol{v} = 0$, then $\boldsymbol{v} \in \mathscr{E}_0^0(\Omega^c)$. Hence, $\boldsymbol{v} = 0$, since $\boldsymbol{v} \in V$. It follows that R is an injective map. Let us show that R is also surjective. Let $\boldsymbol{E} \in \mathscr{E}$ be arbitrary. To prove surjectivity of R we need to show that there exists $\boldsymbol{v} \in V$, such that $R\boldsymbol{v} = \boldsymbol{E}|\Omega$. Let $\boldsymbol{E}_0 \in \mathscr{E}_0^0(\Omega^c)$ be the orthogonal projection of $\boldsymbol{E}$ onto $\mathscr{E}_0^0(\Omega^c)$. Then $\boldsymbol{v} = \boldsymbol{E} - \boldsymbol{E}_0 \in V$ and $R\boldsymbol{v} = \boldsymbol{E}|\Omega$ because $\boldsymbol{E}_0|\Omega = 0$. We have proved that R is a bijection. The subspace V of $\mathscr{E}$ is closed by construction, and $\mathscr{E}(\Omega)$ is closed by assumption. Therefore, by the Banach inverse operator theorem R^{-1}: $\mathscr{E}(\Omega) \to V \subset \mathscr{E}$ is a bounded extension operator. □

Proof of theorem 2.12.

Recall from (2.9) that Γ is a Fourier multiplier operator with continuous and homogeneous of degree zero symbol, while $\phi \in C_0(\mathbb{R}^d)$. Therefore, by the first commutation lemma [14]

$$\mathsf{K}\boldsymbol{f} = \phi\Gamma\boldsymbol{f} - \Gamma(\phi\boldsymbol{f})$$

is a compact operator on $L^2(\mathbb{R}^d; \mathscr{T})$. To complete the proof of the theorem it remains to observe that

$$K_\phi = (1 - \Gamma)\mathsf{K}X_{\mathscr{E}}(B_1), \quad T_\phi = \Gamma\mathsf{K}X_{\mathscr{J}}(B_1),$$

showing that, in view of Lemma 2.13, operators K_ϕ and T_ϕ are compact for any $\phi \subset C_0(B_1)$. Hence, by Lemma 2.10, the compensated compactness property holds.

The theorem is proved now. □

While the closedness of the subspaces $\mathscr{E}(B_1)$ and $\mathscr{J}(B_1)$ is a natural property, currently there are no general methods for checking if spaces $\mathscr{E}(B_1)$ and $\mathscr{J}(B_1)$ are closed. Nonetheless, in contexts of most interest, such as conductivity and elasticity, and by corollary, for problems coupling a number of electric and strain fields, the closedness of the spaces $\mathscr{E}(B_1)$ and $\mathscr{J}(B_1)$ can be established directly. This is done in chapter 13.

There is another characterization of compensated compactness in terms of harmonic functions.

Definition 2.14. *The space* $\mathscr{H}(\Omega) = \overline{\mathscr{E}(\Omega)} \cap \overline{\mathscr{J}(\Omega)}$ *is called the space of* harmonic functions *on* Ω.

Theorem 2.15. *The compensated compactness postulate 2.3 holds if and only if the restriction operator* R: $\mathscr{H}(\Omega_2) \to \mathscr{H}(\Omega_1)$ *is compact, whenever* Ω_1 *and* Ω_2 *are open and* $\Omega_1 \Subset \Omega_2$.

Proof. Let us first assume that compensated compactness holds. In order to prove compactness of the restriction operator R it is enough to show that it maps any weakly convergent sequence $\boldsymbol{h}_n \in \mathscr{H}(\Omega_2)$ into a strongly convergent sequence. It is sufficient to consider only $\boldsymbol{h}_n \rightharpoonup 0$. Let $\phi \in C_0(\Omega_2)$ be such that $\phi(\boldsymbol{x}) = 1$ for all $\boldsymbol{x} \in \Omega_1$ and $\phi(\boldsymbol{x}) \geqslant 0$ for all $\boldsymbol{x} \in \Omega_2$. Then, by compensated compactness property we have

$$\lim_{n\to\infty} \int_{\Omega_2} \phi(\boldsymbol{x})(\boldsymbol{h}_n(\boldsymbol{x}), \boldsymbol{h}_n(\boldsymbol{x}))d\boldsymbol{x} = 0.$$

But then

$$\|R\boldsymbol{h}_n\|^2_{L^2(\Omega_1)} \leqslant \int_{\Omega_2} \phi(\boldsymbol{x})(\boldsymbol{h}_n(\boldsymbol{x}), \boldsymbol{h}_n(\boldsymbol{x}))d\boldsymbol{x},$$

and hence, $R\boldsymbol{h}_n \to 0$ in $L^2(\Omega_1)$.

Let us now prove the converse. We now have two weakly convergent sequences $\boldsymbol{f}_n \in \overline{\mathscr{E}(\Omega)}$ and $\boldsymbol{g}_n \in \overline{\mathscr{J}(\Omega)}$, $\boldsymbol{f}_n \rightharpoonup 0$, $\boldsymbol{g}_n \rightharpoonup 0$. Let $\phi \in C_0(\Omega')$ for some $\Omega' \Subset \Omega$ and let R: $\mathscr{H}(\Omega) \to \mathscr{H}(\Omega')$ be the compact restriction operator. Let

$$\boldsymbol{F}_n = \boldsymbol{f}_n - \Gamma_\Omega \boldsymbol{f}_n \in \mathscr{H}(\Omega), \qquad \boldsymbol{G}_n = \boldsymbol{g}_n - \Pi_\Omega \boldsymbol{g}_n \in \mathscr{H}(\Omega).$$

Then

$$\int_\Omega \phi(\boldsymbol{f}_n, \boldsymbol{g}_n)d\boldsymbol{x} = \int_{\Omega'} \phi\{(R\boldsymbol{F}_n, \boldsymbol{g}_n) + (\Gamma_\Omega \boldsymbol{f}_n, R\boldsymbol{G}_n) + (\Gamma_\Omega \boldsymbol{f}_n, \Pi_\Omega \boldsymbol{g}_n)\}d\boldsymbol{x}.$$

Compensated compactness then follows, if we can prove that

$$\lim_{n\to\infty} \int_\Omega \phi(\Gamma_\Omega \boldsymbol{f}_n, \Pi_\Omega \boldsymbol{g}_n)d\boldsymbol{x} = 0. \tag{2.25}$$

The proof of this statement follows the proof of theorem 2.12. The key observation is that $\boldsymbol{E}_n = X_0(\Omega)\Gamma_\Omega \boldsymbol{f}_n \in \mathscr{E}$, $\boldsymbol{J}_n = X_0(\Omega)\Pi_\Omega \boldsymbol{g}_n \in \mathscr{J}$. We also have $\boldsymbol{E}_n \rightharpoonup 0$ and $\boldsymbol{J}_n \rightharpoonup 0$ in $L^2(\mathbb{R}^d; \mathscr{T})$. But then, using the orthogonality of $\mathscr{E}$ and $\mathscr{J}$, we have

$$\int_\Omega \phi(\Gamma_\Omega \boldsymbol{f}_n, \Pi_\Omega \boldsymbol{g}_n) d\boldsymbol{x} = \int_{\mathbb{R}^d} \phi(\boldsymbol{E}_n, \boldsymbol{J}_n) d\boldsymbol{x} = \int_{\mathbb{R}^d} (\mathsf{K}\boldsymbol{E}_n, \boldsymbol{J}_n) d\boldsymbol{x},$$

where $\mathsf{K}\boldsymbol{f} = \phi(\Gamma \boldsymbol{f}) - \Gamma(\phi \boldsymbol{f})$ is a compact operator by the first commutation lemma [14]. The statement of the theorem now follows. □

2.6 Geometry of local spaces

In this section we build technical tools needed for rigorous analysis of the homogenization problem in the next chapter. It is hoped that mathematically inclined readers will enjoy this section, since it discusses novel mathematical questions that in classical contexts have been obviated by direct access to potentials. In the interest of tracing logical dependencies between different parts of the theory we will assume only postulate 2.1 throughout the entire theoretical development, while any additional assumptions will always be stated explicitly.

2.6.1 Symbol non-degeneracy

Definition 2.16. *We say that the symbol* $\boldsymbol{\Gamma}(\boldsymbol{\eta})$ *is* $\mathscr{E}$-nondegenerate *if* $\boldsymbol{\Gamma}(\boldsymbol{\eta})\boldsymbol{e}_0 = 0$ *for all* $\boldsymbol{\eta} \in \mathbb{S}^{d-1}$ *implies that* $\boldsymbol{e}_0 = 0$, *or equivalently,* $\mathscr{T}_\mathscr{E} = \mathscr{T}$, *where*

$$\mathscr{T}_\mathscr{E} = \sum_{\boldsymbol{\eta} \in \mathbb{S}^{d-1}} \mathscr{E}_{\boldsymbol{\eta}}. \tag{2.26}$$

Similarly, the symbol $\boldsymbol{\Gamma}(\boldsymbol{\eta})$ *is* $\mathscr{J}$-nondegenerate *if* $\boldsymbol{\Gamma}(\boldsymbol{\eta})\boldsymbol{e}_0 = \boldsymbol{e}_0$ *for all* $\boldsymbol{\eta} \in \mathbb{S}^{d-1}$ *implies that* $\boldsymbol{e}_0 = 0$, *or equivalently,* $\mathscr{T}_\mathscr{J} = \mathscr{T}$, *where*

$$\mathscr{T}_\mathscr{J} = \sum_{\boldsymbol{\eta} \in \mathbb{S}^{d-1}} \mathscr{J}_{\boldsymbol{\eta}}. \tag{2.27}$$

We remark that if the spaces $\mathscr{E}$ and $\mathscr{J}$ are given by linear differential constraints (2.23), then $\mathscr{E}$-nondegeneracy is equivalent to the operator $\mathscr{A}$ being *cocanceling* or the operator $\mathscr{B}$ being *canceling*, [15]. Similarly, $\mathscr{J}$ nondegeneracy is equivalent to the operator $\mathscr{A}^*$ being canceling or the operator $\mathscr{B}^*$ being cocanceling.

One can easily verify that the symbols $\boldsymbol{\Gamma}(\boldsymbol{\eta})$ for conductivity and elasticity are both $\mathscr{E}$ and $\mathscr{J}$-nondegenerate. We will soon see that fields in $\mathscr{E}(\Omega)$ take values in $\mathscr{T}_\mathscr{E}$, so that from the point of view of the boundary value problem (2.19), only the restriction of $\mathsf{L}(\boldsymbol{x})$ to $\mathscr{T}_\mathscr{E}$ plays any role. In that sense we can simply replace $\mathscr{T}$ with $\mathscr{T}_\mathscr{E}$ and assume, without loss of generality, that the symbol $\boldsymbol{\Gamma}$ is always $\mathscr{E}$-nondegenerate.

2.6.2 Density theorems

Theorem 2.17. *Let Ω be a measurable subset of $\mathbb{R}^d$. Then the spaces $C^\infty(\overline{\Omega}; \mathcal{T}) \cap \mathcal{E}(\Omega)$ and $C^\infty(\overline{\Omega}; \mathcal{T}) \cap \mathcal{J}(\Omega)$ are dense in $\mathcal{E}(\Omega)$ and $\mathcal{J}(\Omega)$, respectively.*

Proof. Let ρ_ε be a standard convolution kernel. Then $\boldsymbol{E}_\varepsilon = \rho_\varepsilon * \boldsymbol{E} \in C^\infty(\mathbb{R}^d; \mathcal{T}) \cap \mathcal{E}$ for any $\boldsymbol{E} \in \mathcal{E}$. Thus, $\boldsymbol{E}_\varepsilon|\Omega \in C^\infty(\overline{\Omega}; \mathcal{T}) \cap \mathcal{E}(\Omega)$. Moreover, $\boldsymbol{E}_\varepsilon \to \boldsymbol{E}$ in $L^2(\mathbb{R}^d; \mathcal{T})$. Hence, $\boldsymbol{E}_\varepsilon|\Omega \to \boldsymbol{E}|\Omega$ in $L^2(\Omega; \mathcal{T})$. The statement about $\mathcal{J}(\Omega)$ is proved in the same way. □

Next, we show that compensated compactness guarantees the density of smooth, compactly supported functions in $\mathcal{E}$ and $\mathcal{J}$.

Theorem 2.18. *Suppose that the compensated compactness postulate 2.3 holds. Then, the set of smooth compactly supported functions*

$$\mathscr{C}^\infty(\mathcal{E}) = \bigcup_{n=1}^{\infty} \mathcal{E}_0^\infty(B_n) \tag{2.28}$$

is dense in $\mathcal{E}$, where $\mathcal{E}_0^\infty(D) = \mathcal{E} \cap C_0^\infty(D; \mathcal{T})$. The statement with $\mathcal{E}$ and $\mathcal{J}$ interchanged is also true.

Proof. We begin by proving several simple lemmas about weak convergence in Lebesgue spaces.

Lemma 2.19. *Let $\phi \in C(\overline{B_1})$. We define*

$$\psi_n(\boldsymbol{x}) = \begin{cases} \phi\left(\dfrac{\boldsymbol{x}}{n}\right), & \boldsymbol{x} \in B_n, \\ 0, & \boldsymbol{x} \in \mathbb{R}^d \setminus B_n. \end{cases}$$

Then $\psi_n \overset{*}{\rightharpoonup} \phi(0)$, as $n \to \infty$, in $L^\infty(\mathbb{R}^d)$.

Proof. It is obvious that

$$\lim_{n\to\infty} \psi_n(\boldsymbol{x}) = \phi(0)$$

for every $\boldsymbol{x} \in \mathbb{R}^d$. Let $\eta \in L^1(\mathbb{R}^d)$ be arbitrary. Then $|\psi_n(\boldsymbol{x})\eta(\boldsymbol{x})| \leqslant \|\phi\|_{C(\overline{B_1})}|\eta(\boldsymbol{x})|$. Thus, by the Lebesgue dominated convergence theorem

$$\lim_{n\to\infty} \int_{\mathbb{R}^d} \psi_n(\boldsymbol{x})\eta(\boldsymbol{x})d\boldsymbol{x} = \phi(0)\int_{\mathbb{R}^d} \eta(\boldsymbol{x})d\boldsymbol{x}.$$

The lemma is proved. □

Lemma 2.20. *Let $\phi \in C(\overline{B_1})$. We define*

$$\psi_n(\boldsymbol{x}) = \begin{cases} n^{-d/2}\phi\left(\dfrac{\boldsymbol{x}}{n}\right), & \boldsymbol{x} \in B_n, \\ 0, & \boldsymbol{x} \in \mathbb{R}^d \setminus B_n. \end{cases}$$

Then $\psi_n \rightharpoonup 0$, as $n \to \infty$, in $L^2(\mathbb{R}^d)$.

Proof. It is easy to verify that ψ_n is a bounded sequence in $L^2(\mathbb{R}^d)$. It is obvious that

$$\lim_{n\to\infty} \psi_n(\boldsymbol{x}) = 0$$

for every $\boldsymbol{x} \in \mathbb{R}^d$. Let $\eta \in L^1(\mathbb{R}^d) \cap L^2(\mathbb{R}^d)$ be arbitrary. Then $|\psi_n(\boldsymbol{x})\eta(\boldsymbol{x})| \leqslant \|\phi\|_{C(\overline{B_1})} |\eta(\boldsymbol{x})|$. Thus, by the Lebesgue dominated convergence theorem

$$\lim_{n\to\infty} \int_{\mathbb{R}^d} \psi_n(\boldsymbol{x})\eta(\boldsymbol{x}) d\boldsymbol{x} = 0.$$

The statement of lemma follows from the density of $L^1(\mathbb{R}^d) \cap L^2(\mathbb{R}^d)$ in $L^2(\mathbb{R}^d)$ and the boundedness of the sequence ψ_n in $L^2(\mathbb{R}^d)$. □

Lemma 2.21. *Let $f \in L^2(\mathbb{R}^d)$. We define $g_n(z) = n^{d/2} f(nz)$, $z \in B_1$. Then $g_n \rightharpoonup 0$, as $n \to \infty$, in $L^2(B_1)$.*

Proof. It is easy to check that $\|g_n\|_{L^2(B_1)} \leqslant \|f\|_{L^2(\mathbb{R}^d)}$. For every $\phi \in C(\overline{B_1})$ we have

$$(g_n, \phi)_{L^2(B_1)} = n^{-d/2} \int_{B_n} f(\boldsymbol{x})\phi\left(\frac{\boldsymbol{x}}{n}\right) d\boldsymbol{x} \to 0,$$

as $n \to \infty$, by lemma 2.20. The statement of lemma follows from the density of $C(\overline{B_1})$ in $L^2(B_1)$ and the boundedness of the sequence g_n in $L^2(B_1)$. □

We are now ready to prove theorem 2.18. Let $\boldsymbol{E}_0 \in \mathscr{E}$ be orthogonal to all functions in $\mathscr{E}_0^0(B_n)$, $n \geqslant 1$, which denotes functions in $\mathscr{E}_0(B_n)$, extended by 0 to all of $\mathbb{R}^d$. Observe that $\boldsymbol{E}_0|B_n \in \overline{\mathscr{J}(B_n)}$, since $\boldsymbol{E}_0|B_n$ is orthogonal to $\mathscr{E}_0(B_n)$. At the same time, $\boldsymbol{E}_0|B_n \in \mathscr{E}(B_n)$, since $\boldsymbol{E}_0 \in \mathscr{E}$. Then, the sequence $\boldsymbol{E}_n(z) = n^{d/2}\boldsymbol{E}_0(nz)$, $z \in B_1$, satisfies $\boldsymbol{E}_n \in \overline{\mathscr{J}(B_1)} \cap \mathscr{E}(B_1)$ for every $n \geqslant 1$. By lemma 2.21, $\boldsymbol{E}_n \rightharpoonup 0$ in $L^2(B_1; \mathscr{T})$. But then, by theorem 2.9, we must have

$$\lim_{n\to\infty} \int_{B_1} \phi(z)(\boldsymbol{E}_n(z), \boldsymbol{E}_n(z)) dz = 0$$

for every $\phi \in C_0(B_1)$. However,

$$(\phi \boldsymbol{E}_n, \boldsymbol{E}_n)_{L^2(B_1)} = \int_{B_n} \phi\left(\frac{\boldsymbol{x}}{n}\right) |\boldsymbol{E}_0(\boldsymbol{x})|^2 d\boldsymbol{x},$$

and by lemma 2.19

$$\lim_{n\to\infty} \int_{B_n} \phi\left(\frac{\boldsymbol{x}}{n}\right) |\boldsymbol{E}_0(\boldsymbol{x})|^2 d\boldsymbol{x} = \phi(0) \|\boldsymbol{E}_0\|_{L^2(\mathbb{R}^d)}.$$

We conclude that $\boldsymbol{E}_0 = 0$, and the set of functions $\bigcup_{n=1} \mathscr{E}_0^0(B_n)$ is dense in $\mathscr{E}$. We finish the proof by observing that for any $\boldsymbol{e} \in \mathscr{E}_0^0(B_n)$ functions $\boldsymbol{e}_\varepsilon = \rho_\varepsilon * \boldsymbol{e} \in \mathscr{E}_0^\infty(B_{n+1})$, where ρ_ε is a standard convolution kernel, and $\varepsilon \in (0, 1)$. The theorem follows from the fact that $\boldsymbol{e}_\varepsilon \to \boldsymbol{e}$ in $L^2(\mathbb{R}^d; \mathscr{T})$, as $\varepsilon \to 0$. □

When the spaces $\mathscr{E}$ and $\mathscr{J}$ are given in terms of differential constraints, as in postulate 2.4, we can prove a much more detailed density theorem.

Theorem 2.22. *Under assumptions of postulate 2.4 the set of compactly supported functions*

$$\mathscr{E}_c = \{\mathscr{A}^*\mathbf{\Phi}: \mathbf{\Phi} \in C_0^\infty(\mathbb{R}^d; \mathscr{T})\} \subset \mathscr{C}^\infty(\mathscr{E})$$

is dense in $\mathscr{E}$.

Proof. Let $\boldsymbol{E} \in \mathscr{E}$ be arbitrary. Let $\phi_\varepsilon(\boldsymbol{x})$ be given by its Fourier transform

$$\hat{\phi}_\varepsilon(\boldsymbol{\xi}) = \chi_{\hat{B}_\varepsilon^c}(\boldsymbol{\xi})\boldsymbol{A}^{\dagger *}(\boldsymbol{\xi})\hat{\boldsymbol{E}}(\boldsymbol{\xi}).$$

Since $\boldsymbol{A}^\dagger(\boldsymbol{\xi})$ is bounded on the unit sphere and homogeneous of degree $-m$ the function $\sqrt{1 + |\boldsymbol{\xi}|^{2m}}\,\boldsymbol{A}^\dagger(\boldsymbol{\xi})$ is bounded on $\hat{B}_\varepsilon^c$. Therefore, $\sqrt{1 + |\boldsymbol{\xi}|^{2m}}\;\hat{\phi}_\varepsilon(\boldsymbol{\xi}) \in L^2(\mathbb{R}^d; \mathscr{T})$ for any $\varepsilon > 0$. Hence (see, e.g., [1]) $\phi_\varepsilon \in H^m(\mathbb{R}^d; \mathscr{T})$. We also see that

$$\lim_{\varepsilon\to 0} \boldsymbol{A}^*(\boldsymbol{\xi})\hat{\phi}_\varepsilon(\boldsymbol{\xi}) = \lim_{\varepsilon\to 0} \chi_{\hat{B}_\varepsilon^c}(\boldsymbol{\xi})\boldsymbol{\Gamma}(\boldsymbol{\xi})\hat{\boldsymbol{E}}(\boldsymbol{\xi}) = \boldsymbol{\Gamma}(\boldsymbol{\xi})\hat{\boldsymbol{E}}(\boldsymbol{\xi}) = \hat{\boldsymbol{E}}(\boldsymbol{\xi}),$$

where the limit as $\varepsilon \to 0$ is understood in the sense of $L^2(\mathbb{R}^d)$. This shows that the space

$$\mathscr{C}'(\mathscr{E}) = \{\mathscr{A}^*\phi: \phi \in H^m(\mathbb{R}^d; \mathscr{T})\}$$

is dense in $\mathscr{E}$. It is well-known that $C_0^\infty(\mathbb{R}^d)$ is dense in $H^m(\mathbb{R}^d)$. Furthermore, if $C_0^\infty(\mathbb{R}^d; \mathscr{T}) \ni \psi_n \to \phi \in H^m(\mathbb{R}^d; \mathscr{T})$, as $n \to \infty$ in $H^m(\mathbb{R}^d; \mathscr{T})$, then $\mathscr{A}^*\psi_n \to \mathscr{A}^*\phi$, as $n \to \infty$ in $L^2(\mathbb{R}^d; \mathscr{T})$. Thus, $\mathscr{E}_c$ is dense in $\mathscr{C}'(\mathscr{E})$. The L^2 topology comes from a norm. Therefore, $\mathscr{E}_c$ is dense in $\mathscr{E}$. The statement about $\mathscr{J}$ is proved by the observation made in the proof of theorem 2.11 that the subspace $\mathscr{E}$ can be described as

$$\mathscr{E} = \{\boldsymbol{f} \in L^2(\mathbb{R}^d; \mathscr{T}): \mathscr{B}^*\boldsymbol{f} = 0 \text{ in } \mathscr{D}'\}, \tag{2.29}$$

for some linear homogeneous differential operator $\mathscr{B}$ with constant rank property, satisfying $\mathscr{R}(\boldsymbol{B}(\boldsymbol{\xi})) = \ker \boldsymbol{A}(\boldsymbol{\xi})$ for all $\boldsymbol{\xi} \neq 0$. □

Corollary 2.23. *Under the assumptions of theorem 2.22, let* Ω *and* Ω' *be open subsets of* $\mathbb{R}^d$*, such that* $\overline{\Omega} \subset \Omega'$*. Then the sets of functions*

$$\mathscr{E}_0^\infty(\Omega, \Omega') = \{\mathscr{A}^*\mathbf{\Phi}|\Omega: \mathbf{\Phi} \in C_0^\infty(\Omega'; \mathscr{T})\},$$

$$\mathscr{J}_0^\infty(\Omega, \Omega') = \{\mathscr{B}\mathbf{\Phi}|\Omega: \mathbf{\Phi} \in C_0^\infty(\Omega'; \mathscr{T})\}$$

are dense in $\mathscr{E}(\Omega)$ *and* $\mathscr{J}(\Omega)$*, respectively.*

Proof. For any $\boldsymbol{E} \in \mathscr{E}(\Omega)$ there exists $\widetilde{\boldsymbol{E}} \in \mathscr{E}$, such that $\widetilde{\boldsymbol{E}}|\Omega = \boldsymbol{E}$. By theorem 2.22 there exists a sequence $\mathbf{\Phi}_n \in C_0^\infty(\mathbb{R}^d; \mathscr{T})$, such that $\mathscr{A}^*\mathbf{\Phi}_n \to \widetilde{\boldsymbol{E}}$ in $L^2(\mathbb{R}^d; \mathscr{T})$. Let $\eta \in C_0^\infty(\Omega')$ be such that $\eta(\boldsymbol{x}) = 1$ for all $\boldsymbol{x} \in \Omega$. Then

$$\boldsymbol{E}_n = \mathscr{A}^*\mathbf{\Phi}_n|\Omega = \mathscr{A}^*(\eta\mathbf{\Phi}_n)|\Omega,$$

and $E_n \to E$ in $L^2(\Omega; \mathcal{T})$. Moreover, $\eta\Phi \in C_0^\infty(\Omega'; \mathcal{T})$. The statement for $\mathcal{J}(\Omega)$ is proved in the same way. □

Corollary 2.24. *Under the assumptions of corollary 2.23 the sets of functions*

$$\{E|\Omega\colon E \in \mathcal{E}_0(\Omega')\}, \qquad \{J|\Omega\colon E \in \mathcal{J}_0(\Omega')\}.$$

are dense in $\mathcal{E}(\Omega)$ and $\mathcal{J}(\Omega)$, respectively.

2.6.3 Pointwise behavior of the fields †

We now proceed to establish pointwise properties of the space $\mathcal{E}$.

Theorem 2.25. *Assuming only postulate 2.1, the following statements are true.*

(a) *If $E \in \mathcal{E}$ then $E(x) \in \mathcal{T}_{\mathcal{E}}$ for a.e. $x \in \mathbb{R}^d$, where $\mathcal{T}_{\mathcal{E}}$ is defined in (2.26).*

(b) *Assume that Ω is bounded. Then for any $e_0 \in \mathcal{T}_{\mathcal{E}}$ the function $E_0(x) = e_0$ for all $x \in \Omega$ is in $\overline{\mathcal{E}(\Omega)}$.*

(c) *Assume that Ω is an open and bounded subset of $\mathbb{R}^d$ and $x_0 \in \Omega$ is fixed. Then*

$$\mathrm{Span}\{E(x_0)\colon E \in \mathcal{E}_0(\Omega),\ \textit{such that } x_0 \textit{ is a Lebesgue point of } E\}$$

is the same subspace $V_{\mathcal{E}} \subset \mathcal{T}_{\mathcal{E}}$ for all open and bounded subsets $\Omega \subset \mathbb{R}^d$ and all $x_0 \in \Omega$. Moreover,

$$V_{\mathcal{E}} = \mathrm{Span}\{E_1(x_0), E_2(x_0), \dots\}, \quad x_0 \in \Omega_0^* \subset \Omega,$$

where $\{E_j\}$ is any countable dense subset of $\mathcal{E}_0(\Omega)$, all points of Ω_0^* are common Lebesgue points of $\{E_j\}$, and Ω_0^* differs from Ω by a set of measure zero.

(d) *If $\mathcal{C}^\infty(\mathcal{E})$, given by (2.28) is dense in $\mathcal{E}$ then $V_{\mathcal{E}} = \mathcal{T}_{\mathcal{E}}$.*

Proof. Part (a). Let $E \in \mathcal{E}$. Then

$$E(x) = \int_{\mathbb{R}^d} \hat{E}(\xi)e^{2\pi i x\cdot\xi}d\xi \in (\mathcal{T}_{\mathcal{E}} \otimes \mathbb{C}) \bigcap \mathcal{T} = \mathcal{T}_{\mathcal{E}},$$

since $\hat{E}(\xi) \in \mathcal{E}_{\xi/|\xi|} \otimes \mathbb{C}$ for a.e. $\xi \in \mathbb{R}^d$.

Part (b). For any $J \in \mathcal{J}_0(\Omega)$ we have $J_0 = X_0(\Omega)J \in L^1(\mathbb{R}^d; \mathcal{T})$, since Ω is a bounded set. Then $\hat{J}_0(\xi)$ is continuous on $\mathbb{R}^d$ and for any $\varepsilon > 0$ and $\xi \neq 0$ we have

$$0 = \lim_{\varepsilon\to 0}\Gamma\left(\frac{\varepsilon\xi}{|\varepsilon\xi|}\right)\hat{J}_0(\varepsilon\xi) = \Gamma\left(\frac{\xi}{|\xi|}\right)\hat{J}_0(0),$$

since $J_0 \in \mathcal{J}$. Thus,

$$\hat{J}_0(0) \in \bigcap_{|\eta|=1} \mathcal{E}_\eta^\perp = \mathcal{T}_{\mathcal{E}}^\perp.$$

Thus, for all $e_0 \in \mathcal{T}_{\mathcal{E}}$ and any $J \in \mathcal{J}_0(\Omega)$ we have

$$(J, E_0)_{L^2(\Omega)} = \left(e_0, \int_{\mathbb{R}^d} J_0(x)dx\right) = (e_0, \hat{J}_0(0)) = 0,$$

where $E_0(x) = e_0$ for all $x \in \Omega$. Therefore, $E_0 \in (\mathcal{J}_0(\Omega))^{\perp} = \overline{\mathcal{E}(\Omega)}$.

Part (c). Before the statement of part (c) is proved let us temporarily denote[3]

$$V_x(\Omega) = \text{Span}\{E(x)\colon E \in \mathcal{E}_0(\Omega), \text{ so that } x \text{ is a Lebesgue point of } E\}.$$

Let

$$\mathcal{N}(\Omega) = \{f\colon \Omega \to \mathcal{T}\colon \forall E \in \mathcal{E}_0(\Omega)\ (f(x), E(x)) = 0 \text{ for a.e. } x \in \Omega\}. \quad (2.30)$$

Here we do not require $f(x)$ to be a measurable function. Yet, we still identify any two $\mathcal{T}$-valued functions that differ on a set of Lebesgue measure zero. The subspace $\mathcal{N}(\Omega)$ of $\mathcal{T}$-valued functions has the following properties: $\chi_D(x)f \in \mathcal{N}(\Omega)$, for any $D \subset \Omega$, provided $f \in \mathcal{N}(\Omega)$. Also, $f(\alpha x + \beta) \in \mathcal{N}(\Omega)$, when $f \in \mathcal{N}(\Omega)$, provided $\alpha\Omega + \beta \subset \Omega$. Indeed, for any $E \in \mathcal{E}_0(\Omega)$ the function

$$\widetilde{E}(x) = (X_0(\Omega)E)((x - \beta)/\alpha)|\Omega \in \mathcal{E}_0(\Omega).$$

Therefore, $(f(x), \widetilde{E}(x)) = 0$ for a.e. $x \in \Omega$. But then for a.e. $x \in \Omega$ we also have

$$0 = (f(\alpha x + \beta), \widetilde{E}(\alpha x + \beta)) = (f(\alpha x + \beta), E(x)).$$

We conclude that $f(\alpha x + \beta) \in \mathcal{N}(\Omega)$.

Let $\{E_1, E_2, \ldots\}$ be a countable dense subset of $\mathcal{E}_0(\Omega)$. Let $\Omega_0 \subset \Omega$ be the set of all common Lebesgue points of $\{E_1, E_2, \ldots\}$. Let

$$\widetilde{V}_x = \text{Span}\{E_n(x)\colon n \geqslant 1\}, \qquad x \in \Omega_0.$$

We now define the map $\mathcal{P}\colon \Omega_0 \to \text{Sym}(\mathcal{T})$ such that $\mathcal{P}(x)$ is the orthogonal projection onto $\widetilde{V}_x^{\perp}$.

Lemma 2.26. *The function $\mathcal{P}\colon \Omega_0 \to \text{Sym}(\mathcal{T})$ such that $\mathcal{P}(x)$ is the orthogonal projection onto $\widetilde{V}_x^{\perp}$ is Lebesgue measurable.*

Proof. Let $m = \dim \mathcal{T}$. Let $\mathfrak{I}$ be the set of all finite, nonempty subsets of $\mathbb{N}$ of length at most m. The set $\mathfrak{I}$ is countable. For each $I = \{n_1, \ldots, n_k\} \in \mathfrak{I}$, let E_I denote the list $\{E_{n_1}, \ldots, E_{n_k}\}$. Then for every $I \in \mathfrak{I}$, let

$$\Omega_I = \{x \in \Omega_0\colon E_I \text{ is a basis of } \widetilde{V}_x\}.$$

All subsets $\Omega_I \subset \Omega_0$ are Lebesgue measurable because

$$\Omega_I = \widetilde{\Omega}_I \bigcap \bigcap_{n \notin I} \Omega_I^{(n)},$$

[3] Our goal is to prove that $V_x(\Omega)$ depends neither on x, nor on Ω.

where

$$\widetilde{\Omega}_I = \{x \in \Omega_0 \colon \bigwedge E_I(x) \neq 0\}, \quad \Omega_I^{(n)} = \{x \in \Omega_0 \colon E_n(x) \wedge \bigwedge E_I(x) = 0\}.$$

Here

$$\bigwedge E_I(x) = E_{n_1}(x) \wedge \cdots \wedge E_{n_k}(x), \quad I = \{n_1, \ldots, n_k\}.$$

For each $x \in \Omega_I$, let $(e_1(x), \ldots e_k(x))$ be the orthonormal bases of $\widetilde{V}_x$ obtained by means of the Gram-Schmidt orthogonalization from $E_I(x)$. Obviously, each function $e_i \colon \Omega_I \to \mathscr{T}$ is measurable. Then, for every $x \in \Omega_I$

$$\mathscr{P}(x) = \mathsf{I} - \sum_{i=1}^{k} e_i(x) \otimes e_i(x)$$

is a measurable function on Ω_I. Moreover, $\mathscr{P}(x)$ does not depend on the choice of the orthonormal basis of $\widetilde{V}_x$. Therefore, the function $\mathscr{P}(x)$, given by the above construction, whenever $x \in \Omega_I$, is well-defined. Thus, $\mathscr{P}(x)$ is measurable on Ω_0, since $\mathfrak{J}$ is countable and

$$\Omega_0 = \bigcup_{I \in \mathfrak{J}} \Omega_I.$$

□

Let $\Omega_0^* \subset \Omega_0$ be the set of all Lebesgue points of $\mathscr{P}(x)$. Let $\{x_0, x_1\} \subset \Omega_0^*$. Let $v \in \widetilde{V}_{x_0}^{\perp}$. Let $p_v(x) = \mathscr{P}(x)v$. Then $p_v(x)$ is measurable and bounded, $p_v(x_0) = v$, and x_0 is a Lebesgue point of $p_v(x)$, since x_0 is a Lebesgue point of $\mathscr{P}(x)$. We claim that $p_v \in \mathscr{N}(\Omega)$. Indeed, by construction $p_v(x) \in \widetilde{V}_x^{\perp}$ for all $x \in \Omega_0^*$. Then $(p_v(x), E_n(x)) = 0$ for all $n \geqslant 1$ and $x \in \Omega_0^*$. Let $E \in \mathscr{E}_0(\Omega)$ and $E_{n_k} \to E$ in $L^2(\Omega; \mathscr{T})$. But then in $L^1(\Omega)$ we have

$$0 = (p_v(x), E_{n_k}(x)) \to (p_v(x), E(x)) \text{ in } L^1(\Omega; \mathscr{T}).$$

Thus $p_v \in \mathscr{N}(\Omega)$. Now for sufficiently small α we have $\alpha(\Omega - x_1) + x_0 \subset \Omega$. Therefore,

$$\widetilde{p}(x) = p_v(\alpha(x - x_1) + x_0) \in \mathscr{N}(\Omega).$$

Also, since the affine function $x \mapsto \alpha(x - x_1) + x_0$ maps x_1 into x_0, we conclude that x_1 is a Lebesgue point of $\widetilde{p}$. We also have $\widetilde{p}(x_1) = p_v(x_0) = v$. Recalling that x_1 is a Lebesgue point of both $E_n(x)$ and $\widetilde{p}(x)$ and that $\widetilde{p}(x)$ is bounded, we conclude that x_1 is also a Lebesgue point of $(\widetilde{p}(x), E_n(x))$. However, $(\widetilde{p}(x), E_n(x)) = 0$ for a.e. $x \in \Omega$, since $\widetilde{p}(x) \in \mathscr{N}(\Omega)$. Thus,

$$0 = \lim_{r \to 0} ⨍_{B_r(x_1)} (\widetilde{p}(x), E_n(x)) dx = (\widetilde{p}(x_1), E_n(x_1)) = (v, E_n(x_1)).$$

Hence, $v \in \widetilde{V}_{x_1}^{\perp}$. We have proved that $\widetilde{V}_{x_0}^{\perp} \subset \widetilde{V}_{x_1}^{\perp}$ for any $\{x_0, x_1\} \subset \Omega_0^*$. Switching the order of x_0 and x_1 we obtain the reverse inclusion. Thus, the subspaces $\widetilde{V}_x$ are independent of x. Let us denote this common subspace by $\widetilde{V}_\Omega$.

We have just proved that $(\boldsymbol{v}, \boldsymbol{E}_n(\boldsymbol{x})) = 0$ for all $\boldsymbol{v} \in \widetilde{V}_\Omega^\perp$ and all $\boldsymbol{x} \in \Omega_0^*$. Therefore, for any $\boldsymbol{E} \in \mathscr{E}_0(\Omega)$ and any of its Lebesgue points $\boldsymbol{x} \in \Omega$ we have $(\boldsymbol{v}, \boldsymbol{E}(\boldsymbol{x})) = 0$ for any $\boldsymbol{v} \in \widetilde{V}_\Omega^\perp$. Therefore, $\boldsymbol{E}(\boldsymbol{x}) \in \widetilde{V}_\Omega$ for a.e. $\boldsymbol{x} \in \Omega$.

Let us prove now that the spaces $V_{\boldsymbol{x}}(\Omega) = \widetilde{V}_\Omega$ are also independent of Ω. Let Ω' be another open and bounded subset of $\mathbb{R}^d$. Let us fix $\boldsymbol{x}_0' \in \Omega'$ and $\boldsymbol{x}_0 \in \Omega$ arbitrarily. For sufficiently small α we have $\alpha(\Omega - \boldsymbol{x}_0) + \boldsymbol{x}_0' \subset \Omega'$. Let $\mathscr{C}(\boldsymbol{x}_0)$ be the set of all $\boldsymbol{E} \in \mathscr{E}_0(\Omega)$ for which $\boldsymbol{x}_0$ is a Lebesgue point. Then for any $\boldsymbol{E} \in \mathscr{C}(\boldsymbol{x}_0)$ we define

$$\widetilde{\boldsymbol{E}}(\boldsymbol{x}') = \boldsymbol{E}\left(\boldsymbol{x}_0 + \frac{\boldsymbol{x}' - \boldsymbol{x}_0'}{\alpha}\right).$$

Then $\boldsymbol{E}' \in \mathscr{E}_0(\Omega')$ and $\boldsymbol{x}_0'$ is its Lebesgue point. Therefore,

$$\widetilde{V}_\Omega = \mathrm{Span}\{\boldsymbol{E}(\boldsymbol{x}_0)\colon \boldsymbol{E} \in \mathscr{C}(\boldsymbol{x}_0)\} = \mathrm{Span}\{\widetilde{\boldsymbol{E}}(\boldsymbol{x}_0')\colon \boldsymbol{E} \in \mathscr{C}(\boldsymbol{x}_0)\} \subset \widetilde{V}_{\Omega'}.$$

Switching the order of Ω and Ω' we obtain the reverse inclusion. Hence, we have proved that $V_{\boldsymbol{x}}(\Omega) = \widetilde{V}_\Omega = V_{\mathscr{E}}$.

Part (d). Let us show that the equality $V_{\mathscr{E}} = \mathscr{T}_{\mathscr{E}}$ is a consequence of the density of $\mathscr{C}(\mathscr{E})$ in $\mathscr{E}$. Suppose $\boldsymbol{e}_0 \in \mathscr{T}_{\mathscr{E}}$. By part (b) of this theorem there exists $\boldsymbol{E}_n \in \mathscr{E}$ such that $\boldsymbol{E}_n|B_1 \to \boldsymbol{e}_0$, as $n \to \infty$ in $L^2(B_1; \mathscr{T})$. By assumption, each $\boldsymbol{E}_n \in \mathscr{E}$ can be approximated arbitrarily well by functions in $\mathscr{C}(\mathscr{E})$. Since the topology of $L^2(B_1; \mathscr{T})$ comes from a norm, there exist a sequence of balls B_{r_n} and $\boldsymbol{E}_n' \in \mathscr{E}_0(B_{r_n})$, such that $\boldsymbol{E}_n'|B_1 \to \boldsymbol{e}_0$ in $L^2(B_1, \mathscr{T})$. But, by part (c) of this theorem $\boldsymbol{E}_n'|B_1 \in L^2(B_1, V_{\mathscr{E}})$, which is a closed subspace of $L^2(B_1, \mathscr{T})$. Therefore, $\boldsymbol{e}_0 \in L^2(B_1, V_{\mathscr{E}})$. In particular, $\boldsymbol{e}_0 \in V_{\mathscr{E}}$. Hence, $\mathscr{T}_{\mathscr{E}} \subset V_{\mathscr{E}}$. The reverse inclusion was proved in part (c). The theorem is proved now. □

Corollary 2.27. *The subspace $\mathscr{N}(\Omega)$ given by (2.30) consists of all $V_{\mathscr{E}}^\perp$-valued functions, where we do not distinguish between two functions that differ on a set of Lebesgue measure zero.*

Proof. We have already proved that all $V_{\mathscr{E}}^\perp$-valued functions are in $\mathscr{N}(\Omega)$. Now, let $\boldsymbol{f} \in \mathscr{N}(\Omega)$. Let $\boldsymbol{e}_1, \ldots, \boldsymbol{e}_k$ be a fixed orthonormal basis of $V_{\mathscr{E}}$. Then, for all $\boldsymbol{x} \in \Omega_0^*$ there exist numbers $\alpha_i^n(\boldsymbol{x})$, such that

$$\boldsymbol{e}_i = \sum_{n=1}^{\infty} \alpha_i^n(\boldsymbol{x}) \boldsymbol{E}_n(\boldsymbol{x}),$$

where all, but finitely many numbers $\alpha_i^n(\boldsymbol{x})$ are zero for each $\boldsymbol{x} \in \Omega_0^*$. But then, by definition of $\mathscr{N}(\Omega)$, there exists a set $G \subset \Omega_0^*$ of full measure, such that $(\boldsymbol{f}(\boldsymbol{x}), \boldsymbol{E}_n(\boldsymbol{x})) = 0$ for $n \geqslant 1$, and all $\boldsymbol{x} \in G$. But then for every $\boldsymbol{x} \in G$

$$(\boldsymbol{f}(\boldsymbol{x}), \boldsymbol{e}_i) = \sum_{n=1}^{\infty} \alpha_i^n(\boldsymbol{x})(\boldsymbol{f}(\boldsymbol{x}), \boldsymbol{E}_n(\boldsymbol{x})) = 0.$$

But that means that $\boldsymbol{f}(\boldsymbol{x}) \in V_{\mathscr{E}}^\perp$ for all $\boldsymbol{x} \in G$. □

Corollary 2.28. *Combining theorem 2.18 with part (d) of theorem 2.25 we deduce that the compensated compactness property implies* $V_{\mathcal{E}} = \mathcal{T}_{\mathcal{E}}$.

References

[1] Calderón A 1961 Lebesgue spaces of differentiable functions *Proc. Symp. Pure Math.* **4** 33–49

[2] Campbell S L and Meyer C D 2009 *Generalized Inverses of Linear Transformations* (Philadelphia, PA: SIAM)

[3] Dell'Antonio G F, Figari R and Orlandi E 1986 An approach through orthogonal projections to the study of inhomogeneous random media with linear response *Ann. Inst. Henri Poincaré* **44** 1–28

[4] Fonseca I and Müller S 1999 A-quasiconvexity, lower semicontinuity and young measures *Siam J. Math. Anal.* **30** 1355–90

[5] Guerra A 2022 Quasiconvexity, null Lagrangians, and Hardy space integrability under constant rank constraints *Arch. Ration. Mech. Anal.* **245** 279–320

[6] Kohler W and Papanicolaou G C 1982 Bounds for effective conductivity of random media *Macroscopic Properties of Disordered Media* ed R Burridge, S Childress and G Papanicolaou (Berlin: Springer) 111–30

[7] Milton G W 1987 Multicomponent composites, electrical networks and new types of continued fraction I *Commun. Math. Phys.* **111** 281–327

[8] Milton G W 1987 Multicomponent composites, electrical networks and new types of continued fraction II *Commun. Math. Phys.* **111** 329–72

[9] Milton G W 1990 On characterizing the set of possible effective tensors of composites: the variational method and the translation method *Commun. Pure Appl. Math.* **43** 63–125

[10] Milton G W 2002 *The Theory of Composites* Cambridge Monographs on Applied and Computational Mathematics (Cambridge: Cambridge University Press)

[11] Milton G W and Kohn R V 1988 Variational bounds on the effective moduli of anisotropic composites *J. Mech. Phys. Solids* **36** 597–629

[12] Raiţă B 2019 Potentials for $\mathscr{A}$-quasiconvexity *Calc. Var. Partial Differ. Equ.* **58** 1–16

[13] Shipman S P and Welters A T 2013 Resonant electromagnetic scattering in anisotropic layered media *J. Math. Phys.* **54** 103511

[14] Tartar L 1990 H-measures, a new approach for studying homogenisation, oscillations and concentration effects in partial differential equations *Proc. R. Soc. Edinburgh* A **115** 193–230

[15] Van Schaftingen J 2013 Limiting Sobolev inequalities for vector fields and canceling linear differential operators *J. Eur. Math. Soc.* **15** 877–921

IOP Publishing

Composite Materials (Second Edition)
Mathematical theory and exact relations
Yury Grabovsky

Chapter 3

Composite materials

3.1 Mathematical definition of a composite

The intuitive notion of a composite material and its effective properties is clear to most people. However, a rigorous mathematical analysis requires a definition. Let us consider an example of a composite domain $\Omega \subset \mathbb{R}^3$ occupied by two materials A and B. To describe such a medium one needs to specify a subset of Ω occupied by the material A. (The complement will be occupied by the material B.) However, such a description does not draw a clear line between a true composite and a compound object, like a wooden door with metal door knob. This is the reason why in the mathematical theory of composite materials we define a composite as a *limit of a sequence of structures* [5, 19, 21, 22]. With such a definition we imagine that a specific composite material is a member of a family of heterogeneous media labeled by the small parameter ε that may (but does not have to) represent a typical length scale. The effective properties of the 'composite' will then be defined to be the properties of the 'limit material', as $\varepsilon \to 0$. In this book we discuss the questions of existence and properties of such limit materials with full mathematical rigor. In practice, a composite material is identified with a finite (but 'small') value of ε. See, e.g., [11] for a numerical comparison between the finite ε and limit materials for periodic composites, where ε is the ratio of the size of the period cell to the size of the sample.

In this book, by a 'composite material' we mean a family of heterogeneous media, whose tensors of local material properties we will denote by $\mathsf{L}_\varepsilon(\boldsymbol{x})$, $\varepsilon > 0$. In other words, the relation between the fields $\boldsymbol{E}(\boldsymbol{x})$ and $\boldsymbol{J}(\boldsymbol{x})$ will be $\boldsymbol{J}(\boldsymbol{x}) = \mathsf{L}_\varepsilon(\boldsymbol{x})\boldsymbol{E}(\boldsymbol{x})$ for every $\boldsymbol{x} \in \Omega$. For a subset $U \subset \mathrm{End}^+(\mathscr{T})$ of material properties we define

$$U^\Omega = \{\mathsf{L} \in L^\infty(\Omega; \mathrm{End}(\mathscr{T})) \colon \mathsf{L}(\boldsymbol{x}) \in U \text{ for a.e. } \boldsymbol{x} \in \Omega\}. \tag{3.1}$$

In other words, U^Ω denotes the set of all possible heterogeneous media occupying the domain Ω, made with materials from a given set U. In this book we assume that $\mathsf{L}_\varepsilon \in \mathscr{M}(\alpha, \beta)^\Omega$, for some $\beta > \alpha > 0$, all $\varepsilon > 0$, where the sets $\mathscr{M}(\alpha, \beta)$ are defined in (2.18). We can now make rigorous the intuitive picture of a composite. On the

doi:10.1088/978-0-7503-6249-8ch3

micro-scale a composite is described by the family $\mathsf{L}_\varepsilon(\boldsymbol{x})$ that oscillates on smaller and smaller length scales as $\varepsilon \to 0$. Yet, the oscillations of $\mathsf{L}_\varepsilon(\boldsymbol{x})$ describing a specific composite should not be completely random, in the sense that for small ε the *response* of the composite should be indistinguishable from the response of a material, which is locally homogeneous, i.e., its pointwise properties may vary only on the macroscale. The idea is then to look at the limiting behavior of the solutions $\boldsymbol{E}_\varepsilon$ of (2.19), as $\varepsilon \to 0$.

Definition 3.1. *We say that the family* $\{\mathsf{L}_\varepsilon\colon \varepsilon > 0\}$ *H-converges to* $\mathsf{L}_*(\boldsymbol{x})$ *(denoted by* $\mathsf{L}_\varepsilon \overset{H}{\rightharpoonup} \mathsf{L}_*$*) if the family of solutions* $\boldsymbol{E}_\varepsilon$ *of*

$$\boldsymbol{E}_\varepsilon \in \mathscr{E}_0(\Omega), \qquad \boldsymbol{J}_\varepsilon = \mathsf{L}_\varepsilon(\boldsymbol{x})\boldsymbol{E}, \qquad \Gamma_\Omega \boldsymbol{J}_\varepsilon = \boldsymbol{f} \tag{3.2}$$

converges weakly in $L^2(\Omega; \mathscr{T})$ *to the solution* $\boldsymbol{E}_*$ *of*

$$\boldsymbol{E}_* \in \mathscr{E}_0(\Omega), \qquad \boldsymbol{J}_* = \mathsf{L}_*(\boldsymbol{x})\boldsymbol{E}_{**}, \qquad \Gamma_\Omega \boldsymbol{J}_* = \boldsymbol{f}, \tag{3.3}$$

for every $\boldsymbol{f} \in \mathscr{E}_0(\Omega)$, *and additionally* $\boldsymbol{J}_\varepsilon = \mathsf{L}_\varepsilon \boldsymbol{E}_\varepsilon$ *converges weakly in* $L^2(\Omega; \mathscr{T})$ *to* $\boldsymbol{J}_* = \mathsf{L}_*(\boldsymbol{x})\boldsymbol{E}_*$. *The H-limit* $\mathsf{L}_*(\boldsymbol{x})$ *is called the effective tensor of the composite* $\mathsf{L}_\varepsilon(\boldsymbol{x})$.

The notion of H-convergence was first studied by Spagnolo [22] for the conductivity problem. In that specific context it was shown [1] that what we defined as H-convergence can be described in terms of the convergence of Green's functions for the operators $\phi \mapsto \nabla \cdot (\boldsymbol{\sigma}_\varepsilon(\boldsymbol{x})\nabla\phi)$. For that reason the name 'G-convergence' was given to this kind of limit. Later, Murat and Tartar [19] generalized this notion to the case of non-symmetric conductivity tensors $\boldsymbol{\sigma}_\varepsilon$ by adding the requirement of convergence of fluxes. However, as our approach reveals, the need for convergence of fluxes is related not to the lack of symmetry of material tensors $\mathsf{L}_\varepsilon(\boldsymbol{x})$ per se, but to deeper geometric properties of the spaces $\mathscr{E}$ and $\mathscr{J}$. Since our approach to homogenization in this book does not deal with specific partial differential equations and their Green's functions explicitly, and is based on Murat and Tartar's compensated compactness approach [18, 23], we adopt the H-convergence terminology, where 'H' stands for homogenization. It is important to remark that for a physical composite material there is no mathematically sharp boundary between the 'microstructure', represented by the oscillatory nature of $\mathsf{L}_\varepsilon(\boldsymbol{x})$ and the 'macroscale' inhomogeneity, represented by the dependence of the H-limit $\mathsf{L}_*(\boldsymbol{x})$ on $\boldsymbol{x} \in \Omega$. In modeling specific composite materials it is a judgement call to identify the oscillatory regions (i.e., the microstructure) and the non-oscillatory regions (macroscale inhomogeneity). The latter can be modeled mathematically as a pointwise convergent family $\mathsf{L}_\varepsilon(\boldsymbol{x})$, in which case H-limit reduces to the pointwise limit.

Lemma 3.2. *Suppose* $\mathsf{L}_\varepsilon(\boldsymbol{x}) \in \mathscr{M}(\alpha, \beta)$ *for a.e.* $\boldsymbol{x} \in \Omega$ *and* $\mathsf{L}_\varepsilon(\boldsymbol{x}) \to \mathsf{L}_*(\boldsymbol{x})$ *in measure (strong convergence) then* $\mathsf{L}_\varepsilon \overset{H}{\rightharpoonup} \mathsf{L}_*$ *in* Ω.

Proof. The sequence of solutions $\boldsymbol{E}_\varepsilon$ of (2.19) is uniformly bounded in $L^2(\Omega; \mathcal{T})$. Therefore, there is a weakly convergent subsequence $\boldsymbol{E}_{\varepsilon_k} \rightharpoonup \boldsymbol{E}_0$ in $L^2(\Omega; \mathcal{T})$. The uniform boundedness of L_ε and convergence in measure implies that $\mathsf{L}_{\varepsilon_k}\boldsymbol{E}_{\varepsilon_k} \rightharpoonup \mathsf{L}_*\boldsymbol{E}_0$ in $L^2(\Omega; \mathcal{T})$. Thus, the weak limit $\boldsymbol{E}_0 \in \mathscr{E}_0(\Omega)$ of $\boldsymbol{E}_{\varepsilon_k}$ is uniquely determined by the equation $\Gamma_\Omega(\mathsf{L}_*\boldsymbol{E}_0) = \boldsymbol{f}$, and is independent of the chosen weakly convergent subsequence. It follows that the entire family $\boldsymbol{E}_\varepsilon$ weakly converges to $\boldsymbol{E}_0$ in $L^2(\Omega; \mathcal{T})$. Hence, $\mathsf{L}_\varepsilon \overset{H}{\rightharpoonup} \mathsf{L}_*$. □

H-convergence has a remarkable compactness property [19].

Theorem 3.3. (H-compactness). *Assume that the Fourier multiplier symbol* $\boldsymbol{\Gamma}(\boldsymbol{\eta})$ *has the compensated compactness property and is* $\mathscr{E}$*-nondegenerate in the sense of definition 2.16. Suppose* $\mathsf{L}_\varepsilon \in \mathscr{M}(\alpha, \beta)^\Omega$ *is an arbitrary family of microstructures. Then there exists a subsequence* $\varepsilon_k \to 0$ *and* $\mathsf{L}_* \in \mathscr{M}(\alpha, \beta)^\Omega$, *such that* $\mathsf{L}_{\varepsilon_k} \overset{H}{\rightharpoonup} \mathsf{L}_*$.

The proof is quite technical and we postpone it until section 3.3.3. This theorem tells us that up to a subsequence *any* infinite family of microstructures can be homogenized.

3.2 Periodic composites

Periodic composites are the most important and, as we will see, fundamental examples of H-convergent families of heterogeneous media. Let $\{\boldsymbol{p}_1,\dots, \boldsymbol{p}_d\} \subset \mathbb{R}^d$ be a basis of periods, and let

$$Q = \left\{\sum_{i=1}^{d}\lambda_i\boldsymbol{p}_i : 0 < \lambda_i < 1,\ i = 1,\dots, d\right\}$$

be the parallelepiped of periods. For $\mathsf{L} \in \mathscr{M}(\alpha, \beta)^Q$ and $0 < \alpha < \beta$ we define

$$\mathsf{L}_\varepsilon(\boldsymbol{x}) = \mathsf{L}_{\rm per}\left(\frac{\boldsymbol{x}}{\varepsilon}\right), \quad \boldsymbol{x} \in \mathbb{R}^d, \tag{3.4}$$

where $\mathsf{L}_{\rm per}(\boldsymbol{z})$, $\boldsymbol{z} \in \mathbb{R}^d$ is the Q-periodic extension of $\mathsf{L}(\boldsymbol{z})$ from the period cell Q to all of $\mathbb{R}^d$. We observe that tensor fields $\mathsf{L}_\varepsilon(\boldsymbol{x})$ are εQ-periodic. For example, if $A \cup B = Q$ are complementary subsets of Q occupied by materials L_A, L_B, then

$$\mathsf{L}(\boldsymbol{z}) = \chi_A(\boldsymbol{z})\mathsf{L}_A + \chi_B(\boldsymbol{z})\mathsf{L}_B, \quad \boldsymbol{z} \in Q$$

defines a tensor field of local material properties of a two-phase composite material. The family $\{\mathsf{L}_\varepsilon(\boldsymbol{x}) : \boldsymbol{x} \in \Omega\}$, where $\mathsf{L}_\varepsilon(\boldsymbol{x})$ is given by (3.4), is then interpreted as a periodic two-phase composite occupying the domain $\Omega \subset \mathbb{R}^d$.

In general, we will call the function $\mathsf{L}(\boldsymbol{z})$ *the local tensor* of material properties. Here and throughout the entire section we assume that $\mathsf{L} \in \mathscr{M}(\alpha, \beta)^Q$. We will prove that $\mathsf{L}_\varepsilon \overset{H}{\rightharpoonup} \mathsf{L}_*$, where $\mathsf{L}_* \in \mathrm{End}^+(\mathcal{T})$ is constant. The homogeneity of the effective medium L_* is intuitively clear, since the microstructure in the vicinity of any point of Ω is exactly the same periodic microstructure, as in the vicinity of any other point of Ω. In order to give a formula for the H-limit L_* we first need to define periodic versions of subspaces $\mathscr{E}$ and $\mathscr{J}$ of $L^2_{\text{per}}(Q;\mathscr{T})$—the set of Q-periodic locally L^2 vector fields.

Let $\boldsymbol{Q}$ be the invertible $d \times d$ matrix, such that $\boldsymbol{Q}Q = [0, 1]^d$. (The basis vectors $\{\boldsymbol{p}_1, \dots, \boldsymbol{p}_d\}$ form the columns of $\boldsymbol{Q}^{-1}$.) We define

$$\mathscr{E}_{\text{per}} = \left\{ \boldsymbol{E} \in L^2_{\text{per}}(Q;\mathscr{T}) \colon \hat{\boldsymbol{E}}(0) = 0,\ \hat{\boldsymbol{E}}(\boldsymbol{k}) \in \mathscr{E}_{\frac{\boldsymbol{Q}^T\boldsymbol{k}}{|\boldsymbol{Q}^T\boldsymbol{k}|}} \otimes \mathbb{C},\ \boldsymbol{k} \neq 0 \right\}, \tag{3.5}$$

$$\mathscr{J}_{\text{per}} = \left\{ \boldsymbol{J} \in L^2_{\text{per}}(Q;\mathscr{T}) \colon \hat{\boldsymbol{J}}(0) = 0,\ \hat{\boldsymbol{J}}(\boldsymbol{k}) \in \mathscr{J}_{\frac{\boldsymbol{Q}^T\boldsymbol{k}}{|\boldsymbol{Q}^T\boldsymbol{k}|}} \otimes \mathbb{C},\ \boldsymbol{k} \neq 0 \right\}, \tag{3.6}$$

where $\hat{\boldsymbol{J}}(\boldsymbol{k})$ (and $\hat{\boldsymbol{E}}(\boldsymbol{k})$), $\boldsymbol{k} \in \mathbb{Z}^d$, are the Fourier coefficients in the expansion

$$\boldsymbol{J}(\boldsymbol{z}) = \sum_{\boldsymbol{k}\in\mathbb{Z}^d} \hat{\boldsymbol{J}}(\boldsymbol{k}) e^{2\pi i \boldsymbol{Q}\boldsymbol{z}\cdot\boldsymbol{k}}, \tag{3.7}$$

and are given by

$$\hat{\boldsymbol{J}}(\boldsymbol{k}) = ⨍_Q \boldsymbol{J}(\boldsymbol{z}) e^{-2\pi i \boldsymbol{Q}\boldsymbol{z}\cdot\boldsymbol{k}} d\boldsymbol{z}, \quad \boldsymbol{k} \in \mathbb{Z}^d, \tag{3.8}$$

where

$$⨍_Q = \frac{1}{|Q|}\int_Q$$

denotes the average over a period cell.

Let $\mathscr{U}$ be the space of constant vector fields in $L^2_{\text{per}}(Q;\mathscr{T})$. Then we have the orthogonal decomposition

$$L^2_{\text{per}}(Q;\mathscr{T}) = \mathscr{E}_{\text{per}} \oplus \mathscr{J}_{\text{per}} \oplus \mathscr{U}. \tag{3.9}$$

In order to compute the effective tensor of the periodic composite with period cell Q and local tensor $\mathsf{L}(\boldsymbol{z})$ we need to solve the *periodic cell problem*

$$\boldsymbol{E} \in \mathscr{E}_{\text{per}} \oplus \mathscr{U}, \quad \boldsymbol{J}(\boldsymbol{z}) = \mathsf{L}(\boldsymbol{z})\boldsymbol{E}(\boldsymbol{z}), \quad \boldsymbol{J} \in \mathscr{J}_{\text{per}} \oplus \mathscr{U}, \quad \langle \boldsymbol{E} \rangle = \boldsymbol{e}_0, \tag{3.10}$$

where

$$\langle \boldsymbol{E} \rangle = ⨍_Q \boldsymbol{E}(\boldsymbol{z}) d\boldsymbol{z}$$

is a compact notation for the average over a period cell.

Theorem 3.4. *Assume that the compensated compactness postulate 2.3 holds. Then the effective tensor* L_* *of the periodic composite (3.4) is given by its action on the arbitrary field* $\boldsymbol{e}_0 \in \mathscr{T}$

$$\mathsf{L}_*\boldsymbol{e}_0 = ⨍_Q \mathsf{L}(\boldsymbol{z})\boldsymbol{E}(\boldsymbol{z}) d\boldsymbol{z}, \tag{3.11}$$

where $\boldsymbol{E}(\boldsymbol{z})$ *is the unique solution of the periodic cell problem (3.10).*

The proof of this theorem requires development of sophisticated machinery of homogenization and is postponed to section 3.3.

3.2.1 Fiber-reinforced periodic composites

The microstructure of fiber-reinforced periodic composites is assumed to be the same in any horizontal cross-section. Mathematically this means that the local tensor of material properties $\mathsf{L}(z)$ is Q'-periodic, where Q' is a parallelepiped of periods in $\mathbb{R}^{d'}$, for some $d' < d$, and is independent of the remaining $d - d'$ variables $\tilde{z}$. Such periodic composites are also Q-periodic, where $Q = Q' \times [0, 1]^{d-d'}$. Therefore the original periodic homogenization theorem applies, and the effective tensor L_* of such a composite is still determined by the solutions of the original cell problem (3.10). The fibrous nature of such composites manifests itself only through the independence of the local tensor $\mathsf{L}(z)$ of $\tilde{z}\in[0, 1]^{d-d'}$, where we write $z = (z', \tilde{z}) \in Q' \times [0, 1]^{d-d'}$. Let us then show that as a consequence, the solution $\boldsymbol{E}(z)$ of the cell problem (3.10) is always independent of $\tilde{z}$ for any choice of the average field $\boldsymbol{e}_0 \in \mathscr{T}$.

Theorem 3.5. *Suppose that* $\mathsf{L}(z) = \mathsf{L}(z')$ *is* Q'*-periodic and is independent of the* $d - d'$ *remaining coordinates* $\tilde{z}$. *Then the solution* $\boldsymbol{E}(z)$ *of the cell problem (3.10) is also independent of* $\tilde{z}$.

Proof. The orthogonality structure (3.9) implies

$$\int_{Q'}\left(\int_{[0,\,1]^{d-d'}}(\mathsf{L}(z')\boldsymbol{E},\, \boldsymbol{E} - \boldsymbol{e}_0)d\tilde{z}\right)dz' = 0.$$

Applying Parceval's identity we obtain

$$\int_{Q'}\sum_{\tilde{\boldsymbol{k}}\in\mathbb{Z}^{d-d'}\setminus\{0\}}(\mathsf{L}(z')\hat{\boldsymbol{E}}(z', \tilde{\boldsymbol{k}}),\, \hat{\boldsymbol{E}}(z', \tilde{\boldsymbol{k}}))dz' = 0,$$

where

$$\hat{\boldsymbol{E}}(z', \tilde{\boldsymbol{k}}) = \int_{[0,\,1]^{d-d'}}\boldsymbol{E}(z', \tilde{z})e^{-2\pi i\tilde{\boldsymbol{k}}\cdot\tilde{z}}d\tilde{z}.$$

Now, the assumption of positive definiteness of $\mathsf{L}(z')$ implies that $\hat{\boldsymbol{E}}(z', \tilde{\boldsymbol{k}}) = 0$ for a.e. $z' \in Q'$, for all $\tilde{\boldsymbol{k}}\in\mathbb{Z}^{d-d'}\setminus\{0\}$, which is equivalent to the independence of $\boldsymbol{E}(z)$ on $\tilde{z}$. □

Theorem 3.5 permits us to reformulate the cell problem (3.10) for fiber-reinforced composites in the abstract Hilbert space framework that incorporates the fibrous nature of the microgeometry into the structure of underlying subspaces. The key observation that if $\boldsymbol{E}(z)$ depends only on z', then $\hat{\boldsymbol{E}}(\boldsymbol{k}) = 0$ whenever $\boldsymbol{k} = (\boldsymbol{k}', \tilde{\boldsymbol{k}})$ and $\tilde{\boldsymbol{k}}\neq 0$. We can therefore drop all reference to $\tilde{z}$ variables, and regard all Q-periodic fields as Q'-periodic functions. The subspaces $\mathscr{E}_{\text{per}}$ and $\mathscr{J}_{\text{per}}$ of $L^2(Q; \mathscr{T})$ are then replaced by the subspaces $\mathscr{E}'_{\text{per}}$ and $\mathscr{J}'_{\text{per}}$ of $L^2(Q'; \mathscr{T})$, defined by the subspaces

$\mathscr{E}'_{\eta'} = \mathscr{E}_{(\eta',0)}$ and $\mathscr{J}'_{\eta'} = \mathscr{J}_{(\eta',0)}$ of $\mathscr{T}$. Thus, the cell problem for fibrous periodic composites can be regarded as a d'-dimensional problem that is obtained from the original d-dimensional problem by replacing the frequency variable $\boldsymbol{\eta} \in \mathbb{S}^{d-1}$ by $(\boldsymbol{\eta}', 0)$, $\boldsymbol{\eta}' \in \mathbb{S}^{d'-1}$, while retaining the same inner-product space $\mathscr{T}$. If the original symbol $\boldsymbol{\Gamma}(\boldsymbol{\eta})$ satisfied the rotational equivariance postulate 2.5, then the fibrous symbol $\boldsymbol{\Gamma}'(\boldsymbol{\eta}') = \boldsymbol{\Gamma}((\boldsymbol{\eta}', 0))$ also satisfies it, if we regard SO(d') as a subgroup of SO(d) by identifying $\boldsymbol{R}' \in SO(d')$ with $\boldsymbol{R} = \begin{bmatrix} \boldsymbol{R}' & 0 \\ 0 & \boldsymbol{I}_{d-d'} \end{bmatrix} \in \mathrm{SO}(d)$, which rotates within the subspace orthogonal to the fibers, while leaving all fiber directions fixed.

We remark, that the Hilbert space structure modification for fiber-reinforced composites pertains only to the interpretation of the periodic cell problem (3.10). The homogenization theorem leading to (3.10) operates in the original unmodified form, since both the physical domain $\Omega \subset \mathbb{R}^d$, occupied by the composite, and the applied boundary conditions are allowed to depend on $\tilde{\boldsymbol{x}}$ variables.

3.2.2 G-closure

In this book it will generally be assumed that the properties of the constituent materials and their volume fractions are known exactly. However, the properties of composites can also be strongly dependent upon microstructure, which is rarely known exactly. This gives rise to the so-called G-closure problem [12] of describing the set of all possible material moduli that a composite can have, provided that it is made of a given set of materials. In this book we deal exclusively with the situation where the microstructure is not constrained in any way, except possibly by the specified volume fractions of the phases.

Consider a closed subset $U \subset \mathscr{M}(\alpha, \beta)$ representing the set of available materials. We then consider all possible composites made with materials from U. In practical applications the set U can be either finite or infinite, as is the case for polycrystalline composites, where an anisotropic crystallite can occur in a composite in arbitrary orientations.

Definition 3.6. *The G-closure $G(U)$ of a subset $U \subset \mathscr{M}(\alpha, \beta)$ of tensors of material properties is the closure of the set of all effective tensors L_* of periodic composites $\mathsf{L}(\boldsymbol{x}/\varepsilon)$, where $\mathsf{L}(\boldsymbol{z})$ can be any measurable function on Q with values in U.*

Remark 3.7. *Let $U \subset \mathscr{M}(\alpha, \beta)$ be a subset. Then the set $G(U)$ from definition 3.6 is independent of the period cell Q. Indeed, let $\boldsymbol{Q}$ map Q into $[0, 1]^d$ and L^0: $\mathbb{R}^d \to U$ be $[0, 1]^d$-periodic. Then $\mathsf{L}(\boldsymbol{z}) = \mathsf{L}^0(\boldsymbol{Q}\boldsymbol{z})$ is Q-periodic and has values in U. Conversely, every measurable Q-periodic map L: $\mathbb{R}^d \to U$ can be represented in this way, since $\mathsf{L}^0(\boldsymbol{z}) = \mathsf{L}(\boldsymbol{Q}^{-1}\boldsymbol{z})$ is $[0, 1]^d$-periodic and has values in U. Obviously, $\boldsymbol{E}_0(\boldsymbol{z})$ solves the $[0, 1]^d$-periodic cell problem if and only if $\boldsymbol{E}(\boldsymbol{z}) = \boldsymbol{E}_0(\boldsymbol{Q}\boldsymbol{z})$ solves the Q-periodic cell problem. But then formula (3.11) shows that $\mathsf{L}^0_* = \mathsf{L}_*$.*

So far, we presented periodic composites as a class of examples of H-convergent sequences of microstructures. We now present a remarkable theorem formulated by Kohn and Dal Maso and proved by Tartar [26] and Raitums [20] in the context of conductivity. See also [2] for the analogous result in elasticity.

Theorem 3.8. *Assume that the Fourier multiplier symbol* $\boldsymbol{\Gamma}(\eta)$ *has the compensated compactness property and is* $\mathscr{E}$*-nondegenerate in the sense of definition 2.16. Let U be any closed subset of* $\mathscr{M}(\alpha, \beta)$*.*

(i) *Assume that* $\mathsf{L}_\varepsilon \in U^\Omega$ *is such that* $\mathsf{L}_\varepsilon \overset{H}{\rightharpoonup} \mathsf{L}_*$*. Then* $\mathsf{L}_* \in G(U)^\Omega$*.*

(ii) *Assume that* $\mathsf{L}_* \in G(U)^\Omega$*. Then there exists* $\mathsf{L}_\varepsilon \in U^\Omega$*, such that* $\mathsf{L}_\varepsilon \overset{H}{\rightharpoonup} \mathsf{L}_*$*.*

The proof of this result is in section 3.3.4. In essence, it says that there is no loss of generality in studying periodic composites if we are interested in questions about G-closure, like the ones studied in this book. It is natural to expect that G-closure of a set U would be G-closed. This is indeed the case. However, the mathematical proof of this simple statement requires going rather deep into the relationship between G-closures and H-convergence. See corollary 3.38 in section 3.3.4.

We now distinguish two cases of special interest. In the first one the set U is a finite set $U = \{\mathsf{L}_1, \dots \mathsf{L}_r\}$. The tensors in $G(U)$ will be referred to as the effective tensors of r-material, or r-phase composites. In the second case U is the union of finitely many SO(d) orbits

$$U = \bigcup_{j=1}^{r} \{\boldsymbol{R} \cdot \mathsf{L}_j \colon \boldsymbol{R} \in \mathrm{SO}(d)\}. \tag{3.12}$$

In this case the composites made from the materials in U are called polycrystals. Since single crystals $\mathsf{L}_1, \dots, \mathsf{L}_r$ determine the set U we will talk about the polycrystalline G-closure of materials $\mathsf{L}_1, \dots, \mathsf{L}_r$ and write $G^{\mathrm{pc}}(\{\mathsf{L}_1, \dots, \mathsf{L}_r\})$ to denote the G-closure of the set U in (3.12). We note that complexity of computing G-closures analytically skyrockets as the number of materials increases. While, $G(\{\mathsf{L}_1\}) = \mathsf{L}_1$, trivially, the sets $G(\{\mathsf{L}_1, \mathsf{L}_2\})$ and $G^{\mathrm{pc}}(\{\mathsf{L}_1\})$ are known in very few settings, while the sets $G(\{\mathsf{L}_1, \dots, \mathsf{L}_r\})$ and $G^{\mathrm{pc}}(\{\mathsf{L}_1, \dots, \mathsf{L}_{r-1}\})$ would be hopeless to compute analytically for any $r \geqslant 3$.

In the two special cases above we can restrict the set of admissible microstructures so that each material or each crystal L_j is used in specified volume fraction θ_j. Let $\vartheta = (\theta_1, \dots, \theta_r)$, such that $0 \leqslant \theta_j \leqslant 1$ and $\theta_1 + \dots \theta_r = 1$.

Definition 3.9. *Let* $\vartheta = (\theta_1, \dots, \theta_r)$ *be such that* $0 \leqslant \theta_j \leqslant 1$ *and* $\theta_1 + \dots \theta_r = 1$*. The partition of the period cell Q into r disjoint measurable subsets* $A_1, \dots, A_r$ *so that* $|A_j|/|Q| = \theta_j$*,* $j = 1, \dots, r$ *is called a* ϑ*-partition of Q.*

Definition 3.10. *The set of all effective tensors of periodic composites with*

$$\mathsf{L}(z) = \sum_{j=1}^{r} \mathsf{L}_j \chi_{A_j}(z),$$

where $A_1, \ldots, A_r$ *is a* ϑ*-partition of* Q*, is called* G_ϑ*-closure of the set* $U = \{\mathsf{L}_1, \ldots, \mathsf{L}_r\}$.

Definition 3.11. *The polycrystalline* G_ϑ*-closure of the set* $U = \{\mathsf{L}_1, \ldots, \mathsf{L}_r\}$ *is the set of all effective tensors of periodic composites with*

$$\mathsf{L}(z) = \boldsymbol{R}(z) \cdot \sum_{j=1}^{r} \mathsf{L}_j \chi_{A_j}(z),$$

where $A_1, \ldots, A_r$ *is a* ϑ*-partition of* Q *and* $\boldsymbol{R}\colon Q \to \mathrm{SO}(d)$ *is an arbitrary measurable function.*

The knowledge of the G-closure of a set of given materials provides bounds on the possible properties of all composite made of these materials, and can also tell us whether or not a desirable combination of properties is theoretically possible to achieve by mixing materials that possess only partial set of required properties. Very few G-closures have been computed exactly, e.g., [7, 9, 10, 12–14]. In general it is an almost intractable problem [4].

3.2.3 A formula for the effective tensor

So far, the effective tensor of a periodic composite is defined in terms of a solution of a periodic cell problem (3.10). In this section we will derive a formula for L_*, which will be central for the theory of exact relations. Moreover, we will show here that this formula has other remarkable features, such as superior convergence properties for effective tensors of high contrast composites.

If we approximate local material properties by a given homogeneous reference medium $\mathsf{L}_0 \in \mathrm{End}^+(\mathscr{T})$, then the resulting flux field $\boldsymbol{J}$ will be approximated by $\mathsf{L}_0\boldsymbol{E}$. The discrepancy $\boldsymbol{P} = \boldsymbol{J} - \mathsf{L}_0\boldsymbol{E}$ is called the polarization field. We will see that there are benefits if we rewrite the cell problem (3.10) in terms of the polarization field without direct reference to subspaces $\mathscr{E}_{\mathrm{per}}$, and $\mathscr{J}_{\mathrm{per}}$. Instead, we encode the subspaces by a non-orthogonal projection operator Γ' onto $\mathsf{L}_0\mathscr{E}_{\mathrm{per}}$ along $\mathscr{J}_{\mathrm{per}} \oplus \mathscr{U}$. It is easy to see that Γ' is a scale-free Fourier multiplier operator

$$\widehat{\Gamma' \boldsymbol{f}}(\boldsymbol{k}) = \begin{cases} \boldsymbol{\Gamma}'\left(\dfrac{\boldsymbol{Q}^T\boldsymbol{k}}{|\boldsymbol{Q}^T\boldsymbol{k}|}\right)\hat{\boldsymbol{f}}(\boldsymbol{k}), \boldsymbol{k} \in \mathbb{Z}^d \setminus \{0\}, \\ 0, \boldsymbol{k} = 0, \end{cases} \qquad \boldsymbol{Q}Q = [0, 1]^d,$$

where $\boldsymbol{\Gamma}'(\boldsymbol{n})$ is the non-orthogonal projection operator onto the subspace $\mathsf{L}_0\mathscr{E}_{\boldsymbol{n}}$ along the subspace $\mathscr{J}_{\boldsymbol{n}}$. We note that the projection $\boldsymbol{\Gamma}'(\boldsymbol{n})$ is well defined as long as L_0 is positive definite, since in that case, $\mathsf{L}_0\mathscr{E}_{\boldsymbol{n}}$ and $\mathscr{J}_{\boldsymbol{n}}$ have trivial intersection. Indeed, if $\mathsf{L}_0\boldsymbol{e} = \boldsymbol{j}$ for some $\boldsymbol{e} \in \mathscr{E}_{\boldsymbol{n}}$ and $\boldsymbol{j} \in \mathscr{J}_{\boldsymbol{n}}$, then taking the inner product with $\boldsymbol{e}$ we obtain

$$0 = (\boldsymbol{j}, \boldsymbol{e}) = (\mathsf{L}_0\boldsymbol{e}, \boldsymbol{e}),$$

and thus, $\boldsymbol{e} = 0$. Noting that a non-orthogonal projection operator is necessarily non-symmetric, it will be convenient to introduce a related symmetric operator

$$\Gamma_0 = \mathsf{L}_0^{-1}\Gamma'. \tag{3.13}$$

Lemma 3.12. *Let*

$$\boldsymbol{\Gamma}_0(\boldsymbol{n}) = \mathsf{L}_0^{-1}\boldsymbol{\Gamma}'(\boldsymbol{n}). \tag{3.14}$$

Then $\boldsymbol{\Gamma}_0(\boldsymbol{n}) \in \mathrm{Sym}(\mathscr{T})$, *provided* $\mathsf{L}_0 \in \mathrm{Sym}^+(\mathscr{T})$.

Proof. For $\boldsymbol{u}_1, \boldsymbol{u}_2 \subset \mathscr{T}$ we write

$$\boldsymbol{u}_1 = \boldsymbol{j}_1 + \mathsf{L}_0\boldsymbol{e}_1, \qquad \boldsymbol{u}_2 = \boldsymbol{j}_2 + \mathsf{L}_0\boldsymbol{e}_2, \qquad \{\boldsymbol{e}_1, \boldsymbol{e}_2\} \subset \mathscr{E}_{\boldsymbol{n}}, \quad \{\boldsymbol{j}_1, \boldsymbol{j}_2\} \subset \mathscr{J}_{\boldsymbol{n}}.$$

Then $\boldsymbol{\Gamma}_0\boldsymbol{u}_1 = \boldsymbol{e}_1$, $\boldsymbol{\Gamma}_0\boldsymbol{u}_2 = \boldsymbol{e}_2$, so that

$$(\boldsymbol{\Gamma}_0(\boldsymbol{n})\boldsymbol{u}_1, \boldsymbol{u}_2) = (\boldsymbol{e}_1, \mathsf{L}_0\boldsymbol{e}_2) = (\mathsf{L}_0\boldsymbol{e}_1, \boldsymbol{e}_2) = (\boldsymbol{u}_1, \boldsymbol{\Gamma}_0(\boldsymbol{n})\boldsymbol{u}_2).$$

□

We can now restate the cell problem (3.10) as a single operator equation in terms of the polarization field. If $\boldsymbol{E} \in \mathscr{E}_{\mathrm{per}} \oplus \mathscr{U}$ solves the cell problem (3.10), then

$$\Gamma_0\boldsymbol{P} = \Gamma_0\boldsymbol{J} - \Gamma_0\mathsf{L}_0\boldsymbol{E} = -(\boldsymbol{E} - \boldsymbol{e}_0) = \boldsymbol{e}_0 - (\mathsf{L} - \mathsf{L}_0)^{-1}\boldsymbol{P}. \tag{3.15}$$

Hence, we obtain,

$$(\Gamma_0 + (\mathsf{L} - \mathsf{L}_0)^{-1})\boldsymbol{P} = \boldsymbol{e}_0 \qquad \boldsymbol{P} \in L^2_{\mathrm{per}}(Q; \mathscr{T}). \tag{3.16}$$

Conversely, if $\boldsymbol{P} \in L^2_{\mathrm{per}}(Q; \mathscr{T})$ solves (3.16), then $\boldsymbol{E} = (\mathsf{L} - \mathsf{L}_0)^{-1}\boldsymbol{P}$ satisfies $\boldsymbol{E} = \boldsymbol{e}_0 - \Gamma_0\boldsymbol{P}$. It follows that $\boldsymbol{E} \in \mathscr{E}_{\mathrm{per}} \oplus \mathscr{U}$ and $\langle \boldsymbol{E} \rangle = \boldsymbol{e}_0$. We also see that

$$\boldsymbol{J} = \boldsymbol{P} + \boldsymbol{L}_0\boldsymbol{E} = (\boldsymbol{P} - \Gamma'\boldsymbol{P}) + \mathsf{L}_0\Gamma_0\boldsymbol{P} + \mathsf{L}_0\boldsymbol{E} = (\boldsymbol{P} - \Gamma'\boldsymbol{P}) + \mathsf{L}_0\boldsymbol{e}_0.$$

This shows that $\boldsymbol{J} \in \mathscr{J}_{\mathrm{per}} \oplus \mathscr{U}$, so that $\boldsymbol{E} = (\mathsf{L} - \mathsf{L}_0)^{-1}\boldsymbol{P}$ solves the cell problem (3.10).

We note that the relation between the intensity field $\boldsymbol{E}$ and the polarization field $\boldsymbol{P}$ may become singular, if $\mathsf{L}(z) - \mathsf{L}_0$ is not invertible. This problem can be resolved using the idea of preconditioning, i.e., rewriting the same linear equation in an equivalent form, that behaves in a more stable manner with respect to its parameters. The idea is to replace the polarization field $\boldsymbol{P}$ with a different linear function of $\boldsymbol{E}$ and $\boldsymbol{J}$:

$$\boldsymbol{S} = \mathsf{A}\boldsymbol{E} + \mathsf{M}\boldsymbol{J} = (\mathsf{A} + \mathsf{M}\mathsf{L})\boldsymbol{E}, \tag{3.17}$$

where the values of constant linear operators $\{\mathsf{A}, \mathsf{M}\} \subset \mathrm{End}(\mathscr{T})$ will be determined in the course of our analysis. The goal is to chose operators A and M, so that $\mathsf{A} + \mathsf{M}\mathsf{L}_*$ and $\mathsf{A} + \mathsf{M}\mathsf{L}(z)$ are *stably invertible* on Q.

Definition 3.13. *We say that* $\mathsf{F} \in L^\infty(Q; \mathrm{End}(\mathscr{T}))$ *is stably invertible on* Q *if* $\mathsf{F}^{-1}(\boldsymbol{z})$ *exists for almost all* $\boldsymbol{z} \in Q$ *and* $\mathsf{F}^{-1} \in L^\infty(Q; \mathrm{End}(\mathscr{T}))$.

Taking the average of (3.17) over a period cell we obtain

$$\langle \boldsymbol{S} \rangle = (\mathsf{A} + \mathsf{M}\mathsf{L}_*)\boldsymbol{e}_0. \tag{3.18}$$

Using (3.17), we can express the polarization field in terms of $\boldsymbol{S}$:

$$\boldsymbol{P} = (\mathsf{L} - \mathsf{L}_0)(\mathsf{A} + \mathsf{M}\mathsf{L})^{-1}\boldsymbol{S}.$$

We can also express the average of the polarization field in terms of $\langle \boldsymbol{S} \rangle$, using (3.18):

$$\langle \boldsymbol{P} \rangle = (\mathsf{L}_* - \mathsf{L}_0)\boldsymbol{e}_0 = (\mathsf{L}_* - \mathsf{L}_0)(\mathsf{A} + \mathsf{M}\mathsf{L}_*)^{-1}\langle \boldsymbol{S} \rangle$$

This suggests defining a transformation

$$W(\mathsf{L}) = (\mathsf{L} - \mathsf{L}_0)(\mathsf{A} + \mathsf{M}\mathsf{L})^{-1},$$

so that in terms of $W(\mathsf{L})$ we have

$$\boldsymbol{P} = W(\mathsf{L})\boldsymbol{S}, \qquad \langle \boldsymbol{P} \rangle = W(\mathsf{L}_*)\langle \boldsymbol{S} \rangle. \tag{3.19}$$

We would like the W-transformation to preserve symmetry. In other words, if L and L_0 are symmetric, we want $W(\mathsf{L})$ to be symmetric as well. This means that we must have

$$(\mathsf{L} - \mathsf{L}_0)(\mathsf{A} + \mathsf{M}\mathsf{L})^{-1} = (\mathsf{A}^T + \mathsf{L}\mathsf{M}^T)^{-1}(\mathsf{L} - \mathsf{L}_0).$$

for every $\mathsf{L} \in \mathrm{Sym}(\mathscr{T})$. Equivalently,

$$(\mathsf{A}^T + \mathsf{L}\mathsf{M}^T)(\mathsf{L} - \mathsf{L}_0) = (\mathsf{L} - \mathsf{L}_0)(\mathsf{A} + \mathsf{M}\mathsf{L}).$$

It follows that $\mathsf{M}^T = \mathsf{M}$ and

$$\mathsf{L}(\mathsf{A} + \mathsf{M}\mathsf{L}_0) = (\mathsf{A} + \mathsf{M}\mathsf{L}_0)^T\mathsf{L}, \quad \forall \mathsf{L} \in \mathrm{Sym}(\mathscr{T}).$$

It follows that $\mathsf{A} + \mathsf{M}\mathsf{L}_0 = \alpha\mathsf{I}$ for some $\alpha \in \mathbb{R}$. Conversely, if $\mathsf{M} \in \mathrm{Sym}(\mathscr{T})$ and

$$\mathsf{A} = \alpha\mathsf{I} - \mathsf{M}\mathsf{L}_0, \tag{3.20}$$

then

$$W(\mathsf{L}) = (\mathsf{L} - \mathsf{L}_0)(\alpha\mathsf{I} + \mathsf{M}(\mathsf{L} - \mathsf{L}_0))^{-1} \in \mathrm{Sym}(\mathscr{T})$$

for every $\mathsf{L} \in \mathrm{Sym}(\mathscr{T})$. We will set the value of A according to (3.20), even if L is not necessarily symmetric, in which case M will not be assumed to be symmetric either. We also observe that

$$W(\mathsf{L}) = \frac{1}{\alpha}(\mathsf{L} - \mathsf{L}_0)(\mathsf{I} + (\mathsf{M}/\alpha)(\mathsf{L} - \mathsf{L}_0))^{-1}.$$

This shows that replacing M by M/α and W by W/α we can set $\alpha = 1$ without loss of generality. From now on we will use the notation

$$W_{\mathsf{M}}(\mathsf{L}) = (\mathsf{L} - \mathsf{L}_0)(\mathsf{I} + \mathsf{M}(\mathsf{L} - \mathsf{L}_0))^{-1}. \tag{3.21}$$

In order to rewrite the cell problem in terms of the field $\boldsymbol{S}$ we observe that formula (3.17) now reads $\boldsymbol{S} = (\mathsf{I} + \mathsf{M}(\mathsf{L} - \mathsf{L}_0))\boldsymbol{E}$, so that

$$\boldsymbol{E} = (\mathsf{I} + \mathsf{M}(\mathsf{L} - \mathsf{L}_0))^{-1}\boldsymbol{S} = (\mathsf{I} - \mathsf{M}W_{\mathsf{M}}(L))\boldsymbol{S} \tag{3.22}$$

The cell problem (3.10) can then be rewritten in terms of the field $\boldsymbol{S}$ using the calculation (3.15): $\Gamma_0\boldsymbol{P} = \boldsymbol{e}_0 - \boldsymbol{E}$, and formulas (3.19) and (3.22)

$$(\mathsf{I} + (\Gamma_0 - \mathsf{M})W_{\mathsf{M}}(\mathsf{L}))\boldsymbol{S} = \boldsymbol{e}_0. \tag{3.23}$$

The new form (3.23) of the cell problem suggests that it will be convenient to introduce a Fourier multiplier operator Λ_{M}, given by

$$\widehat{\Lambda_{\mathsf{M}}\boldsymbol{f}}(\boldsymbol{k}) = \begin{cases} \left(\mathsf{M} - \Gamma_0\left(\dfrac{Q^T\boldsymbol{k}}{|Q^T\boldsymbol{k}|}\right)\right)\hat{\boldsymbol{f}}(\boldsymbol{k}), \boldsymbol{k} \in \mathbb{Z}^d\backslash\{0\}, \\ 0, \boldsymbol{k} = 0, \end{cases} \qquad QQ = [0, 1]^d. \tag{3.24}$$

With this notation (3.23) can be written as

$$(\mathsf{I} - \Lambda_{\mathsf{M}}W_{\mathsf{M}}(\mathsf{L}))\boldsymbol{S} = \langle\boldsymbol{S}\rangle, \tag{3.25}$$

where the relation between $\boldsymbol{e}_0$ and $\langle\boldsymbol{S}\rangle$ is easy to obtain taking the average of (3.23), using (3.19):

$$\boldsymbol{e}_0 = (\mathsf{I} - \mathsf{M}W_{\mathsf{M}}(\mathsf{L}_*))\langle\boldsymbol{S}\rangle = (\mathsf{I} + \mathsf{M}(\mathsf{L}_* - \mathsf{L}_0))^{-1}\langle\boldsymbol{S}\rangle.$$

From (3.19), we obtain

$$W_{\mathsf{M}}(\mathsf{L}_*)\langle\boldsymbol{S}\rangle = \langle W_{\mathsf{M}}(\mathsf{L})\boldsymbol{S}\rangle = \langle W_{\mathsf{M}}(\mathsf{L})(\mathsf{I} - \Lambda_{\mathsf{M}}W_{\mathsf{M}}(\mathsf{L}))^{-1}\langle\boldsymbol{S}\rangle\rangle.$$

Since $\langle\boldsymbol{S}\rangle$ can be an arbitrary vector in $\mathscr{T}$, we obtain the desired formula for the effective tensor of a composite as a formula for its W-transform in terms of the W-transform of the local tensor:

$$W_{\mathsf{M}}(\mathsf{L}_*) = \langle W_{\mathsf{M}}(\mathsf{L})(\mathsf{I} - \Lambda_{\mathsf{M}}W_{\mathsf{M}}(\mathsf{L}))^{-1}\rangle. \tag{3.26}$$

In order to make our derivation of (3.26) rigorous we need to choose the 'translation tensor' M in such a way that the operators $\mathsf{I} + \mathsf{M}(\mathsf{L}_* - \mathsf{L}_0)$ and $\mathsf{I} + \mathsf{M}(\mathsf{L}(z) - \mathsf{L}_0)$ are stably invertible on Q. We will then need to prove that the operator $\mathfrak{T} = \mathsf{I} - \Lambda_{\mathsf{M}}W_{\mathsf{M}}(\mathsf{L})$ is invertible on $L^2_{\mathrm{per}}(Q, \mathscr{T})$.

We note that it is possible to avoid having to check stable invertibility of $\mathsf{I} + \mathsf{M}(\mathsf{L}(z) - \mathsf{L}_0)$ and $\mathsf{I} + \mathsf{M}(\mathsf{L}_* - \mathsf{L}_0)$ for each specific periodic composite by choosing M with the *universal invertibility* property.

Definition 3.14. *We say that* $\mathsf{M} \in \operatorname{End}(\mathscr{T})$ *has the universal invertibility property (with respect to a given* $\mathsf{L}_0 \in \operatorname{End}^+(\mathscr{T})$*) if* $\mathsf{F} = \mathsf{I} + \mathsf{M}(\mathsf{L} - \mathsf{L}_0)$ *is invertible for any positive definite operator* $\mathsf{L} \in \operatorname{End}^+(\mathscr{T})$.

Lemma 3.15. $\mathsf{M} \in \operatorname{End}(\mathscr{T})$ *has the universal invertibility property if and only if* $\mathscr{R}(\mathsf{M}) = \mathscr{R}(\mathsf{M}^T)$, *where* $\mathscr{R}(\mathsf{M})$ *is the range of* M, *and*

$$(\mathsf{M}^{-1}\boldsymbol{u}, \boldsymbol{u}) \geqslant (\mathsf{L}_0\boldsymbol{u}, \boldsymbol{u}), \qquad \forall \boldsymbol{u} \in \mathscr{R}(\mathsf{M}). \tag{3.27}$$

Here M^{-1} *is understood as the inverse of the bijective map* $\mathsf{M}\colon \mathscr{R}(\mathsf{M}) \to \mathscr{R}(\mathsf{M})$.

The proof of this lemma is a bit technical. It can be found in section 3.2.5. We remark that if $\mathsf{M}\boldsymbol{v} = \boldsymbol{u}$ for some $\boldsymbol{v} \in \mathscr{T}$, then $\boldsymbol{v} = \mathsf{M}^{-1}\boldsymbol{u} + \boldsymbol{w}$, where $\boldsymbol{w} \in \ker(\mathsf{M})$, but then, since $\ker(\mathsf{M}) = \mathscr{R}(\mathsf{M}^T)^\perp = \mathscr{R}(\mathsf{M})^\perp$, we have $(\boldsymbol{v}, \boldsymbol{u}) = (\mathsf{M}^{-1}\boldsymbol{u}, \boldsymbol{u})$. Hence, $\mathsf{M}^{-1}\boldsymbol{u}$ in the left-hand side of (3.27) can also be interpreted as *any* preimage of $\boldsymbol{u}$.

Corollary 3.16. *For any unit vector* $\boldsymbol{n}$ *the operator* $\mathsf{M} = \boldsymbol{\Gamma}_0(\boldsymbol{n})$ *has the universal invertibility property.*

Proof. We have $\ker(\boldsymbol{\Gamma}_0(\boldsymbol{n})) = \mathscr{J}_{\boldsymbol{n}}$ and $\mathscr{R}(\boldsymbol{\Gamma}_0(\boldsymbol{n})) = \mathscr{E}_{\boldsymbol{n}} = \ker(\boldsymbol{\Gamma}_0(\boldsymbol{n}))^\perp = \mathscr{R}(\boldsymbol{\Gamma}_0(\boldsymbol{n})^T)$. Also, from the equality $\boldsymbol{\Gamma}_0(\boldsymbol{n})\mathsf{L}_0\boldsymbol{u} = \boldsymbol{u}$ that holds for all $\boldsymbol{u} \in \mathscr{E}_{\boldsymbol{n}}$, we conclude that $\mathsf{L}_0\boldsymbol{u}$ is a preimage of $\boldsymbol{u}$ under the map $\boldsymbol{\Gamma}_0(\boldsymbol{n})$. Thus, $((\boldsymbol{\Gamma}_0(\boldsymbol{n}))^{-1}\boldsymbol{u}, \boldsymbol{u}) = (\mathsf{L}_0\boldsymbol{u}, \boldsymbol{u})$ for all $\boldsymbol{u} \in \mathscr{E}_{\boldsymbol{n}}$. Therefore $\boldsymbol{\Gamma}_0(\boldsymbol{n})$ satisfies (3.27) with equality. □

Lemma 3.17. *Suppose that the operators* $\mathsf{I} + \mathsf{M}(\mathsf{L}_* - \mathsf{L}_0)$ *and* $\mathsf{I} + \mathsf{M}(\mathsf{L}(z) - \mathsf{L}_0)$ *are stably invertible on* Q. *Then the operator* $\mathfrak{T} = \mathsf{I} - \Lambda_{\mathsf{M}} W_{\mathsf{M}}(\mathsf{L})$ *is invertible on* $L^2_{\text{per}}(Q, \mathscr{T})$.

Proof. By our assumptions the operator $\mathfrak{T}$ is bounded. It remains to prove that for every $\boldsymbol{f} \in L^2_{\text{per}}(Q, \mathscr{T})$ there exists unique $\boldsymbol{u} \in L^2_{\text{per}}(Q, \mathscr{T})$ such that $\mathfrak{T}\boldsymbol{u} = \boldsymbol{f}$. Then, by the Banach invertibility theorem, this would imply that $\mathfrak{T}^{-1}$ is a bounded operator on $L^2_{\text{per}}(Q, \mathscr{T})$. Let us examine the equation $\mathfrak{T}\boldsymbol{u} = \boldsymbol{f}$. Using the definition of Λ_{M} and W_{M} we can rewrite it as

$$\Gamma_0(W_{\mathsf{M}}(\mathsf{L})\boldsymbol{u}) = \boldsymbol{f} \quad \mathsf{M}\langle W_{\mathsf{M}}(\mathsf{L})\boldsymbol{u}\rangle - (\mathsf{I} + \mathsf{M}(\mathsf{L} - \mathsf{L}_0))^{-1}\boldsymbol{u} \tag{3.28}$$

Since the left-hand side of (3.28) always belongs to $\mathscr{E}_{\text{per}}$, we will denote the right-hand side of (3.28) by $\boldsymbol{e}$. If we denote the constant field $\mathsf{M}\langle W_{\mathsf{M}}(\mathsf{L})\boldsymbol{u}\rangle$ by $\boldsymbol{e}_0$, we obtain the relation between $\boldsymbol{u}$ and $\boldsymbol{e}$

$$\boldsymbol{u} = (\mathsf{I} + \mathsf{M}(\mathsf{L} - \mathsf{L}_0))(\boldsymbol{f} - \boldsymbol{e} - \boldsymbol{e}_0). \tag{3.29}$$

Multiplying equation (3.28) on the left by L_0 we obtain

$$\Gamma'(W_{\mathsf{M}}(\mathsf{L})\boldsymbol{u}) = \mathsf{L}_0(\boldsymbol{f} - \boldsymbol{e}_0) - \mathsf{L}_0(\mathsf{I} + \mathsf{M}(\mathsf{L} - \mathsf{L}_0))^{-1}\boldsymbol{u}$$

Thus,

$$W_{\mathsf{M}}(\mathsf{L})\boldsymbol{u} - \mathsf{L}_0(\boldsymbol{f} - \boldsymbol{e}_0) + \mathsf{L}_0(\mathsf{I} + \mathsf{M}(\mathsf{L} - \mathsf{L}_0))^{-1}\boldsymbol{u} = W_{\mathsf{M}}(\mathsf{L})\boldsymbol{u} - \Gamma'(W_{\mathsf{M}}(\mathsf{L})\boldsymbol{u}) \in \mathscr{J}_{\text{per}} \oplus \mathscr{U},$$

or, equivalently,

$$\mathsf{L}(\mathsf{I} + \mathsf{M}(\mathsf{L} - \mathsf{L}_0))^{-1}\boldsymbol{u} - \mathsf{L}_0\boldsymbol{f} \in \mathscr{J}_{\text{per}} \oplus \mathscr{U}.$$

If Γ_{per} denotes the orthogonal projection onto $\mathscr{E}_{\text{per}}$ in $L^2_{\text{per}}(Q, \mathscr{T})$, then we have

$$\Gamma_{\text{per}}(\mathsf{L}(\mathsf{I} + \mathsf{M}(\mathsf{L} - \mathsf{L}_0))^{-1}\boldsymbol{u}) = \Gamma_{\text{per}}(\mathsf{L}_0\boldsymbol{f}). \tag{3.30}$$

Eliminating $\boldsymbol{u}$, using (3.29) we rewrite (3.30) as

$$\Gamma_{\text{per}}(\mathsf{L}(\boldsymbol{e} + \boldsymbol{e}_0)) = \Gamma_{\text{per}}((\mathsf{L} - \mathsf{L}_0)\boldsymbol{f}), \qquad \boldsymbol{e} \in \mathscr{E}_{\text{per}}. \tag{3.31}$$

Conversely, it is easy to verify by reversing the arguments that the system of equations (3.31), (3.29), together with

$$\boldsymbol{e}_0 = \mathsf{M}\langle W_{\mathsf{M}}(\mathsf{L})\boldsymbol{u}\rangle \tag{3.32}$$

is equivalent to (3.28).

We note that by analogy with the proof of theorem 2.8 the Lax-Milgram lemma guarantees that the operator $\Gamma_{\text{per}}(\mathsf{L}\boldsymbol{e})$ is invertible on $\mathscr{E}_{\text{per}}$. Hence, the solution $\boldsymbol{e}$ of (3.31) can be represented as $\boldsymbol{e} + \boldsymbol{e}_0 = \tilde{\boldsymbol{e}} + \boldsymbol{E}$, where $\tilde{\boldsymbol{e}} \in \mathscr{E}_{\text{per}}$ is the unique solution of

$$\Gamma_{\text{per}}(\mathsf{L}\tilde{\boldsymbol{e}}) = \Gamma_{\text{per}}((\mathsf{L} - \mathsf{L}_0)\boldsymbol{f}),$$

while $\boldsymbol{E} \in \mathscr{E}_{\text{per}} \oplus \mathscr{U}$ is the unique solution of the periodic cell problem (3.10). The solution $\boldsymbol{u}$ of $\mathfrak{T}\boldsymbol{u} = \boldsymbol{f}$ is then given by (3.29), establishing existence and uniqueness, if we show that $\boldsymbol{e}_0 \in \mathscr{U}$ can be determined uniquely.

In order to determine $\boldsymbol{e}_0$ we use (3.29) to compute

$$W_{\mathsf{M}}(\mathsf{L})\boldsymbol{u} = (\mathsf{L} - \mathsf{L}_0)(\boldsymbol{f} - \boldsymbol{e} - \boldsymbol{e}_0), \tag{3.33}$$

Averaging (3.33) over the period cell, multiplying by M, and using (3.32) we obtain

$$\boldsymbol{e}_0 = \mathsf{M}\langle(\mathsf{L} - \mathsf{L}_0)(\boldsymbol{f} - \boldsymbol{e} - \boldsymbol{e}_0)\rangle,$$

Replacing $\boldsymbol{e} + \boldsymbol{e}_0$ with $\tilde{\boldsymbol{e}} + \boldsymbol{E}$ we obtain

$$\boldsymbol{e}_0 = \mathsf{M}\langle(\mathsf{L} - \mathsf{L}_0)(\boldsymbol{f} - \tilde{\boldsymbol{e}})\rangle - \mathsf{M}(\mathsf{L}_* - \mathsf{L}_0)\boldsymbol{e}_0,$$

where we took into account that $\boldsymbol{E}$ solves (3.10). This gives

$$\boldsymbol{e}_0 = (\mathsf{I} + \mathsf{M}(\mathsf{L}_* - \mathsf{L}_0))^{-1}\mathsf{M}\langle(\mathsf{L} - \mathsf{L}_0)(\boldsymbol{f} - \tilde{\boldsymbol{e}})\rangle. \tag{3.34}$$

Conversely, it is easy to verify that (3.34) implies (3.32). Thus, for every $\boldsymbol{f} \in L^2_{\text{per}}(Q, \mathscr{T})$ we have exhibited the unique solution $\boldsymbol{u}$ of $\mathfrak{T}\boldsymbol{u} = \boldsymbol{f}$. The lemma is proved now. □

We summarize our results in the form of a theorem.

Theorem 3.18. *Let* L_* *be the effective tensor of the composite with the local tensor* $\mathsf{L}(z)$, $z \in Q$. *Let* $\mathsf{L}_0 \in \mathrm{End}^+(\mathscr{T})$ *and* $\mathsf{M} \in \mathrm{End}(\mathscr{T})$ *be such that the operators* $\mathsf{I} + \mathsf{M}(\mathsf{L}_* - \mathsf{L}_0)$ *and* $\mathsf{I} + \mathsf{M}(\mathsf{L}(z) - \mathsf{L}_0)$ *are stably invertible on* Q *in the sense of definition 3.13. Then the effective tensor* L_* *can be determined by (3.26).*

Remark 3.19. *Substituting (3.20) into (3.17) with* $\alpha = 1$, *we obtain* $\boldsymbol{S} = \boldsymbol{E} + \mathsf{M}\boldsymbol{P}$. *Hence,* $\boldsymbol{S}$ *can be interpreted as a modified total field, which is the sum of the applied field* $\boldsymbol{E}$ *and the modified polarization field* $\mathsf{M}\boldsymbol{P}$. *Formula (3.19) gives the physical meaning of* $W_{\mathsf{M}}(\mathsf{L})$ *as the tensor in the linear constitutive relation between the polarization field* $\boldsymbol{P}$ *and the modified total field* $\boldsymbol{S}$.

We reiterate that the effective tensor L_* of the periodic composite does not depend on the reference medium L_0 and the translation tensor M. These objects can be regarded as preconditioners for the solution of the cell problem (3.10). In particular, it was shown in [6, 15] (see also [17], section 14.9) that for $\mathsf{L} \in \mathscr{M}(\alpha, \beta)^{\Omega}$ choosing $\mathsf{L}_0 = \sqrt{\alpha\beta}\,\mathsf{I}$ and $\mathsf{M} = \mathsf{L}_0^{-1}/2$, we can make the operator norm of $\Lambda_{\mathsf{M}} W_{\mathsf{M}}(\mathsf{L})$ no larger than $(\sqrt{c} - 1)/(\sqrt{c} + 1)$, where $c = \beta/\alpha$ is the contrast ratio. Hence, formula (3.26) can be written in the form of the Neumann series

$$W_{\mathsf{M}}(\mathsf{L}_*) = \sum_{n=0}^{\infty} \langle W_{\mathsf{M}}(\mathsf{L})(\Lambda_{\mathsf{M}} W_{\mathsf{M}}(\mathsf{L}))^n \rangle, \tag{3.35}$$

converging geometrically fast.

3.2.4 Lamination formula

Let us apply formula (3.26) to a laminar microstructure, where $\mathsf{L}(z) = \mathsf{L}(z_1)$ and the period cell is chosen to be $Q = [0, 1]^d$. Then $W_{\mathsf{M}}(\mathsf{L})$ is also a function of z_1 only. In that case it is easy to see that

$$\Lambda_{\mathsf{M}} W_{\mathsf{M}}(\mathsf{L}) = (\mathsf{M} - \boldsymbol{\Gamma}_0(\boldsymbol{e}_1))(W_{\mathsf{M}}(\mathsf{L}) - \langle W_{\mathsf{M}}(\mathsf{L}) \rangle).$$

This suggests the choice $\mathsf{M} = \boldsymbol{\Gamma}_0(\boldsymbol{e}_1)$, (which by corollary 3.16 has the universal invertibility property) reducing (3.26) to a surprisingly simple formula: $W_{\mathsf{M}}(\mathsf{L}_*) = \langle W_{\mathsf{M}}(\mathsf{L}) \rangle$. In general, if the lamination normal is a unit vector $\boldsymbol{n}$ (see figure 3.1), then we choose $\mathsf{M} = \boldsymbol{\Gamma}_0(\boldsymbol{n})$. The result is summarized as a theorem.

Theorem 3.20. *Let* $\mathsf{L}(z) = \tilde{\mathsf{L}}(z \cdot \boldsymbol{n})$ *and let* L_0 *be an arbitrary positive definite reference medium. Then*

$$\langle W_{\boldsymbol{n}}(\mathsf{L}) \rangle = W_{\boldsymbol{n}}(\mathsf{L}_*), \tag{3.36}$$

where

$$W_{\boldsymbol{n}}(\mathsf{L}) = W_{\boldsymbol{\Gamma}_0(\boldsymbol{n})}(\mathsf{L}) \tag{3.37}$$

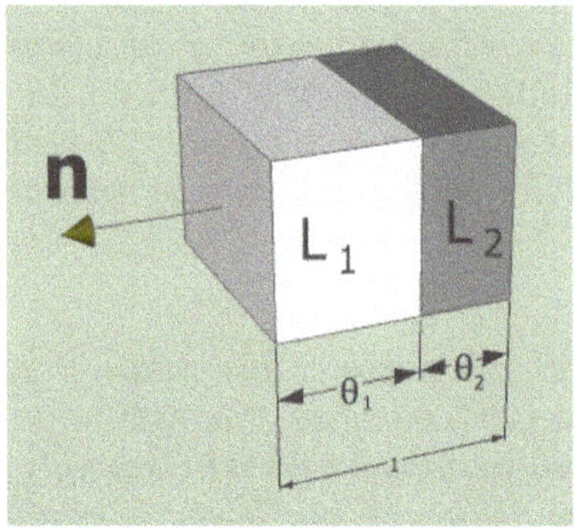

Figure 3.1. The period cell of the laminate.

Corollary 3.21. *A G-closed set is an image of a convex subset of* $\mathrm{End}(\mathscr{T})$ *under the analytic diffeomorphism* W_n^{-1}.

Proof. The statement follows from the fact that if the set G is G-closed and $\{\mathsf{L}_1, \mathsf{L}_2\} \subset G$ then $\mathsf{L}_* \in G$, where L_* is the effective tensor of the laminate made with materials L_1 and L_2 taken in volume fractions θ_1 and $\theta_2 = 1 - \theta_1$ with lamination normal $\boldsymbol{n}$. By theorem 3.20 we obtain

$$W_n(\mathsf{L}_*) = \theta_1 W_n(\mathsf{L}_1) + \theta_2 W_n(\mathsf{L}_2). \tag{3.38}$$

Hence, $W_n(G)$ is convex for all $\boldsymbol{n} \in \mathbb{S}^{d-1}$. The map $W_{\boldsymbol{n}}$ is an analytic diffeomorphism by lemma 3.24 in section 3.2.5, since, by corollary 3.16, $\mathsf{M} = \boldsymbol{\Gamma}_0(\boldsymbol{n})$ has the universal invertibility property. □

Formula (3.38) was first derived by Milton [16] and independently by Zhikov [27]. Other linear lamination formulas were derived by Backus [3] and Tartar [24] for elasticity.

Corollary 3.21 implies substantial regularity of the set $G(U)$ *regardless of regularity of* U, which itself can be any measurable subset of $\mathscr{M}(\alpha, \beta)$, defined in (2.18). Specifically, the set $G(U)$ must be an analytic submanifold of $\mathrm{End}^+(\mathscr{T})$ with Lipschitz boundary. We remark that if $W_n(G)$ is a convex set for all $\boldsymbol{n} \in \mathbb{S}^2$ then the set G is L-closed (or stable under lamination). In almost all cases the G-closure and L-closure coincide. However, there is an example of Milton of a set that is L-closed but not G-closed [17], section 31.9. Another example is given in this book in chapter 16. Francfort and Milton [8] gave an algorithm for constructing general polycrystalline G-closures in the context of 2D conductivity using corollary 3.21.

3.2.5 Properties of the transformation $W_{\mathsf{M}}(\mathsf{L})$†

In this section we study properties of the transformation $W_{\mathsf{M}}(\mathsf{L})$, where M has the universal invertibility property (see definition 3.14). We first prove the universal invertibility criterion lemma 3.15.

Proof of lemma 3.15. We begin the proof with a technical lemma.

Lemma 3.22. *Let* $\{\boldsymbol{a}, \boldsymbol{b}\} \subset \mathscr{T}\setminus\{0\}$. *Then there exists* $\mathsf{L} \in \mathrm{Sym}^+(\mathscr{T})$, *such that* $\mathsf{L}\boldsymbol{a} = \boldsymbol{b}$ *if and only if* $(\boldsymbol{a}, \boldsymbol{b}) > 0$

Proof. The necessity of the inequality $(\boldsymbol{a}, \boldsymbol{b}) > 0$ is obvious. Now suppose that it is satisfied. If $\boldsymbol{b} = \lambda\boldsymbol{a}$, then $\lambda > 0$ and $\mathsf{L} = \lambda\mathsf{I}$ does the job. Now assume that $\boldsymbol{a}$ and $\boldsymbol{b}$ are linearly independent. We then define $\mathsf{L} = \mathsf{I}$ on the orthogonal complement of $\mathscr{S} = \mathrm{Span}\{\boldsymbol{a}, \boldsymbol{b}\}$ and define $\mathsf{L}\boldsymbol{b} = \boldsymbol{p} \in \mathscr{S}$, where $\boldsymbol{p}$ will be chosen in such a way that L is symmetric and positive definite on $\mathscr{S}$ (and therefore on $\mathscr{T}$). We note that symmetry of L implies that $(\boldsymbol{p}, \boldsymbol{a}) = |\boldsymbol{b}|^2$. In that case the number $\beta = (\boldsymbol{p}, \boldsymbol{b})$ determines vector $\boldsymbol{p} \in \mathscr{S}$ and, hence, $\mathsf{L} \in \mathrm{Sym}(\mathscr{T})$ uniquely. We require that $(\mathsf{L}(x\boldsymbol{a} + y\boldsymbol{b}), (x\boldsymbol{a} + y\boldsymbol{b})) > 0$ for all $(x, y) \neq (0, 0)$. We compute

$$(\mathsf{L}(x\boldsymbol{a} + y\boldsymbol{b}), (x\boldsymbol{a} + y\boldsymbol{b})) = x^2(\boldsymbol{a}, \boldsymbol{b}) + 2xy|\boldsymbol{b}|^2 + y^2\beta.$$

Hence, positivity of L on $\mathscr{S} = \mathrm{Span}\{\boldsymbol{a}, \boldsymbol{b}\}$ is equivalent to $|\boldsymbol{b}|^4 < \beta(\boldsymbol{a}, \boldsymbol{b})$. Hence, any choice of $\beta > |\boldsymbol{b}|^4/(\boldsymbol{a}, \boldsymbol{b})$ will result in a positive definite $\mathsf{L} \in \mathrm{Sym}(\mathscr{T})$, satisfying $\mathsf{L}\boldsymbol{a} = \boldsymbol{b}$. □

To prove lemma 3.15 we observe that $\mathscr{R}(\mathsf{M})$ is an invariant subspace of $\mathsf{F} = \mathsf{I} + \mathsf{M}(\mathsf{L} - \mathsf{L}_0)$ and that F is singular on $\mathscr{T}$ if and only if it is singular on $\mathscr{R}(\mathsf{M})$. Indeed, if $\mathsf{F}\boldsymbol{t} = 0$ for some $\boldsymbol{t} \in \mathscr{T}\setminus\{0\}$, then $\boldsymbol{t} = -\mathsf{M}(\mathsf{L} - \mathsf{L}_0)\boldsymbol{t} \in \mathscr{R}(\mathsf{M})$. Conversely, if $\boldsymbol{u} \in \mathscr{R}(\mathsf{M})\setminus\{0\}$ is such that $\mathsf{F}\boldsymbol{u} = 0$, then F is certainly singular on $\mathscr{T}$. In turn, F: $\mathscr{R}(\mathsf{M}) \to \mathscr{R}(\mathsf{M})$ is singular if and only if we can choose $\boldsymbol{u} \in \mathscr{R}(\mathsf{M})\setminus\{0\}$ such that $(\mathsf{L} - \mathsf{L}_0)\boldsymbol{u} = -\boldsymbol{v}$, where $\mathsf{M}\boldsymbol{v} = \boldsymbol{u}$. By lemma 3.22 we can always choose a positive definite operator L, such that $\mathsf{L}\boldsymbol{u} = \mathsf{L}_0\boldsymbol{u} - \boldsymbol{v}$, provided $(\mathsf{L}_0\boldsymbol{u} - \boldsymbol{v}, \boldsymbol{u}) > 0$. Hence, in order to ensure that F is non-singular we must require that $(\mathsf{L}_0\boldsymbol{u} - \boldsymbol{v}, \boldsymbol{u}) \leqslant 0$ for every $\boldsymbol{u} \in \mathscr{R}(\mathsf{M})$ and every $\boldsymbol{v} \in \mathscr{T}$, satisfying $\boldsymbol{u} = \mathsf{M}\boldsymbol{v}$. Recall that $\ker(\mathsf{M})^\perp = \mathscr{R}(\mathsf{M}^T)$ and therefore M: $\mathscr{R}(\mathsf{M}^T) \to \mathscr{R}(\mathsf{M})$ is an invertible transformation. Hence, there exists a unique $\boldsymbol{v}_0 \in \mathscr{R}(\mathsf{M}^T)$, such that $\mathsf{M}\boldsymbol{v}_0 = \boldsymbol{u}$. We denote $\boldsymbol{v}_0 = \mathsf{M}^{-1}\boldsymbol{u}$. Thus, if $\mathsf{M}\boldsymbol{v} = \boldsymbol{u}$, then $\boldsymbol{v} = \boldsymbol{v}_0 + \boldsymbol{w}$, where $\boldsymbol{w} \in \ker(\mathsf{M})$ can be arbitrary. Our conclusion is that F is non-singular for every positive definite L if and only if

$$(\mathsf{L}_0\boldsymbol{u} - \mathsf{M}^{-1}\boldsymbol{u} - \boldsymbol{w}, \boldsymbol{u}) \leqslant 0, \qquad \forall\,\boldsymbol{u} \in \mathscr{R}(\mathsf{M}), \forall\,\boldsymbol{w} \in \ker(\mathsf{M}). \tag{3.39}$$

In particular, we must have $(\boldsymbol{u}, \boldsymbol{w}) = 0$ for all $\boldsymbol{u} \in \mathscr{R}(\mathsf{M})$ and all $\boldsymbol{w} \in \ker(\mathsf{M})$. Thus, $\mathscr{R}(\mathsf{M}) - (\ker(\mathsf{M}))^\perp = \mathscr{R}(\mathsf{M}^T)$, and inequality (3.39) reduces to (3.27). □

Lemma 3.23. *Suppose that* $\mathsf{M} \in \mathrm{End}(\mathscr{T})$ *has the universal invertibility property. Then the map* $W_\mathsf{M}(\mathsf{L})$, *given by (3.21), is injective. Moreover,*

$$W_\mathsf{M}(\mathsf{L}) = [\mathsf{I} + (\mathsf{L} - \mathsf{L}_0)\mathsf{M}]^{-1}(\mathsf{L} - \mathsf{L}_0). \tag{3.40}$$

Proof. To prove formula (3.40) we observe that

$$(\mathsf{L} - \mathsf{L}_0)(\mathsf{I} + \mathsf{M}(\mathsf{L} - \mathsf{L}_0)) = (\mathsf{I} + (\mathsf{L} - \mathsf{L}_0)\mathsf{M})(\mathsf{L} - \mathsf{L}_0). \tag{3.41}$$

The operator $\tilde{\mathsf{F}} = \mathsf{I} + (\mathsf{L} - \mathsf{L}_0)\mathsf{M}$ is invertible if and only if -1 is not an eigenvalue of $(\mathsf{L} - \mathsf{L}_0)\mathsf{M}$. Recall that AB and BA have the same spectrum for any two operators

$\{A, B\} \subset \mathrm{End}(\mathscr{T})$. Therefore, $\tilde{F}$ is invertible if and only if $F = I + M(L - L_0)$ is invertible. Thus, if M has the universal invertibility property then $\tilde{F}$ is invertible and multiplying (3.41) by $\tilde{F}^{-1}$ on the left and F^{-1} on the right we obtain (3.40).

Now we will show that W_M is injective, i.e., W_M maps distinct operators L into distinct operators $K = W_M(L)$. Suppose $W_M(L_1) = W_M(L_2)$. Using representation (3.21) for $W_M(L_1)$ and (3.40) for $W_M(L_2)$ we obtain

$$(I + (L_2 - L_0)M)(L_1 - L_0) = (L_2 - L_0)(I + M(L_1 - L_0)).$$

Opening parentheses we obtain $L_1 = L_2$. □

Lemma 3.24. *Suppose that* $M \in \mathrm{End}(\mathscr{T})$ *has the universal invertibility property. Then the map* $W_M(L)$*, given by (3.21), is an analytic diffeomorphism between* $\mathrm{End}^+(\mathscr{T})$ *and* $\mathscr{R} = W_M(\mathrm{End}^+(\mathscr{T})) \subset \mathrm{End}(\mathscr{T})$*. Moreover, for any* $K \in \mathscr{R}$ *there is a unique* $L \in \mathrm{End}^+(\mathscr{T})$*, given by*

$$L = W_M^{-1}(K) = L_0 + (I - KM)^{-1}K = L_0 + K(I - MK)^{-1}. \tag{3.42}$$

In particular, the operators $I - KM$ *and* $I - MK$ *are invertible.*

Proof. Now suppose $K \in \mathscr{R}$. Then, as we have just shown, there exists a unique $L \in \mathrm{End}^+(\mathscr{T})$, such that $K = W_M(L)$. It is an easy matrix algebra exercise to verify that

$$I - KM = \tilde{F}^{-1}, \qquad I - MK = F^{-1}, \tag{3.43}$$

proving the invertibility of $I - KM$ and $I - MK$. Multiplying the first equation in (3.43) on the right and the second equation on the left by $L - L_0$ we obtain

$$(I - KM)(L - L_0) = K, \qquad (L - L_0)(I - MK) = K.$$

Solving for L we obtain (3.42). Since both formulas (3.37) and (3.42) define analytic (rational) maps, which are defined and injective on $\mathrm{End}^+(\mathscr{T})$ and $\mathscr{R}$, respectively, we may conclude that W_M is an analytic diffeomorphism. □

Corollary 3.25. *Suppose that* $\{M, M'\} \subset \mathrm{End}(\mathscr{T})$ *have the universal invertibility property. Suppose* $K \in \mathscr{R} = W_M(\mathrm{End}^+(\mathscr{T}))$*. Then the operators* $I - K(M - M')$ *and hence* $I - (M - M')K$ *are invertible.*

Proof. Since M' has the universal invertibility property then $I + (L - L_0)M'$ must be invertible for every $L \in \mathrm{End}^+(\mathscr{T})$. Hence, $I + (W_M^{-1}(K) - L_0)M'$ must be invertible. Substituting the first expression for $W_M^{-1}(K)$ from (3.42) we conclude that

$$I + (I - KM)^{-1}KM' = (I - KM)^{-1}(I - (M - M')K)$$

must be invertible. The invertibility of $I - (M - M')K$ now follows. □

3.3 Properties of H-convergence †

In this section we show that the mathematical concept of H-limit can adequately describe the effective behavior of composites. The following theorem shows that the macroscopic response of a composite is uniquely defined by the microstructure.

Theorem 3.26. (Uniqueness of H-limits). *Assume that $V_{\mathscr{E}} = \mathscr{T}$ (This assumption holds if the symbol $\boldsymbol{\Gamma}(\boldsymbol{\eta})$ is $\mathscr{E}$-nondegenerate and satisfies the compensated compactness property). Suppose that for every $\varepsilon > 0$ $\mathsf{L}_\varepsilon(\boldsymbol{x}) \in \mathscr{M}(\alpha, \beta)$ for a.e. $\boldsymbol{x} \in \Omega$. If $\mathsf{L}_\varepsilon \overset{H}{\rightharpoonup} \mathsf{L}_*$ and $\mathsf{L}_\varepsilon \overset{H}{\rightharpoonup} \mathsf{L}'_*$, as $\varepsilon \to 0$, then $\mathsf{L}_*(\boldsymbol{x}) = \mathsf{L}'_*(\boldsymbol{x})$ for a.e. $\boldsymbol{x} \in \Omega$.*

Proof. Let $\boldsymbol{E} \in \mathscr{E}_0(\Omega)$ be arbitrary. Let $\boldsymbol{f} = \mathsf{T}_{\mathsf{L}_*}\boldsymbol{E}$, where the operator T_L is defined in (2.20). Then, since $\mathsf{L}_\varepsilon \overset{H}{\rightharpoonup} \mathsf{L}_*$ we have

$$\boldsymbol{E}_\varepsilon = \mathsf{T}_{\mathsf{L}_\varepsilon}^{-1}\boldsymbol{f} \rightharpoonup \mathsf{T}_{\mathsf{L}_*}^{-1}\boldsymbol{f} = \boldsymbol{E}, \qquad \mathsf{L}_\varepsilon \boldsymbol{E}_\varepsilon \rightharpoonup \mathsf{L}_*\boldsymbol{E}.$$

But at the same time $\mathsf{L}_\varepsilon \boldsymbol{E}_\varepsilon \rightharpoonup \mathsf{L}'_*\boldsymbol{E}$. Thus, for any $\boldsymbol{E} \in \mathscr{E}_0(\Omega)$ we have $\mathsf{L}_*\boldsymbol{E} = \mathsf{L}'_*\boldsymbol{E}$. Then for any $\boldsymbol{U}_0 \in \mathscr{T}$ we have

$$(\boldsymbol{E}, (\mathsf{L}_* - \mathsf{L}'_*)^T \boldsymbol{U}_0) = 0 \text{ for a.e. } \boldsymbol{x} \in \Omega \text{ and all } \boldsymbol{E} \in \mathscr{E}_0(\Omega).$$

Thus, $(\mathsf{L}_* - \mathsf{L}'_*)^T \boldsymbol{U}_0 \in \mathscr{N}(\Omega)$, where $\mathscr{N}(\Omega)$ is defined in (2.30). By corollary 2.27 the assumption $V_{\mathscr{E}} = \mathscr{T}$ implies that $\mathscr{N}(\Omega) = \{0\}$, and the theorem is proved. □

3.3.1 Proof of the periodic homogenization theorem

We start by establishing the relation between the periodic and local versions of the spaces $\mathscr{E}$ and $\mathscr{J}$.

Lemma 3.27. *Suppose $\boldsymbol{E} \in \mathscr{E}_{\text{per}}$ and $\boldsymbol{J} \in \mathscr{J}_{\text{per}}$. Then for any bounded measurable subset $\Omega \subset \mathbb{R}^d$*

$$\lim_{\varepsilon \to 0} \left\| \Pi_\Omega \boldsymbol{E}\left(\frac{\boldsymbol{x}}{\varepsilon}\right) \right\|_{L^2(\Omega)} = \lim_{\varepsilon \to 0} \left\| \Gamma_\Omega \boldsymbol{J}\left(\frac{\boldsymbol{x}}{\varepsilon}\right) \right\|_{L^2(\Omega)} = 0,$$

where Γ_Ω and Π_Ω are the orthogonal projections onto $\mathscr{E}_0(\Omega)$ and $\mathscr{J}_0(\Omega)$, respectively.

Proof. The proof follows [25], example 2.1. It is enough to prove only one of the statements in the lemma. The other is obtained by renaming $\mathscr{E}$ into $\mathscr{J}$ and vice versa. By a property of orthogonal projections

$$\left\| \Gamma_\Omega \boldsymbol{J}\left(\frac{\boldsymbol{x}}{\varepsilon}\right) \right\|_{L^2(\Omega)} = \inf_{\boldsymbol{j} \in \mathscr{J}(\Omega)} \left\| \boldsymbol{J}\left(\frac{\boldsymbol{x}}{\varepsilon}\right) - \boldsymbol{j} \right\|_{L^2(\Omega)}. \tag{3.44}$$

The idea is to construct a test function $\boldsymbol{j} = \boldsymbol{j}_\varepsilon \in \mathscr{J}(\Omega)$ in (3.44), so that $\boldsymbol{J}(\boldsymbol{x}/\varepsilon) - \boldsymbol{j}_\varepsilon \to 0$, as $\varepsilon \to 0$, in $L^2(\Omega; \mathscr{T})$. Since Ω is bounded, there exists $a > 0$, such that $\Omega \subset [-a/4, a/4]^d$. Let $\phi(\boldsymbol{x})$ be given by its Fourier transform:

$$\hat{\phi}(\boldsymbol{\xi}) = a^d \chi_{[-\frac{1}{a},\, \frac{1}{a}]^d}(\boldsymbol{\xi}), \qquad \phi(\boldsymbol{x}) = \prod_{j=1}^{d} \frac{a \sin(2\pi x_j/a)}{\pi x_j}.$$

It is easy to see that $(4/\pi)^d \leqslant \phi(\boldsymbol{x}) \leqslant 2^d$ on $[-a/4,\, a/4]^d \supset \Omega$. Let $\psi \in C_0(\mathbb{R}^d)$ be such that $\psi(\boldsymbol{x}) = 1/\phi(\boldsymbol{x})$ for all $\boldsymbol{x} \in \Omega$. We now define

$$\boldsymbol{j}_\varepsilon(\boldsymbol{x}) = \boldsymbol{J}\left(\frac{\boldsymbol{x}}{\varepsilon}\right)\phi(\boldsymbol{x})\psi(\boldsymbol{x}) - \Gamma\left(\boldsymbol{J}\left(\frac{\boldsymbol{x}}{\varepsilon}\right)\phi(\boldsymbol{x})\psi(\boldsymbol{x})\right) \in \mathscr{J},$$

and therefore,

$$\boldsymbol{j}_\varepsilon(\boldsymbol{x})|\Omega = \boldsymbol{J}\left(\frac{\boldsymbol{x}}{\varepsilon}\right) - \Gamma\left(\boldsymbol{J}\left(\frac{\boldsymbol{x}}{\varepsilon}\right)\phi(\boldsymbol{x})\psi(\boldsymbol{x})\right)\Bigg|\,\Omega \in \mathscr{J}(\Omega).$$

Using this as a test function in (3.44) we obtain

$$\left\|\Gamma_\Omega \boldsymbol{J}\left(\frac{\boldsymbol{x}}{\varepsilon}\right)\right\|_{L^2(\Omega)} \leqslant \left\|\Gamma\left(\boldsymbol{J}\left(\frac{\boldsymbol{x}}{\varepsilon}\right)\phi(\boldsymbol{x})\psi(\boldsymbol{x})\right)\right\|_{L^2(\Omega)} \leqslant \left\|\Gamma\left(\boldsymbol{J}\left(\frac{\boldsymbol{x}}{\varepsilon}\right)\phi(\boldsymbol{x})\psi(\boldsymbol{x})\right)\right\|_{L^2(\mathbb{R}^d)}.$$

Let $\mathsf{K} = \Gamma\psi - \psi\Gamma$. By the first commutation lemma, [25], K is a compact operator on $L^2(\mathbb{R}^d;\, \mathscr{T})$. This would allow us to conclude that $\mathsf{K}(\phi(\boldsymbol{x})\boldsymbol{J}(\boldsymbol{x}/\varepsilon)) \to 0$ in the norm of $L^2(\mathbb{R}^d;\, \mathscr{T})$, if we can show that $\phi(\boldsymbol{x})\boldsymbol{J}(\boldsymbol{x}/\varepsilon) \rightharpoonup 0$, as $\varepsilon \to 0$, in the weak topology of $L^2(\mathbb{R}^d;\, \mathscr{T})$. This is a consequence of the following lemma.

Lemma 3.28. *Suppose that $\phi \in L^2(\mathbb{R}^d)$, such that $\hat{\phi}$ has a compact support. Let $f \in L^2_{\text{per}}(Q)$. Let $g_\varepsilon(\boldsymbol{x}) = \phi(\boldsymbol{x})f(\boldsymbol{x}/\varepsilon)$. Then $g_\varepsilon \rightharpoonup \phi(\boldsymbol{x})\langle f\rangle$, as $\varepsilon \to 0$, in the weak topology of $L^2(\mathbb{R}^d)$.*

Proof. The first observation is that it is sufficient to prove the lemma only for the case $\langle f\rangle = 0$. Let us first show that g_ε is bounded in $L^2(\mathbb{R}^d)$. Using the expansion

$$f(\boldsymbol{z}) = \sum_{\boldsymbol{k}\in\mathbb{Z}^d} \hat{f}(\boldsymbol{k}) e^{2\pi i \boldsymbol{Q}\boldsymbol{z}\cdot\boldsymbol{k}}$$

we compute

$$\hat{g}_\varepsilon(\boldsymbol{\xi}) = \sum_{\boldsymbol{k}\in\mathbb{Z}^d} \hat{f}(\boldsymbol{k})\hat{\phi}\left(\boldsymbol{\xi} - \frac{\boldsymbol{Q}^T\boldsymbol{k}}{\varepsilon}\right). \tag{3.45}$$

Then,

$$\|g_\varepsilon\|^2_{L^2(\mathbb{R}^d)} = \|\hat{g}_\varepsilon\|^2_{L^2(\mathbb{R}^d)} = \sum_{\boldsymbol{k}\in\mathbb{Z}^d} \int_{\mathbb{R}^d} \left|\hat{f}(\boldsymbol{k})\hat{\phi}\left(\boldsymbol{\xi} - \frac{\boldsymbol{Q}^T\boldsymbol{k}}{\varepsilon}\right)\right|^2 d\boldsymbol{\xi}$$

because for all sufficiently small $\varepsilon > 0$ functions $\hat{\phi}(\boldsymbol{\xi} - \boldsymbol{Q}^T\boldsymbol{k}/\varepsilon)$ and $\hat{\phi}(\boldsymbol{\xi} - \boldsymbol{Q}^T\boldsymbol{m}/\varepsilon)$ have disjoint support, whenever $\boldsymbol{m} \neq \boldsymbol{k}$. But then

$$\|g_\varepsilon\|^2_{L^2(\mathbb{R}^d)} = \sum_{\boldsymbol{k}\in\mathbb{Z}^d} |\hat{f}(\boldsymbol{k})|^2 \|\hat{\phi}\|^2_{L^2(\mathbb{R}^d)} = \det \boldsymbol{Q}\|f\|^2_{L^2(Q)}\|\phi\|^2_{L^2(\mathbb{R}^d)}$$

for all sufficiently small $\varepsilon > 0$. In particular, g_ε is bounded in $L^2(\mathbb{R}^d)$.

It is well-known that the set of functions with compact support is dense in $L^2(\mathbb{R}^d)$. Therefore, the set of functions, whose Fourier transform has compact support, is also dense in $L^2(\mathbb{R}^d)$. Let ψ be such a function. Then, by Plancherel's formula and (3.45), we obtain

$$(g_\varepsilon, \psi)_{L^2(\mathbb{R}^d)} = \sum_{k \in \mathbb{Z}^d \setminus \{0\}} \hat{f}(\boldsymbol{k}) \int_{\mathbb{R}^d} \hat{\phi}\left(\boldsymbol{\xi} - \frac{\boldsymbol{Q}^T \boldsymbol{k}}{\varepsilon}\right) \overline{\hat{\psi}(\boldsymbol{\xi})} d\boldsymbol{\xi} = 0,$$

for all sufficiently small $\varepsilon > 0$, because functions $\hat{\phi}(\boldsymbol{\xi} - \boldsymbol{Q}^T\boldsymbol{k}/\varepsilon)$ and $\hat{\psi}(\boldsymbol{\xi})$ have disjoint support for any $\boldsymbol{k} \in \mathbb{Z}^d \setminus \{0\}$. We conclude that $g_\varepsilon \rightharpoonup 0$, as $\varepsilon \to 0$, in the weak topology of $L^2(\mathbb{R}^d)$. □

Writing

$$\Gamma\left(\boldsymbol{J}\left(\frac{\boldsymbol{x}}{\varepsilon}\right)\phi(\boldsymbol{x})\psi(\boldsymbol{x})\right) = \mathsf{K}\left(\boldsymbol{J}\left(\frac{\boldsymbol{x}}{\varepsilon}\right)\phi(\boldsymbol{x})\right) + \psi(\boldsymbol{x})\Gamma\left(\boldsymbol{J}\left(\frac{\boldsymbol{x}}{\varepsilon}\right)\phi(\boldsymbol{x})\right),$$

we see that by compactness of K and lemma 3.28, the first term goes to 0 in the norm of $L^2(\mathbb{R}^d; \mathscr{T})$. Hence, in order to prove the lemma it remains to show that $\Gamma(\phi(\boldsymbol{x})\boldsymbol{J}(\boldsymbol{x}/\varepsilon)) \to 0$, as $\varepsilon \to 0$ in the norm of $L^2(\mathbb{R}^d; \mathscr{T})$. Replacing $\boldsymbol{J}(z)$ by its Fourier expansion (3.7) and applying the Plancherel identity we obtain

$$\left\| \Gamma\left(\boldsymbol{J}\left(\frac{\boldsymbol{x}}{\varepsilon}\right)\phi(\boldsymbol{x})\right) \right\|_{L^2(\mathbb{R}^d)}^2 = \int_{\mathbb{R}^d} \left| \boldsymbol{\Gamma}\left(\frac{\boldsymbol{\xi}}{|\boldsymbol{\xi}|}\right) \sum_{\boldsymbol{k} \in \mathbb{Z}^d \setminus \{0\}} \hat{\boldsymbol{J}}(\boldsymbol{k})\hat{\phi}\left(\boldsymbol{\xi} - \frac{\boldsymbol{Q}^T\boldsymbol{k}}{\varepsilon}\right) \right|^2 d\boldsymbol{\xi}.$$

Recall that for all sufficiently small $\varepsilon > 0$ functions $\hat{\phi}(\boldsymbol{\xi} - \boldsymbol{Q}^T\boldsymbol{k}/\varepsilon)$ and $\hat{\phi}(\boldsymbol{\xi} - \boldsymbol{Q}^T\boldsymbol{m}/\varepsilon)$ have disjoint support, whenever $\boldsymbol{m} \neq \boldsymbol{k}$. Therefore,

$$\int_{\mathbb{R}^d} \left| \boldsymbol{\Gamma}\left(\frac{\boldsymbol{\xi}}{|\boldsymbol{\xi}|}\right) \sum_{\boldsymbol{k} \in \mathbb{Z}^d \setminus \{0\}} \hat{\boldsymbol{J}}(\boldsymbol{k})\hat{\phi}\left(\boldsymbol{\xi} - \frac{\boldsymbol{Q}^T\boldsymbol{k}}{\varepsilon}\right) \right|^2 d\boldsymbol{\xi} =$$

$$\sum_{\boldsymbol{k} \in \mathbb{Z}^d \setminus \{0\}} \int_{\mathbb{R}^d} \left| \boldsymbol{\Gamma}\left(\frac{\boldsymbol{\xi}}{|\boldsymbol{\xi}|}\right) \hat{\boldsymbol{J}}(\boldsymbol{k})\hat{\phi}\left(\boldsymbol{\xi} - \frac{\boldsymbol{Q}^T\boldsymbol{k}}{\varepsilon}\right) \right|^2 d\boldsymbol{\xi}.$$

Changing variables $\boldsymbol{\xi}' = \boldsymbol{\xi} - \boldsymbol{Q}^T\boldsymbol{k}/\varepsilon$ in the integral, we obtain

$$\left\| \Gamma\left(\boldsymbol{J}\left(\frac{\boldsymbol{x}}{\varepsilon}\right)\phi(\boldsymbol{x})\right) \right\|_{L^2(\mathbb{R}^d)}^2 = \sum_{\boldsymbol{k} \in \mathbb{Z}^d \setminus \{0\}} \int_{\mathbb{R}^d} \left| \boldsymbol{\Gamma}\left(\frac{\boldsymbol{Q}^T\boldsymbol{k} + \varepsilon\boldsymbol{\xi}'}{|\boldsymbol{Q}^T\boldsymbol{k} + \varepsilon\boldsymbol{\xi}'|}\right) \hat{\boldsymbol{J}}(\boldsymbol{k})\hat{\phi}(\boldsymbol{\xi}') \right|^2 d\boldsymbol{\xi}'.$$

Since $\boldsymbol{\Gamma}(\eta)$ is projection-valued we have

$$\left| \boldsymbol{\Gamma}\left(\frac{\boldsymbol{Q}^T\boldsymbol{k} + \varepsilon\boldsymbol{\xi}'}{|\boldsymbol{Q}^T\boldsymbol{k} + \varepsilon\boldsymbol{\xi}'|}\right) \hat{\boldsymbol{J}}(\boldsymbol{k})\hat{\phi}(\boldsymbol{\xi}') \right| \leqslant |\hat{\boldsymbol{J}}(\boldsymbol{k})\hat{\phi}(\boldsymbol{\xi}')|.$$

Observing that

$$\sum_{\boldsymbol{k} \in \mathbb{Z}^d \setminus \{0\}} \int_{\mathbb{R}^d} |\hat{\boldsymbol{J}}(\boldsymbol{k})\hat{\phi}(\boldsymbol{\xi}')|^2 d\boldsymbol{\xi}' = \det \boldsymbol{Q} \|\boldsymbol{J}\|_{L^2_{\text{per}}(Q)}^2 \|\phi\|_{L^2(\mathbb{R}^d)}^2 < +\infty,$$

we conclude that the Lebesgue dominated convergence theorem is applicable and hence

$$\lim_{\varepsilon\to 0}\sum_{k\in\mathbb{Z}^d\setminus\{0\}}\int_{\mathbb{R}^d}\left|\Gamma\left(\frac{Q^Tk+\varepsilon\xi'}{|Q^Tk+\varepsilon\xi'|}\right)\hat{J}(k)\hat{\phi}(\xi')\right|^2 d\xi'=$$

$$\sum_{k\in\mathbb{Z}^d\setminus\{0\}}\int_{\mathbb{R}^d}\left|\Gamma\left(\frac{Q^Tk}{|Q^Tk|}\right)\hat{J}(k)\hat{\phi}(\xi')\right|^2 d\xi'=0.$$

□

For a given $e_0 \in \mathscr{U}$ let $E \in \mathscr{E}_{\text{per}}$ be the unique solution of the cell problem

$$E \in \mathscr{E}_{\text{per}}, \qquad J(z) = \mathsf{L}^T(z)(E(z) + e_0), \qquad J \in \mathscr{J}_{\text{per}} \oplus \mathscr{U}. \tag{3.46}$$

We note that the cell problem (3.46) is slightly different from the cell problem (3.10). At the end of this section we will establish their equivalence. At the moment, however, the periodic cell problem in the form (3.46) will be more convenient.

Theorem 3.29. (Periodic homogenization). *Assume that the compensated compactness postulate 2.3 holds. Let $\mathsf{L}(z) \in L^\infty(\mathbb{R}^d; \mathrm{End}(\mathscr{T}))$ be Q-periodic, satisfying $\mathsf{L}(z) \in \mathscr{M}(\alpha, \beta)$ for all $z \in Q$. Let Ω be an open and bounded domain in $\mathbb{R}^d$, and let $\mathsf{L}_\varepsilon(x) = \mathsf{L}(x/\varepsilon)$, $x \in \Omega$. Then L_ε H-converges to the constant tensor field $\mathsf{L}_* \in \mathrm{End}(\mathscr{T})$ on Ω, defined by its action on an arbitrary vector $e_0 \in \mathscr{T}$ by*

$$\mathsf{L}_*^T e_0 = \fint_Q \mathsf{L}^T(z)(E(z) + e_0)dz, \tag{3.47}$$

where $E(z)$ solves the cell problem (3.46).

Proof. Here we present the elegant proof of Murat and Tartar [19] based on the compensated compactness property. We fix an arbitrary $f \in \mathscr{E}_0(\Omega)$ and let $E_\varepsilon = \mathsf{T}^{-1}_{\mathsf{L}(x/\varepsilon)} f$. The sequences E_ε and $J_\varepsilon = \mathsf{L}(x/\varepsilon)E_\varepsilon$ are uniformly bounded in $L^2(\Omega; \mathscr{T})$ because of (2.22). By the Banach-Alaoglu theorem there is a subsequence ε_k such that $E_{\varepsilon_k} \rightharpoonup E_*$ and $J_{\varepsilon_k} \rightharpoonup J_*$ in $L^2(\Omega; \mathscr{T})$. We observe that $\Gamma_\Omega J_{\varepsilon_k} = f$ and $\Pi_\Omega E_{\varepsilon_k} = 0$. Observe that

$$\left(E\left(\frac{x}{\varepsilon_k}\right) + e_0, J_{\varepsilon_k}\right) = \left(\mathsf{L}^T\left(\frac{x}{\varepsilon_k}\right)\left(E\left(\frac{x}{\varepsilon_k}\right) + e_0\right), E_{\varepsilon_k}\right).$$

By lemma 3.27 and the compensated compactness property we obtain

$$(e_0, J_*) = (\mathsf{L}_*^T e_0, E_*) = (e_0, \mathsf{L}_* E_*)$$

for any $e_0 \in \mathscr{T}$. Thus, $J_* = \mathsf{L}_* E_*$, and therefore, $E_* = \mathsf{T}^{-1}_{\mathsf{L}_*} f$. Observe that the weak limits E_* and J_* are independent of the choice of the subsequence ε_k. We conclude that the original families E_ε and J_ε have weak limits E_* and J_*, respectively. Hence, $\mathsf{L}(x/\varepsilon)$ H-converges to L_*, and the theorem is proved. □

Corollary 3.30. *Let* L_* *be the effective tensor of the periodic composite with period cell* Q *and local tensor* $\mathsf{L}(\boldsymbol{x})$. *Then* L_* *is given by (3.11).*

Proof. By theorem 3.29 $\mathsf{L}^T(\boldsymbol{x}/\varepsilon) \overset{H}{\rightharpoonup} \mathsf{K}_*$ and $\mathsf{K}_*^T\boldsymbol{e}_0 = \langle \mathsf{L}(\boldsymbol{z})(\boldsymbol{E}(\boldsymbol{z}) + \boldsymbol{e}_0)\rangle$, for all $\boldsymbol{e}_0 \in \mathscr{T}$, where $\boldsymbol{E}(\boldsymbol{z})$ solves the cell problem (3.10). Let $\boldsymbol{E}' \in \mathscr{E}_{\text{per}}$ be the solution of the cell problem (3.46) with $\boldsymbol{e}_0$ replaced by $\boldsymbol{e}_0'$. We use the observation that for any $\tilde{\boldsymbol{E}} \in \mathscr{E}_{\text{per}} \oplus \mathscr{U}$ and any $\tilde{\boldsymbol{J}} \in \mathscr{J}_{\text{per}} \oplus \mathscr{U}$ we have $\langle(\tilde{\boldsymbol{E}}, \tilde{\boldsymbol{J}})\rangle = (\langle\tilde{\boldsymbol{E}}\rangle, \langle\tilde{\boldsymbol{J}}\rangle)$. Therefore,

$$(\boldsymbol{e}_0, \mathsf{L}_*^T\boldsymbol{e}_0') = \langle(\boldsymbol{E} + \boldsymbol{e}_0, \mathsf{L}^T(\boldsymbol{E}' + \boldsymbol{e}_0'))\rangle = \langle(\mathsf{L}(\boldsymbol{E} + \boldsymbol{e}_0), \boldsymbol{E}' + \boldsymbol{e}_0')\rangle = (\mathsf{K}_*^T\boldsymbol{e}_0, \boldsymbol{e}_0').$$

We conclude that $\mathsf{L}_* = \mathsf{K}_*^T$. □

3.3.2 Basic properties of H-convergence

In this section we prove a number of fundamental properties of H-convergence, given by definition 3.1.

Theorem 3.31. *Assume that*

(i) $\Gamma(\eta)$ *is* $\mathscr{E}$*-nondegenerate*
(ii) $\Gamma(\eta)$ *satisfies the compensated compactness property*
(iii) $\mathsf{L}_\varepsilon \in \mathscr{M}(\alpha, \beta)^\Omega$
(iv) $\mathsf{L}_\varepsilon \overset{H}{\rightharpoonup} \mathsf{L}_* \in \mathscr{M}(\alpha, \beta)^\Omega$, *as* $\varepsilon \to 0$

Then the following statements are true.

(a) $\mathsf{L}_\varepsilon^T \overset{H}{\rightharpoonup} \mathsf{L}_*^T$, *as* $\varepsilon \to 0$ *on* Ω.
(b) *Boundary conditions do not affect H-limits. Suppose* $\omega \subset \Omega$, $\boldsymbol{E}_\varepsilon \in \overline{\mathscr{E}(\omega)}$, $\boldsymbol{E}_\varepsilon \rightharpoonup \boldsymbol{E}_*$ *in* $L^2(\omega; \mathscr{T})$, $\boldsymbol{g} \in L^2(\omega; \mathscr{T})$ *and* $\Gamma_\omega(\mathsf{L}_\varepsilon(\boldsymbol{g} + \boldsymbol{E}_\varepsilon)) \to \boldsymbol{f}$ *in* $L^2(\omega; \mathscr{T})$. *Then* $\mathsf{L}_\varepsilon(\boldsymbol{g} + \boldsymbol{E}_\varepsilon) \rightharpoonup \mathsf{L}_*(\boldsymbol{g} + \boldsymbol{E}_*)$ *in* $L^2(\omega; \mathscr{T})$.
(c) *H-limit is local: if* $\omega \subset \Omega$, *then* $\mathsf{L}_\varepsilon \overset{H}{\rightharpoonup} \mathsf{L}_*$ *in* ω.
(d) *Suppose* Ω *is a disjoint union of a finite number of sets* ω_i, $i = 1,\ldots, N$ *in the sense that* $\omega_i \cap \omega_j = \varnothing$, *if* $i \neq j$ *and*

$$\left|\Omega \setminus \bigcup_{i=1}^{N} \omega_i\right| = 0.$$

Suppose that $\mathsf{L}_\varepsilon \in \mathscr{M}(\alpha, \beta)^\Omega$ *and* $\mathsf{L}_\varepsilon|\omega_i \overset{H}{\rightharpoonup} \mathsf{L}_i$ *in* ω_i, $i = 1,\ldots, N$. *Then* $\mathsf{L}_\varepsilon \overset{H}{\rightharpoonup} \mathsf{L}_*$ *in* Ω, *where* $\mathsf{L}_*(\boldsymbol{x}) = \mathsf{L}_i(\boldsymbol{x})$, *if* $\boldsymbol{x} \in \omega_i$.

Proof. Part (a). Let $\{\boldsymbol{f}, \boldsymbol{g}\} \subset \mathscr{E}_0(\Omega)$ be arbitrary. Let $\boldsymbol{E}_\varepsilon' = \mathsf{T}_{\mathsf{L}_\varepsilon^T}^{-1}\boldsymbol{f}$ and $\boldsymbol{E}_\varepsilon = \mathsf{T}_{\mathsf{L}_\varepsilon}^{-1}\boldsymbol{g}$. Then

$$(\boldsymbol{E}_\varepsilon', \boldsymbol{g})_{L^2(\Omega)} = (\boldsymbol{E}_\varepsilon', \mathsf{T}_{\mathsf{L}_\varepsilon}\boldsymbol{E}_\varepsilon)_{L^2(\Omega)} = (\mathsf{T}_{\mathsf{L}_\varepsilon^T}\boldsymbol{E}_\varepsilon', \boldsymbol{E}_\varepsilon)_{L^2(\Omega)} = (\boldsymbol{f}, \boldsymbol{E}_\varepsilon)_{L^2(\Omega)}.$$

By definition of H-convergence

$$\lim_{\varepsilon\to 0}(\boldsymbol{E}'_\varepsilon, \boldsymbol{g})_{L^2(\Omega)} = \lim_{\varepsilon\to 0}(\boldsymbol{f}, \boldsymbol{E}_\varepsilon)_{L^2(\Omega)} = (\boldsymbol{f}, \mathsf{T}_{\mathsf{L}_*}^{-1}\boldsymbol{g})_{L^2(\Omega)} = (\mathsf{T}_{\mathsf{L}_*^T}^{-1}\boldsymbol{f}, \boldsymbol{g})_{L^2(\Omega)}.$$

Hence, $\boldsymbol{E}'_\varepsilon \rightharpoonup \boldsymbol{E}_* = \mathsf{T}_{\mathsf{L}_*^T}^{-1}\boldsymbol{f}$. Now we extract a weakly convergent subsequence $\boldsymbol{J}_{\varepsilon_k}$ from $\boldsymbol{J}_\varepsilon = \mathsf{L}_\varepsilon^T\boldsymbol{E}'_\varepsilon$ and apply the compensated compactness property. On the one hand we have

$$(\mathsf{L}_{\varepsilon_k}^T\boldsymbol{E}'_{\varepsilon_k}, \boldsymbol{E}_{\varepsilon_k}) \overset{*}{\rightharpoonup} (\boldsymbol{J}_*, \mathsf{T}_{\mathsf{L}_*}^{-1}\boldsymbol{g}).$$

On the other hand

$$(\mathsf{L}_{\varepsilon_k}^T\boldsymbol{E}'_{\varepsilon_k}, \boldsymbol{E}_{\varepsilon_k}) = (\boldsymbol{E}'_{\varepsilon_k}, \mathsf{L}_{\varepsilon_k}\boldsymbol{E}_{\varepsilon_k}) \overset{*}{\rightharpoonup} (\boldsymbol{E}_*, \mathsf{L}_*\mathsf{T}_{\mathsf{L}_*}^{-1}\boldsymbol{g}).$$

We conclude that for a.e. $\boldsymbol{x} \in \Omega$

$$(\boldsymbol{J}_* - \mathsf{L}_*^T\boldsymbol{E}_*, \mathsf{T}_{\mathsf{L}_*}^{-1}\boldsymbol{g}) = 0 \text{ for any } \boldsymbol{g} \in \mathscr{E}_0(\Omega).$$

Since $\mathsf{T}_{\mathsf{L}_*}: \mathscr{E}_0(\Omega) \to \mathscr{E}_0(\Omega)$ is an isomorphism, we conclude by (2.30) that $\boldsymbol{J}_* - \mathsf{L}_*^T\boldsymbol{E}_* \in \mathscr{N}(\Omega) = \{0\}$, by corollaries 2.27 and 2.28 and the $\mathscr{E}$-nondegeneracy. Thus, the weak limit $\boldsymbol{J}_* = \mathsf{L}_*^T\boldsymbol{E}_*$ is independent of the weakly convergent subsequence of $\boldsymbol{J}_\varepsilon$, and therefore, the entire family $\boldsymbol{J}_\varepsilon \rightharpoonup \mathsf{L}_*^T\boldsymbol{E}_*$. Together with $\boldsymbol{E}_* = \mathsf{T}_{\mathsf{L}_*^T}^{-1}\boldsymbol{f}$, this proves that $\mathsf{L}_\varepsilon^T \overset{H}{\rightharpoonup} \mathsf{L}_*^T$.

Part (b). Fix any $\boldsymbol{E}_0 \in \mathscr{E}_0(\Omega)$ and let $\tilde{\boldsymbol{E}}_\varepsilon \in \mathscr{E}_0(\Omega)$ be the solution of

$$\Gamma_\Omega(\mathsf{L}_\varepsilon^T\tilde{\boldsymbol{E}}_\varepsilon) = \Gamma_\Omega(\mathsf{L}_*^T\boldsymbol{E}_0).$$

By part (a) $\mathsf{L}_\varepsilon^T \overset{H}{\rightharpoonup} \mathsf{L}_*^T$ and therefore, $\tilde{\boldsymbol{E}}_\varepsilon \rightharpoonup \boldsymbol{E}_0$ and $\mathsf{L}_\varepsilon^T\tilde{\boldsymbol{E}}_\varepsilon \rightharpoonup \mathsf{L}_*^T\boldsymbol{E}_0$, as $\varepsilon \to 0$. We now want to apply the compensated compactness theorem on ω. We have $X_0(\Omega)\tilde{\boldsymbol{E}}_\varepsilon \in \mathscr{E}$, and hence $\tilde{\boldsymbol{E}}_\varepsilon|\omega \in \mathscr{E}(\omega)$. Observe that $\mathsf{L}_\varepsilon^T\tilde{\boldsymbol{E}}_\varepsilon - \mathsf{L}_*^T\boldsymbol{E}_0 \in \overline{\mathscr{J}(\Omega)}$. Therefore, $(\mathsf{L}_\varepsilon^T\tilde{\boldsymbol{E}}_\varepsilon - \mathsf{L}_*^T\boldsymbol{E}_0)|\omega \in \overline{\mathscr{J}(\omega)}$, and thus

$$\Gamma_\omega(\mathsf{L}_\varepsilon(\boldsymbol{g} + \boldsymbol{E}_\varepsilon)) \to \boldsymbol{f}, \quad \Pi_\omega(\tilde{\boldsymbol{E}}_\varepsilon|\omega) = 0,$$

$$\Gamma_\omega(\mathsf{L}_\varepsilon^T\tilde{\boldsymbol{E}}_\varepsilon|\omega) = \Gamma_\omega(\mathsf{L}_*^T\boldsymbol{E}_0|\omega), \quad \Pi_\omega(\boldsymbol{g} + \boldsymbol{E}_\varepsilon) = \Pi_\omega\boldsymbol{g}.$$

The family $\mathsf{L}_\varepsilon(\boldsymbol{g} + \boldsymbol{E}_\varepsilon)$ is bounded in $L^2(\omega; \mathscr{T})$, since $\boldsymbol{E}_\varepsilon \rightharpoonup \boldsymbol{E}_*$, and therefore, there exists a sequence $\varepsilon_k \to 0$ such that $\mathsf{L}_{\varepsilon_k}(\boldsymbol{g} + \boldsymbol{E}_{\varepsilon_k}) \rightharpoonup \boldsymbol{q}$ in $L^2(\omega; \mathscr{T})$ By the compensated compactness theorem we have

$$(\mathsf{L}_{\varepsilon_k}(\boldsymbol{g} + \boldsymbol{E}_{\varepsilon_k}), \tilde{\boldsymbol{E}}_{\varepsilon_k}|\omega) \overset{*}{\rightharpoonup} (\boldsymbol{q}, \boldsymbol{E}_0|\omega).$$

But

$$(\mathsf{L}_{\varepsilon_k}(\boldsymbol{g} + \boldsymbol{E}_{\varepsilon_k}), \tilde{\boldsymbol{E}}_{\varepsilon_k}|\omega) = (\boldsymbol{g} + \boldsymbol{E}_{\varepsilon_k}, \mathsf{L}_{\varepsilon_k}^T(\tilde{\boldsymbol{E}}_{\varepsilon_k}|\omega)),$$

and by the compensated compactness theorem

$$(\boldsymbol{g} + \boldsymbol{E}_{\varepsilon_k}, \mathsf{L}_{\varepsilon_k}^T(\tilde{\boldsymbol{E}}_{\varepsilon_k}|\omega)) \overset{*}{\rightharpoonup} (\boldsymbol{g} + \boldsymbol{E}_*, \mathsf{L}_*^T\boldsymbol{E}_0|\omega).$$

Thus, we have proved that

$$(\boldsymbol{q} - \mathsf{L}_*(\boldsymbol{g} + \boldsymbol{E}_*), \boldsymbol{E}_0|\omega) = 0 \text{ for a.e. } \boldsymbol{x} \in \omega.$$

Equivalently,

$$(X_0(\omega)(\boldsymbol{q} - \mathsf{L}_*(\boldsymbol{g} + \boldsymbol{E}_*)), \boldsymbol{E}_0) = 0 \text{ for a.e. } \boldsymbol{x} \in \Omega.$$

We conclude that $X_0(\omega)(\boldsymbol{q} - \mathsf{L}_*(\boldsymbol{g} + \boldsymbol{E}_*)) \in \mathscr{N}(\Omega) = \{0\}$. Thus, the weak limit $\boldsymbol{q} = \mathsf{L}_*(\boldsymbol{g} + \boldsymbol{E}_*)$ is independent of the weakly convergent subsequence of $\mathsf{L}_\varepsilon(\boldsymbol{g} + \boldsymbol{E}_\varepsilon)$ on ω, and therefore, $\mathsf{L}_\varepsilon(\boldsymbol{g} + \boldsymbol{E}_\varepsilon) \rightharpoonup \mathsf{L}_*(\boldsymbol{g} + \boldsymbol{E}_*)$, as $\varepsilon \to 0$ in $L^2(\omega; \mathscr{T})$.

Part (c). Let $\boldsymbol{f} \in \mathscr{E}_0(\omega)$ and $\boldsymbol{E}_\varepsilon \in \mathscr{E}_0(\omega)$ be the unique solution of $\Gamma_\omega(\mathsf{L}_\varepsilon \boldsymbol{E}_\varepsilon) = \boldsymbol{f}$. Then, the family $\boldsymbol{E}_\varepsilon = \mathsf{T}_{\mathsf{L}_\varepsilon}^{-1}\boldsymbol{f}$ is bounded, since due to (2.22) we have

$$\|\boldsymbol{E}_\varepsilon\| \leqslant \|\mathsf{T}_{\mathsf{L}_\varepsilon}^{-1}\| \|\boldsymbol{f}\| \leqslant \frac{1}{\alpha}\|\boldsymbol{f}\|.$$

It follows that we can extract a weakly convergent subsequence $\boldsymbol{E}_{\varepsilon_k} \rightharpoonup \boldsymbol{E}_* \in \mathscr{E}_0(\omega)$, as $k \to \infty$ in $L^2(\omega; \mathscr{T})$. By part (b) we conclude that $\mathsf{L}_{\varepsilon_k}\boldsymbol{E}_{\varepsilon_k} \rightharpoonup \mathsf{L}_*\boldsymbol{E}_*$, as $k \to \infty$ in $L^2(\omega; \mathscr{T})$. Therefore, $\boldsymbol{E}_* \in \mathscr{E}_0(\omega)$ satisfies $\Gamma_\omega(\mathsf{L}_*\boldsymbol{E}_*) = \boldsymbol{f}$. Hence, the weak limit $\boldsymbol{E}_*$ of a subsequence $\boldsymbol{E}_{\varepsilon_k}$ is uniquely determined by $\boldsymbol{f}$ and L_* and hence, the entire family $\boldsymbol{E}_\varepsilon$ converges weakly to $\boldsymbol{E}_*$ in $L^2(\omega; \mathscr{T})$. Thus, $\mathsf{L}_\varepsilon \overset{H}{\rightharpoonup} \mathsf{L}_*$ in ω.

Part (d). Let $\boldsymbol{f} \in \mathscr{E}_0(\Omega)$ and $\boldsymbol{E}_\varepsilon \in \mathscr{E}_0(\Omega)$ be the unique solution of $\Gamma_\Omega(\mathsf{L}_\varepsilon \boldsymbol{E}_\varepsilon) = \boldsymbol{f}$. The family $\boldsymbol{E}_\varepsilon$ is bounded in $L^2(\Omega; \mathscr{T})$. Therefore, we can extract a weakly convergent subsequence $\boldsymbol{E}_{\varepsilon_k} \rightharpoonup \boldsymbol{E}_0$ as $k \to \infty$ in $L^2(\Omega; \mathscr{T})$. Observe that $\Gamma_{\omega_i}(\mathsf{L}_{\varepsilon_k}\boldsymbol{E}_{\varepsilon_k}|\omega_i) = \Gamma_{\omega_i}(\boldsymbol{f}|\omega_i)$. Indeed, $\mathsf{L}_{\varepsilon_k}\boldsymbol{E}_{\varepsilon_k} - \boldsymbol{f} \in \overline{\mathscr{J}(\Omega)}$, and therefore, $(\mathsf{L}_{\varepsilon_k}\boldsymbol{E}_{\varepsilon_k} - \boldsymbol{f})|\omega_i \in \overline{\mathscr{J}(\omega_i)}$. By part (b), $\mathsf{L}_{\varepsilon_k}\boldsymbol{E}_{\varepsilon_k}|\omega_i \rightharpoonup \mathsf{L}_i\boldsymbol{E}_0$ in $L^2(\omega_i; \mathscr{T})$. Therefore, $\mathsf{L}_{\varepsilon_k}\boldsymbol{E}_{\varepsilon_k} \rightharpoonup \mathsf{L}_*\boldsymbol{E}_0$ in $L^2(\Omega; \mathscr{T})$. But then $\boldsymbol{E}_0$ satisfies $\Gamma_\Omega(\mathsf{L}_*\boldsymbol{E}_0) = \boldsymbol{f}$. This determines $\boldsymbol{E}_0$ uniquely, since $\mathsf{L}_* \in \mathscr{M}(\alpha, \beta)^\Omega$. It follows that the entire family $\boldsymbol{E}_\varepsilon$ weakly converges to $\boldsymbol{E}_0$ in $L^2(\Omega; \mathscr{T})$, since the limit of every weakly convergent subsequence must be $\boldsymbol{E}_0$. Hence, $\mathsf{L}_\varepsilon \overset{H}{\rightharpoonup} \mathsf{L}_*$. □

3.3.3 H-compactness

Now that the basic properties of H-convergence are established we are ready to prove theorem 3.3 describing the remarkable compactness property of H-convergence. To be precise, we are interested in proving the sequential H-compactness of $\mathscr{M}(\alpha, \beta)^\Omega$. Using the notation $\mathscr{B}(X, Y)$ for the space all bounded linear maps between two Banach spaces X and Y, and $\mathscr{B}(X)$ denoting $\mathscr{B}(X, X)$, we note that by definition, H-topology on $\mathscr{M}(\alpha, \beta)^\Omega$ is a pull-back topology of the weak topology on $\mathscr{B}(\mathscr{E}_0(\Omega), \mathscr{E}_0(\Omega) \times L^2(\Omega; \mathscr{T}))$ via

$$\Phi: \mathscr{M}(\alpha, \beta)^\Omega \to \mathscr{B}(\mathscr{E}_0(\Omega), \mathscr{E}_0(\Omega) \times L^2(\Omega; \mathscr{T})), \quad \Phi(L)\boldsymbol{f} = \begin{bmatrix} \mathsf{T}_{\mathsf{L}}^{-1}\boldsymbol{f} \\ \mathsf{L}\mathsf{T}_{\mathsf{L}}^{-1}\boldsymbol{f} \end{bmatrix}$$

that maps $\mathscr{M}(\alpha, \beta)^\Omega$ into a norm-bounded subset of $\mathscr{B}(\mathscr{E}_0(\Omega), \mathscr{E}_0(\Omega) \times L^2(\Omega; \mathscr{T}))$, on which the weak topology is metrizable, since both $\mathscr{E}_0(\Omega)$ and $\mathscr{E}_0(\Omega) \times L^2(\Omega; \mathscr{T})$

are separable. Hence, H-topology on $\mathscr{M}(\alpha, \beta)^{\Omega}$ is also metrizable. In this case, there is no difference between topological compactness and sequential compactness. With that said, we care only about the sequential compactness property. We are now ready to prove theorem 3.3.

Proof of theorem 3.3. We start the proof with several lemmas.

Lemma 3.32. *Suppose* $\mathsf{L}_\varepsilon \in \mathscr{M}(\alpha, \beta)^{\Omega}$ *and* $\mathsf{A} \in \mathscr{B}(\mathscr{E}_0(\Omega))$ *is such that* $\mathsf{T}_{\mathsf{L}_\varepsilon}^{-1} \rightharpoonup \mathsf{A}$ *in the weak topology of* $\mathscr{B}(\mathscr{E}_0(\Omega))$. *Then for any* $\boldsymbol{f} \in \mathscr{E}_0(\Omega)$ *we have*

$$\|\mathsf{A}\| \leqslant \frac{1}{\alpha}, \qquad (\mathsf{A}\boldsymbol{f}, \boldsymbol{f})_{L^2(\Omega)} \geqslant \frac{\alpha}{\beta^2}\|\boldsymbol{f}\|^2.$$

In particular, $\mathsf{A} \in \mathscr{B}(\mathscr{E}_0(\Omega))$.

Proof. The set

$$C(\alpha, \beta) = \{\mathsf{A} \in \mathscr{B}(\mathscr{E}_0(\Omega)) \colon \|\mathsf{A}\| \leqslant \frac{1}{\alpha}, \forall \boldsymbol{f} \in \mathscr{E}_0(\Omega)\ (\mathsf{A}\boldsymbol{f}, \boldsymbol{f})_{L^2(\Omega)} \geqslant \frac{\alpha}{\beta^2}\|\boldsymbol{f}\|^2\}$$

is a closed, convex, and bounded subset of $\mathscr{B}(\mathscr{E}_0(\Omega))$. It is therefore, weakly compact and sequentially weakly compact, since $\mathscr{E}_0(\Omega)$ is separable. The lemma will then follow if we show that $\mathsf{T}_{\mathsf{L}}^{-1} \in C(\alpha, \beta)$ for any $\mathsf{L} \in \mathscr{M}(\alpha, \beta)^{\Omega}$. Let us show that inequalities (2.21), (2.22) imply this statement. Indeed, the inequality $\|T_{\mathsf{L}}^{-1}\| \leqslant \alpha^{-1}$ is (2.22). From $\|\mathsf{T}_{\mathsf{L}}\| \leqslant \beta$ we have $\|\mathsf{T}_{\mathsf{L}}\boldsymbol{E}\|_{L^2(\Omega)} \leqslant \beta\|\boldsymbol{E}\|_{L^2(\Omega)}$. Setting $\boldsymbol{E} = \mathsf{T}_{\mathsf{L}}^{-1}\boldsymbol{f}$, we obtain

$$\|\mathsf{T}_{\mathsf{L}}^{-1}\boldsymbol{f}\|_{L^2(\Omega)} \geqslant \frac{1}{\beta}\|\boldsymbol{f}\|_{L^2(\Omega)}.$$

Hence,

$$(\mathsf{T}_{\mathsf{L}}\boldsymbol{E}, \boldsymbol{E})_{L^2(\Omega)} = (\mathsf{L}\boldsymbol{E}, \boldsymbol{E})_{L^2(\Omega)} \geqslant \alpha\|\boldsymbol{E}\|^2_{L^2(\Omega)}$$

implies

$$(\mathsf{T}_{\mathsf{L}}^{-1}\boldsymbol{f}, \boldsymbol{f})^2_{L^2(\Omega)} \geqslant \alpha\|\mathsf{T}_{\mathsf{L}}^{-1}\boldsymbol{f}\|^2_{L^2(\Omega)} \geqslant \frac{\alpha}{\beta^2}\|\boldsymbol{f}\|^2_{L^2(\Omega)}.$$

The lemma is proved. □

Lemma 3.33. *Suppose* $V_{\mathscr{E}} = \mathscr{T}_{\mathscr{E}} = \mathscr{T}$. *Let* $\{\mathsf{A}, \mathsf{B}\} \in \mathscr{B}(\mathscr{E}_0(\Omega); L^2(\Omega; \mathscr{T}))$. *Suppose that for every* $\{\boldsymbol{f}_1, \boldsymbol{f}_2\} \subset \mathscr{E}_0(\Omega)$ *we have*

$$((\mathsf{A}\boldsymbol{f}_1)(\boldsymbol{x}), \boldsymbol{f}_2(\boldsymbol{x})) = (\boldsymbol{f}_1(\boldsymbol{x}), (\mathsf{B}\boldsymbol{f}_2)(\boldsymbol{x})) \text{ for a.e. } \boldsymbol{x} \in \Omega. \tag{3.48}$$

Then there exists a unique measurable function $\mathsf{L}\colon \Omega \to \mathrm{End}(\mathscr{T})$, *such that*

$$(\mathsf{A}\boldsymbol{f})(\boldsymbol{x}) = \mathsf{L}(\boldsymbol{x})\boldsymbol{f}(\boldsymbol{x}), \quad (\mathsf{B}\boldsymbol{f})(\boldsymbol{x}) = \mathsf{L}^T(\boldsymbol{x})\boldsymbol{f}(\boldsymbol{x}), \text{ for a.e. } \boldsymbol{x} \in \Omega,$$

for every $\boldsymbol{f} \in \mathscr{E}_0(\Omega)$.

Proof. The difficulty in proving the lemma lies in the fact that the null-sets in Ω on which (3.48) fails may depend on $\boldsymbol{f}_1$ and $\boldsymbol{f}_2$. Let $\{\boldsymbol{E}_j: j \geqslant 1\}$ be a countable dense subset of $\mathscr{E}_0(\Omega)$. By theorem 2.25(c) there exists a null-set (set of measure zero) $\mathscr{N}_0$, such that $\Omega\backslash\mathscr{N}_0$ consists of the common Lebesgue points of $\boldsymbol{E}_j(\boldsymbol{x})$, such that

$$\mathscr{T} = V_{\mathscr{E}} = \mathrm{Span}\{\boldsymbol{E}_j(\boldsymbol{x}): j \geqslant 1\}, \quad \forall \boldsymbol{x} \in \Omega\backslash\mathscr{N}_0.$$

Let $\{\boldsymbol{e}_k: k = 1,\ldots, \dim \mathscr{T}\}$ be an orthonormal basis of $\mathscr{T}$. Then for any $\boldsymbol{x} \in \Omega\backslash\mathscr{N}_0$, there are real numbers $\{\alpha_j^{(k)}(\boldsymbol{x}): j \geqslant 1,\ k = 1,\ldots, \dim \mathscr{T}\}$ such that

$$\boldsymbol{e}_k = \sum_{j=1}^{\infty}\alpha_j^{(k)}(\boldsymbol{x})\boldsymbol{E}_j(\boldsymbol{x}),$$

where only finitely many coefficients $\alpha_j^{(k)}(\boldsymbol{x})$ are nonzero for each k and $\boldsymbol{x} \in \Omega\backslash\mathscr{N}_0$. The numbers $\alpha_j^k(\boldsymbol{x})$ are non-unique, but for each $\boldsymbol{x} \in \Omega\backslash\mathscr{N}_0$ their choices are made and fixed for the duration of the proof. Now let us pick an arbitrary $\boldsymbol{f} \in \mathscr{E}_0(\Omega)$ and let $\mathscr{N}_{\boldsymbol{f}} \supset \mathscr{N}_0$ be a null-set, such that all points of $\Omega\backslash\mathscr{N}_{\boldsymbol{f}}$ are the common Lebesgue points of $(\mathsf{B}\boldsymbol{E}_j)(\boldsymbol{x})$, $\boldsymbol{f}(\boldsymbol{x})$, and $(\mathsf{A}\boldsymbol{f})(\boldsymbol{x})$. Then,

$$((\mathsf{A}\boldsymbol{f})(\boldsymbol{x}),\ \boldsymbol{E}_j(\boldsymbol{x})) = (\boldsymbol{f}(\boldsymbol{x}),\ (\mathsf{B}\boldsymbol{E}_j)(\boldsymbol{x})), \quad \forall \boldsymbol{x} \in \Omega\backslash\mathscr{N}_{\boldsymbol{f}}.$$

Then

$$((\mathsf{A}\boldsymbol{f})(\boldsymbol{x}),\ \boldsymbol{e}_k) = (\boldsymbol{f}(\boldsymbol{x}),\ \boldsymbol{a}_k(\boldsymbol{x})), \quad \forall \boldsymbol{x} \in \Omega\backslash\mathscr{N}_{\boldsymbol{f}},\ \ k = 1,\ldots, \dim \mathscr{T},$$

where

$$\boldsymbol{a}_k(\boldsymbol{x}) = \sum_{j=1}^{\infty}\alpha_j^{(k)}(\boldsymbol{x})(\mathsf{B}\boldsymbol{E}_j)(\boldsymbol{x}).$$

Writing $\boldsymbol{a}_k(\boldsymbol{x}) = L_k^s(\boldsymbol{x})\boldsymbol{e}_s$ (using the Einstein summation convention), we obtain

$$((\mathsf{A}\boldsymbol{f})(\boldsymbol{x}),\ \boldsymbol{e}_k) = L_k^s(\boldsymbol{x})(\boldsymbol{f}(\boldsymbol{x}),\ \boldsymbol{e}_s).$$

In other words $L_k^s(\boldsymbol{x})$ are the orthogonal components of $\mathsf{L}(\boldsymbol{x}) \in \mathrm{End}(\mathscr{T})$, such that $(\mathsf{A}\boldsymbol{f})(\boldsymbol{x}) = \mathsf{L}(\boldsymbol{x})\boldsymbol{f}(\boldsymbol{x})$ for a.e. $\boldsymbol{x} \in \Omega$. It is important to observe that the numbers $L_k^s(\boldsymbol{x})$, $\boldsymbol{x} \in \Omega\backslash\mathscr{N}_0$ are independent of the choice of $\boldsymbol{f} \in \mathscr{E}_0(\Omega)$. Repeating the same argument, reversing the roles of A and B we obtain

$$((\mathsf{B}\boldsymbol{f})(\boldsymbol{x}),\ \boldsymbol{e}_k) = (\boldsymbol{f}(\boldsymbol{x}),\ \boldsymbol{b}_k(\boldsymbol{x})),$$

where

$$\boldsymbol{b}_k(\boldsymbol{x}) = \sum_{j=1}^{\infty}\alpha_j^{(k)}(\boldsymbol{x})(\mathsf{A}\boldsymbol{E}_j)(\boldsymbol{x}) = \sum_{j=1}^{\infty}\alpha_j^{(k)}(\boldsymbol{x})\mathsf{L}(\boldsymbol{x})\boldsymbol{E}_j(\boldsymbol{x}) = \mathsf{L}(\boldsymbol{x})\boldsymbol{e}_k.$$

Thus,

$$((\mathsf{B}\boldsymbol{f})(\boldsymbol{x}),\ \boldsymbol{e}_k) = (\boldsymbol{f}(\boldsymbol{x}),\ \mathsf{L}(\boldsymbol{x})\boldsymbol{e}_k) = (\mathsf{L}^T(\boldsymbol{x})\boldsymbol{f}(\boldsymbol{x}),\ \boldsymbol{e}_k).$$

Hence, $(\mathsf{B}\boldsymbol{f})(\boldsymbol{x}) = \mathsf{L}^T(\boldsymbol{x})\boldsymbol{f}(\boldsymbol{x})$ for a.e. $\boldsymbol{x} \in \Omega$. The uniqueness of $\mathsf{L}(\boldsymbol{x})$ follows from our standard argument. If there are two operators $\mathsf{L}(\boldsymbol{x})$ and $\mathsf{L}'(\boldsymbol{x})$, such that $(\mathsf{A}\boldsymbol{f})(\boldsymbol{x}) = \mathsf{L}(\boldsymbol{x})\boldsymbol{f}(\boldsymbol{x}) = \mathsf{L}'(\boldsymbol{x})\boldsymbol{f}(\boldsymbol{x})$, then $(\mathsf{L}(\boldsymbol{x}) - \mathsf{L}'(\boldsymbol{x}))\boldsymbol{f}(\boldsymbol{x}) = 0$, and for any $\boldsymbol{e}_k$ we have $(\boldsymbol{f}(\boldsymbol{x}), (\mathsf{L} - \mathsf{L}')^T\boldsymbol{e}_k) = 0$ for every $\boldsymbol{f} \in \mathscr{E}_0(\Omega)$. Thus the functions $\boldsymbol{c}_k(\boldsymbol{x}) = (\mathsf{L}(\boldsymbol{x}) - \mathsf{L}'(\boldsymbol{x}))^T\boldsymbol{e}_k$ are in $\mathscr{N}(\Omega) = \{0\}$, by corollary 2.27. It follows that $\mathsf{L}(\boldsymbol{x}) = \mathsf{L}'(\boldsymbol{x})$ for a.e. $\boldsymbol{x} \in \Omega$. Since functions $\mathsf{L}(\boldsymbol{x})\boldsymbol{f}(\boldsymbol{x})$ and $\mathsf{L}^T(\boldsymbol{x})\boldsymbol{f}(\boldsymbol{x})$ are in $L^2(\Omega; \mathscr{T})$ for every $\boldsymbol{f} \in \mathscr{E}_0(\Omega)$, we conclude that $\mathsf{L}(\boldsymbol{x})$ must have measurable components. This statement is proved in the lemma below, which adapts the ideas from the proof of lemma 2.26 to the current context. □

Lemma 3.34. *Let Ω_0^* be the set of full measure containing all common Lebesgue points of a countable dense subset $D = \{\boldsymbol{E}_j\colon j \geqslant 1\} \subset \mathscr{E}_0(\Omega)$ as in theorem 2.25(c). Suppose that $\mathsf{L}\colon \Omega_0^* \to \mathrm{End}(\mathscr{T})$ is such that $\mathsf{L}(\boldsymbol{x})\boldsymbol{f}(\boldsymbol{x})$ is a measurable function for every $\boldsymbol{f} \in \mathscr{E}_0(\Omega)$. Then $\mathsf{L}(\boldsymbol{x})$ is a measurable function.*

Proof. Let $m = \dim \mathscr{T}$. We observe that the set of all m-tuples $(\boldsymbol{E}_{j_1}, \ldots, \boldsymbol{E}_{j_m}) \subset D^m$ is a countable set. Let $\{S_k\colon k \geqslant 1\}$ be an enumeration of this set. For $k \geqslant 1$ we define

$$\Omega_k = \{\boldsymbol{x} \in \Omega_0^*\colon S_k \text{ is a basis of } \mathscr{T}\}.$$

Each set Ω_k is a measurable subset of Ω_0^* because

$$\Omega_k = \{\boldsymbol{x} \in \Omega_0^*\colon \boldsymbol{E}_1(\boldsymbol{x}) \wedge \cdots \wedge \boldsymbol{E}_m(\boldsymbol{x}) \neq 0\},$$

where $S_k = (\boldsymbol{E}_1, \ldots, \boldsymbol{E}_m) \in D^m$. Since $D(\boldsymbol{x}) = \{\boldsymbol{E}_j(\boldsymbol{x})\colon j \geqslant 1\}$ is a spanning set of $\mathscr{T}$ for all $\boldsymbol{x} \in \Omega_0^*$, we conclude that

$$\Omega_0^* = \bigcup_{k=1}^{\infty} \Omega_k.$$

Fixing an orthonormal basis $\{\boldsymbol{e}_1, \ldots, \boldsymbol{e}_m\}$ of $\mathscr{T}$ we observe that for every $\boldsymbol{x} \in \Omega_k$ the numbers $\alpha_i^j(\boldsymbol{x})$, such that

$$\boldsymbol{e}_i = \sum_{j=1}^{m} \alpha_i^j(\boldsymbol{x})\boldsymbol{E}_j(\boldsymbol{x})$$

are uniquely defined and are rational functions of the basis vectors $\boldsymbol{E}_1, \ldots, \boldsymbol{E}_m$. Therefore, $\alpha_i^j(\boldsymbol{x})$ are measurable functions on Ω_k. It follows that

$$\mathsf{L}(\boldsymbol{x})\boldsymbol{e}_i = \sum_{j=1}^{m} \alpha_i^j(\boldsymbol{x})\mathsf{L}(\boldsymbol{x})\boldsymbol{E}_j(\boldsymbol{x}), \quad i = 1, \ldots, m$$

are measurable functions on Ω_k. But then $\mathsf{L}(\boldsymbol{x})\boldsymbol{e}_i$ must be measurable on Ω_0^*, since the sets Ω_k cover Ω_0^*. □

We can now finish the proof of theorem 3.3. Let ε_k be a subsequence for which the sequences of operators $\mathsf{T}_{\mathsf{L}_{\varepsilon_k}}^{-1}$, $\mathsf{T}_{\mathsf{L}_{\varepsilon_k}^T}^{-1}$, $\mathsf{L}_{\varepsilon_k}\mathsf{T}_{\mathsf{L}_{\varepsilon_k}}^{-1}$, and $\mathsf{L}_{\varepsilon_k}^T\mathsf{T}_{\mathsf{L}_{\varepsilon_k}^T}^{-1}$ converge weakly in $\mathscr{B}(\mathscr{E}_0(\Omega))$

and $\mathscr{B}(\mathscr{E}_0(\Omega), L^2(\Omega; \mathscr{T}))$, respectively. Such a subsequence exists, since the Hilbert spaces $L^2(\Omega; \mathscr{T})$ and $\mathscr{E}_0(\Omega)$ are separable. Thus,

$$\mathsf{T}_{\mathsf{L}_{\varepsilon_k}}^{-1} \rightharpoonup \mathsf{A}_1, \quad \mathsf{T}_{\mathsf{L}_{\varepsilon_k}^T}^{-1} \rightharpoonup \mathsf{A}_2, \quad \mathsf{L}_{\varepsilon_k}\mathsf{T}_{\mathsf{L}_{\varepsilon_k}}^{-1} \rightharpoonup \mathsf{A}_3, \quad \mathsf{L}_{\varepsilon_k}^T\mathsf{T}_{\mathsf{L}_{\varepsilon_k}^T}^{-1} \rightharpoonup \mathsf{A}_4, \quad k \to \infty$$

in $\mathscr{B}(\mathscr{E}_0(\Omega))$ and $\mathscr{B}(\mathscr{E}_0(\Omega), L^2(\Omega; \mathscr{T}))$, respectively. We now fix arbitrary fields $\{\boldsymbol{f}_1, \boldsymbol{f}_2\} \subset \mathscr{E}_0(\Omega)$ and let $\{\boldsymbol{E}_k^{(1)}, \boldsymbol{E}_k^{(1)}\} \subset \mathscr{E}_0(\Omega)$ be the unique solutions of

$$\Gamma_\Omega(\mathsf{L}_{\varepsilon_k}\boldsymbol{E}_k^{(1)}) = \mathsf{A}_1^{-1}\boldsymbol{f}_1, \qquad \Gamma_\Omega(\mathsf{L}_{\varepsilon_k}^T\boldsymbol{E}_k^{(2)}) = \mathsf{A}_2^{-1}\boldsymbol{f}_2$$

Here, the existence of bounded operators A_1^{-1} and A_2^{-1} is guaranteed by lemma 3.32. We observe that

$$\boldsymbol{E}_k^{(1)} = \mathsf{T}_{\mathsf{L}_{\varepsilon_k}}^{-1}\mathsf{A}_1^{-1}\boldsymbol{f}_1 \rightharpoonup \boldsymbol{f}_1, \qquad \boldsymbol{E}_k^{(2)} = \mathsf{T}_{\mathsf{L}_{\varepsilon_k}^T}^{-1}\mathsf{A}_2^{-1}\boldsymbol{f}_2 \rightharpoonup \boldsymbol{f}_2.$$

$$\mathsf{L}_{\varepsilon_k}\boldsymbol{E}_k^{(1)} = \mathsf{L}_{\varepsilon_k}\mathsf{T}_{\mathsf{L}_{\varepsilon_k}}^{-1}\mathsf{A}_1^{-1}\boldsymbol{f}_1 \rightharpoonup \mathsf{A}_3\mathsf{A}_1^{-1}\boldsymbol{f}_1, \quad \mathsf{L}_{\varepsilon_k}^T\boldsymbol{E}_k^{(2)} = \mathsf{L}_{\varepsilon_k}^T\mathsf{T}_{\mathsf{L}_{\varepsilon_k}^T}^{-1}\mathsf{A}_2^{-1}\boldsymbol{f}_2 \rightharpoonup \mathsf{A}_4\mathsf{A}_2^{-1}\boldsymbol{f}_2$$

in $L^2(\Omega; \mathscr{T})$.

We now apply the compensated compactness theorem on Ω, noting that

$$\Gamma_\Omega(\mathsf{L}_{\varepsilon_k}\boldsymbol{E}_k^{(1)}) = \Gamma_\Omega(\mathsf{A}_1^{-1}\boldsymbol{f}_1), \qquad \Pi_\Omega\boldsymbol{E}_k^{(1)} = 0,$$

$$\Gamma_\Omega(\mathsf{L}_{\varepsilon_k}^T\boldsymbol{E}_k^{(2)}) = \Gamma_\Omega(\mathsf{A}_2^{-1}\boldsymbol{f}_2), \qquad \Pi_\Omega\boldsymbol{E}_k^{(2)} = 0.$$

Thus, by the compensated compactness theorem

$$(\mathsf{L}_{\varepsilon_k}\boldsymbol{E}_k^{(1)}, \boldsymbol{E}_k^{(2)}) \overset{*}{\rightharpoonup} (\mathsf{A}_3\mathsf{A}_1^{-1}\boldsymbol{f}_1, \boldsymbol{f}_2)$$

in the sense of measures on Ω. However,

$$(\mathsf{L}_{\varepsilon_k}\boldsymbol{E}_k^{(1)}, \boldsymbol{E}_k^{(2)}) = (\boldsymbol{E}_k^{(1)}, \mathsf{L}_{\varepsilon_k}^T\boldsymbol{E}_k^{(2)}).$$

Applying the compensated compactness theorem again we obtain

$$(\boldsymbol{E}_k^{(1)}, \mathsf{L}_{\varepsilon_k}^T\boldsymbol{E}_k^{(2)}) \overset{*}{\rightharpoonup} (\boldsymbol{f}_1, \mathsf{A}_4\mathsf{A}_2^{-1}\boldsymbol{f}_2).$$

Thus,

$$((\mathsf{A}_3\mathsf{A}_1^{-1}\boldsymbol{f}_1)(\boldsymbol{x}), \boldsymbol{f}_2(\boldsymbol{x})) = (\boldsymbol{f}_1(\boldsymbol{x}), (\mathsf{A}_4\mathsf{A}_2^{-1}\boldsymbol{f}_2)(\boldsymbol{x})) \quad \forall\{\boldsymbol{f}_1, \boldsymbol{f}_2\} \subset \mathscr{E}_0(\Omega).$$

for a.e. $\boldsymbol{x} \in \Omega$. Under our assumptions $V_{\mathscr{E}} = \mathscr{T}$, in view of corollary 2.28. Thus, by lemma 3.33 there exists a uniquely defined measurable function $\mathsf{L}_*: \Omega \to \mathrm{End}(\mathscr{T})$ such that

$$(\mathsf{A}_3\mathsf{A}_1^{-1}\boldsymbol{f}_1)(\boldsymbol{x}) = \mathsf{L}_*(\boldsymbol{x}), \qquad (\mathsf{A}_4\mathsf{A}_2^{-1}\boldsymbol{f}_2)(\boldsymbol{x}) = \mathsf{L}_*(\boldsymbol{x})^T\boldsymbol{f}_2(\boldsymbol{x})$$

for a.e. $\boldsymbol{x} \in \Omega$.

Our task is to show that $\mathsf{L}_* \in \mathscr{M}(\alpha, \beta)^\Omega$ and that $\mathsf{L}_{\varepsilon_k} \overset{H}{\rightharpoonup} \mathsf{L}_*$. We note that by construction of L_* via lemma 3.33 we have

$$\mathsf{L}_{\varepsilon_k}\boldsymbol{E}_k^{(1)} \rightharpoonup \mathsf{L}_*\boldsymbol{f}_1, \qquad \mathsf{L}_{\varepsilon_k}^T\boldsymbol{E}_k^{(2)} \rightharpoonup \mathsf{L}_*^T\boldsymbol{f}_2.$$

in $L^2(\Omega; \mathcal{T})$. In addition,

$$(\mathsf{L}_{\varepsilon_k} \boldsymbol{E}_k^{(1)}, \boldsymbol{E}_k^{(1)}) \geqslant \alpha |\boldsymbol{E}_k^{(1)}|^2 \geqslant \alpha(2(\boldsymbol{E}_k^{(1)}, \boldsymbol{f}_1) - |\boldsymbol{f}_1|^2).$$

Passing to the limit as $k \to \infty$ and using the compensated compactness theorem we obtain

$$(\mathsf{L}_* \boldsymbol{f}_1, \boldsymbol{f}_1) \geqslant \alpha |\boldsymbol{f}_1|^2 \text{ for all } \boldsymbol{f}_1 \in \mathscr{E}_0(\Omega). \tag{3.49}$$

Let us show that

$$(\mathsf{L}_*(\boldsymbol{x})\boldsymbol{e}, \boldsymbol{e}) \geqslant \alpha |\boldsymbol{e}|^2. \tag{3.50}$$

Once again, the difficulty is that inequality (3.49) may fail on null-sets that depend on $\boldsymbol{f}_1$, $\boldsymbol{f}_2$. Theorem 2.25(c) provides the tool to deal with this issue.

Let $\{\boldsymbol{E}_j : j \geqslant 1\}$ be a countable dense subset of $\mathscr{E}_0(\Omega)$. According to theorem 2.25(c) there exists a null-set $\mathscr{N}_0$, such that all points in $\Omega \setminus \mathscr{N}_0$ are the Lebesgue points of $\boldsymbol{E}_j(\boldsymbol{x})$, and

$$V_{\mathscr{E}} = \text{Span}\{\boldsymbol{E}_1(\boldsymbol{x}), \boldsymbol{E}_2(\boldsymbol{x}), \ldots\}, \quad \forall \boldsymbol{x} \in \Omega \setminus \mathscr{N}_0.$$

Let $\mathscr{N}_* \supset \mathscr{N}_0$ be a null-set, such that all points in $\Omega \setminus \mathscr{N}_*$ are the Lebesgue points of $\mathsf{L}_*(\boldsymbol{x})\boldsymbol{E}_j(\boldsymbol{x})$. Such a set exists because $\mathsf{L}_* \boldsymbol{f} \in L^2(\Omega; \mathcal{T})$ for any $\boldsymbol{f} \in \mathscr{E}_0(\Omega)$. Let $\boldsymbol{x}_0 \in \Omega \setminus \mathscr{N}_*$ be arbitrary, and let $\boldsymbol{e} \in \mathcal{T} = V_{\mathscr{E}}$ be arbitrary. Then there exist real numbers $\{\alpha_j : j \geqslant 1\}$, such that only finitely many of them are nonzero and

$$\sum_{j=1}^{\infty} \alpha_j \boldsymbol{E}_j(\boldsymbol{x}_0) = \boldsymbol{e}.$$

Since $\boldsymbol{x}_0$ is a Lebesgue point of $\boldsymbol{E}_j(\boldsymbol{x})$ and $\mathsf{L}_* \boldsymbol{E}_j(\boldsymbol{x})$, then it is also a Lebesgue point of $\boldsymbol{E}_0(\boldsymbol{x})$ and $\mathsf{L}_* \boldsymbol{E}_0(\boldsymbol{x})$, where

$$\boldsymbol{E}_0(\boldsymbol{x}) = \sum_{j=1}^{\infty} \alpha_j \boldsymbol{E}_j(\boldsymbol{x}) \in \mathscr{E}_0(\Omega).$$

Then we can pass to the limit in the inequality

$$(\mathsf{L}_*(\boldsymbol{x}_0 + r\boldsymbol{z})\boldsymbol{E}_0(\boldsymbol{x}_0 + r\boldsymbol{z}), \boldsymbol{E}_0(\boldsymbol{x}_0 + r\boldsymbol{z})) \geqslant \alpha |\boldsymbol{E}_0(\boldsymbol{x}_0 + r\boldsymbol{z})|^2 \text{ for a.e. } \boldsymbol{z} \in B_1,$$

whose both sides converge in $L^1(B_1)$, obtaining

$$(\mathsf{L}_*(\boldsymbol{x}_0)\boldsymbol{E}_0(\boldsymbol{x}_0), \boldsymbol{E}_0(\boldsymbol{x}_0)) \geqslant \alpha |\boldsymbol{E}_0(\boldsymbol{x}_0)|^2.$$

The desired inequality (3.50) follows from $\boldsymbol{E}_0(\boldsymbol{x}_0) = \boldsymbol{e}$.

To prove the other inequality in the definition of the set $\mathscr{M}(\alpha, \beta)$ we let

$$\boldsymbol{J}_k = \mathsf{L}_{\varepsilon_k} \boldsymbol{E}_k^{(1)}.$$

Recall that $\boldsymbol{J}_k \rightharpoonup \mathsf{L}_* \boldsymbol{f}_1$ in $L^2(\Omega; \mathcal{T})$. We have

$$(\mathsf{L}_{\varepsilon_k} \boldsymbol{E}_k^{(1)}, \boldsymbol{E}_k^{(1)}) = (\boldsymbol{J}_k, \mathsf{L}_{\varepsilon_k}^{-1} \boldsymbol{J}_k) \geqslant \frac{1}{\beta} |\boldsymbol{J}_k|^2 \geqslant \frac{1}{\beta}(2(\boldsymbol{J}_k, \mathsf{L}_* \boldsymbol{f}_1) - |\mathsf{L}_* \boldsymbol{f}_1|^2).$$

Passing to the limit in the above inequality, using compensated compactness, we obtain

$$(\mathsf{L}_*\boldsymbol{f}_1, \boldsymbol{f}_1) \geqslant \frac{1}{\beta}|\mathsf{L}_*\boldsymbol{f}_1|^2 \text{ for all } \boldsymbol{f}_1 \in \mathscr{E}_0(\Omega).$$

As before, we conclude that for a.e. $\boldsymbol{x} \in \Omega$ and every $\boldsymbol{e} \in \mathscr{T}$ we have

$$(\mathsf{L}_*(\boldsymbol{x})\boldsymbol{e}, \boldsymbol{e}) \geqslant \frac{1}{\beta}|\mathsf{L}_*\boldsymbol{e}|^2,$$

or equivalently,

$$(\mathsf{L}_*^{-1}(\boldsymbol{x})\boldsymbol{j}, \boldsymbol{j}) \geqslant \frac{1}{\beta}|\boldsymbol{j}|^2, \text{ for all } \boldsymbol{j} \in \mathscr{T}.$$

Hence, $\mathsf{L}_* \in \mathscr{M}(\alpha, \beta)^{\Omega}$.

To complete the proof of the compactness theorem we need to show that $\mathsf{L}_{\varepsilon_k} \overset{H}{\rightharpoonup} \mathsf{L}_*$, as $k \to \infty$. For that purpose we fix an arbitrary $\boldsymbol{f} \in \mathscr{E}_0(\Omega)$ and define $\boldsymbol{E}_k \in \mathscr{E}_0(\Omega)$ as the unique solution of

$$\Gamma_\Omega(\mathsf{L}_{\varepsilon_k}\boldsymbol{E}_k) = \boldsymbol{f}.$$

In other words, $\boldsymbol{E}_k = \mathsf{T}_{\mathsf{L}_{\varepsilon_k}}^{-1}\boldsymbol{f}$ and $\boldsymbol{J}_k = \mathsf{L}_{\varepsilon_k}\boldsymbol{E}_k = \mathsf{L}_{\varepsilon_k}\mathsf{T}_{\mathsf{L}_{\varepsilon_k}}^{-1}\boldsymbol{f}$. Then, by the definition of the subsequence ε_k we have $\boldsymbol{E}_k \rightharpoonup \mathsf{A}_1\boldsymbol{f} = \boldsymbol{E}_0$, and $\boldsymbol{J}_k \rightharpoonup \mathsf{A}_3\boldsymbol{f} = \boldsymbol{J}_0$. But that means that

$$\boldsymbol{J}_0(\boldsymbol{x}) = (\mathsf{A}_3\mathsf{A}_1^{-1}\mathsf{A}_1\boldsymbol{f})(\boldsymbol{x}) = (\mathsf{A}_3\mathsf{A}_1^{-1}\boldsymbol{E}_0)(\boldsymbol{x}) = \mathsf{L}_*(\boldsymbol{x})\boldsymbol{E}_0(\boldsymbol{x}).$$

And since $\Gamma_\Omega\boldsymbol{J}_k = \boldsymbol{f}$ we conclude that $\Gamma_\Omega\boldsymbol{J}_0 = \boldsymbol{f}$. Hence $\boldsymbol{E}_0 \in \mathscr{E}_0(\Omega)$ is the unique solution of $\Gamma_\Omega(\mathsf{L}_*\boldsymbol{E}_0) = \boldsymbol{f}$, concluding the proof of the H-convergence $\mathsf{L}_{\varepsilon_k} \overset{H}{\rightharpoonup} \mathsf{L}_*$. □

Corollary 3.35. *The sets* $\mathscr{M}(\alpha, \beta)$ *are G-closed for all* $0 < \alpha < \beta$.

The corollary is an immediate consequence of theorems 3.29 and 3.3.

3.3.4 H-convergence and G-closure

In this book we regarded periodic homogenization as one specific example of application of the more general H-convergence theory (see section 3.3.1). However, similarity of results established for periodic composites and using H-convergence suggested that there should be a deeper connection between the two theories. This connection was first conjectured by Kohn and Dal Maso and proved by Tartar [26] and generalized by Raitums [20] for monotone constitutive relations. See also [1]. Specifically, under our standard assumptions on the symbol $\Gamma(\boldsymbol{\eta})$, the set of all H-limits of composites made with materials in U are described by all measurable functions with values in $G(U)$. This was formulated in theorem 3.8, which we are now ready to prove. This is the deepest result of the theory. We start by establishing a relation between periodic and general composites.

Lemma 3.36. *Let Q be a parallelepiped of periods. Let $\boldsymbol{J} \in \mathscr{J}_{\rm per}$ and $\boldsymbol{E} \in \mathscr{E}_{\rm per}$, where $\mathscr{E}_{\rm per}$ and $\mathscr{J}_{\rm per}$ are defined in (3.5), (3.6). Then $\boldsymbol{J}|Q \in \overline{\mathscr{J}(Q)}$ and $\boldsymbol{E}|Q \in \overline{\mathscr{E}(Q)}$. In other words, $\Gamma_Q \boldsymbol{J} = \Pi_Q \boldsymbol{E} = 0$, where Γ_Q and Π_Q are the orthogonal projections onto $\mathscr{E}_0(Q)$ and $\mathscr{J}_0(Q)$, respectively.*

Proof. $\boldsymbol{J} \in \mathscr{J}_{\rm per}$ means that the coefficients $\hat{\boldsymbol{J}}(\boldsymbol{k})$ of its Fourier series (3.7) satisfy

$$\boldsymbol{J}(\boldsymbol{x}) = \sum_{\boldsymbol{k}\in\mathbb{Z}^d\setminus\{0\}} \hat{\boldsymbol{J}}(\boldsymbol{k})e^{2\pi i Q\boldsymbol{x}\cdot\boldsymbol{k}}, \quad \boldsymbol{\Gamma}\left(\frac{\boldsymbol{Q}^T\boldsymbol{k}}{|\boldsymbol{Q}^T\boldsymbol{k}|}\right)\hat{\boldsymbol{J}}(\boldsymbol{k}) = 0, \quad \boldsymbol{k} \in \mathbb{Z}\setminus\{0\}. \tag{3.51}$$

Let $\boldsymbol{E} \in \mathscr{E}_0(Q)$ be arbitrary. By the Parceval identity

$$\int_Q (\boldsymbol{J}(\boldsymbol{x}), \boldsymbol{E}(\boldsymbol{x}))d\boldsymbol{x} = \sum_{\boldsymbol{k}\in\mathbb{Z}^d\setminus\{0\}} (\hat{\boldsymbol{J}}(\boldsymbol{k}), \boldsymbol{E}_k), \tag{3.52}$$

where

$$\boldsymbol{E}_k = \int_Q \boldsymbol{E}(\boldsymbol{x})e^{-2\pi i Q\boldsymbol{x}\cdot\boldsymbol{k}}d\boldsymbol{x}, \quad \boldsymbol{k} \in \mathbb{Z}^d.$$

Now we observe that

$$\boldsymbol{E}_k = \int_{\mathbb{R}^d} (X_0(Q)\boldsymbol{E})(\boldsymbol{x})e^{-2\pi i Q\boldsymbol{x}\cdot\boldsymbol{k}}d\boldsymbol{x} = \widehat{X_0(Q)\boldsymbol{E}}(\boldsymbol{Q}^T\boldsymbol{k}).$$

By assumption $X_0(Q)\boldsymbol{E} \in \mathscr{E}$ and therefore,

$$\boldsymbol{\Gamma}\left(\frac{\boldsymbol{Q}^T\boldsymbol{k}}{|\boldsymbol{Q}^T\boldsymbol{k}|}\right)\boldsymbol{E}_k = \boldsymbol{\Gamma}\left(\frac{\boldsymbol{Q}^T\boldsymbol{k}}{|\boldsymbol{Q}^T\boldsymbol{k}|}\right)\widehat{X_0(Q)\boldsymbol{E}}(\boldsymbol{Q}^T\boldsymbol{k}) = \widehat{X_0(Q)\boldsymbol{E}}(\boldsymbol{Q}^T\boldsymbol{k}) = \boldsymbol{E}_k. \tag{3.53}$$

Hence, from (3.52) and (3.53) we obtain

$$\begin{aligned}(\boldsymbol{J}, \boldsymbol{E})_{L^2_{\rm per}(Q)} &= \sum_{\boldsymbol{k}\in\mathbb{Z}^d\setminus\{0\}} (\hat{\boldsymbol{J}}(\boldsymbol{k}), \boldsymbol{E}_k) = \sum_{\boldsymbol{k}\in\mathbb{Z}^d\setminus\{0\}} \left(\hat{\boldsymbol{J}}(\boldsymbol{k}), \boldsymbol{\Gamma}\left(\frac{\boldsymbol{Q}^T\boldsymbol{k}}{|\boldsymbol{Q}^T\boldsymbol{k}|}\right)\boldsymbol{E}_k\right)\\ &= \sum_{\boldsymbol{k}\in\mathbb{Z}^d\setminus\{0\}} \left(\boldsymbol{\Gamma}\left(\frac{\boldsymbol{Q}^T\boldsymbol{k}}{|\boldsymbol{Q}^T\boldsymbol{k}|}\right)\hat{\boldsymbol{J}}(\boldsymbol{k}), \boldsymbol{E}_k\right) = 0,\end{aligned}$$

where the last equality is due to (3.51). Thus, we have shown that $\boldsymbol{J} \in \mathscr{E}_0(Q)^\perp = \overline{\mathscr{J}(Q)}$. The statement about $\mathscr{E}_{\rm per}$ is proved in the same way. □

It is natural to use composite materials as constituents of a new periodic composite. This 'second generation' composite should still be regarded as a composite made with the same materials with which the original composites were made. The following iterated homogenization lemma shows that the notion of H-convergence supports this intuition mathematically.

Lemma 3.37. (Iterated homogenization). *Assume that the symbol* $\Gamma(\eta)$ *and the family of local tensors* L_ε *satisfy all conditions of theorem 3.31 with* Ω *replaced by the parallelepiped of periods* Q. *For every* $\varepsilon > 0$ *let* $\mathsf{K}_*(\varepsilon)$ *be the effective tensor of a periodic composite with period cell* Q *and the local tensor* $\mathsf{L}_\varepsilon(\boldsymbol{x})$. *Let* K_* *be the effective tensor of a periodic composite with period cell* Q *and the local tensor* $\mathsf{L}_*(\boldsymbol{x})$. *Then*

$$\lim_{\varepsilon\to 0} \mathsf{K}_*(\varepsilon) = \mathsf{K}_*.$$

In other words

$$\lim_{\varepsilon\to 0} \operatorname{Hlim}_{\delta\to 0} \mathsf{L}_\varepsilon\left(\frac{\boldsymbol{x}}{\delta}\right) = \operatorname{Hlim}_{\delta\to 0} \operatorname{Hlim}_{\varepsilon\to 0} \mathsf{L}_\varepsilon\left(\frac{\boldsymbol{x}}{\delta}\right) = \mathsf{K}_*.$$

Proof. According to theorem 3.29

$$\mathsf{K}_*^T(\varepsilon)\boldsymbol{e}_0 = \fint_Q \mathsf{L}_\varepsilon^T(\boldsymbol{z})(\boldsymbol{E}_\varepsilon(\boldsymbol{z}) + \boldsymbol{e}_0)d\boldsymbol{z}, \tag{3.54}$$

where $\boldsymbol{E}_\varepsilon$ is the unique solution of the periodic cell problem

$$\boldsymbol{E}_\varepsilon \in \mathscr{E}_{\text{per}}, \qquad \boldsymbol{J}_\varepsilon(\boldsymbol{z}) = \mathsf{L}_\varepsilon^T(\boldsymbol{z})(\boldsymbol{E}_\varepsilon(\boldsymbol{z}) + \boldsymbol{e}_0), \qquad \boldsymbol{J}_\varepsilon \in \mathscr{J}_{\text{per}} \oplus \mathscr{U}.$$

Let ε_k be a subsequence such that $\boldsymbol{E}_{\varepsilon_k} \rightharpoonup \boldsymbol{E}_*$ in $L^2_{\text{per}}(Q, \mathscr{T})$ and $\mathsf{K}_*^T(\varepsilon_k)\boldsymbol{e}_0 \to \boldsymbol{j}_*$ in $\mathscr{T}$. By lemma 3.36 we have $\boldsymbol{E}_{\varepsilon_k} \in \overline{\mathscr{E}(Q)}$ and

$$\Gamma_Q\left(\mathsf{L}_{\varepsilon_k}^T(\boldsymbol{z})(\boldsymbol{E}_{\varepsilon_k}(\boldsymbol{z}) + \boldsymbol{e}_0)\right) = \Gamma_Q(\mathsf{K}_*^T(\varepsilon_k)\boldsymbol{e}_0) \to \Gamma_Q(\boldsymbol{j}_*).$$

By part (a) of theorem 3.31 we have $\mathsf{L}_\varepsilon^T \overset{H}{\rightharpoonup} \mathsf{L}_*^T$ and then part (b) of theorem 3.31 applies: $\mathsf{L}_{\varepsilon_k}^T(\boldsymbol{E}_{\varepsilon_k} + \boldsymbol{e}_0) \rightharpoonup \mathsf{L}_*^T(\boldsymbol{E}_* + \boldsymbol{e}_0)$. Hence, the weak limit $\boldsymbol{E}_*$ is the solution of the cell problem

$$\boldsymbol{E}_* \in \mathscr{E}_{\text{per}}, \quad \boldsymbol{J}_* = \mathsf{L}_*^T(\boldsymbol{E}_* + \boldsymbol{e}_0), \quad \boldsymbol{J}_* \in \mathscr{J}_{\text{per}} \oplus \mathscr{U}. \tag{3.55}$$

The solution $\boldsymbol{E}_*$ of (3.55) is unique, since $\mathsf{L}_* \in \mathscr{M}(\alpha, \beta)^Q$ by theorem 3.3. We conclude that the entire family $\boldsymbol{E}_\varepsilon$ converges weakly in $L^2_{\text{per}}(Q; \mathscr{T})$ to $\boldsymbol{E}_*$. Thus, according to (3.54)

$$\lim_{\varepsilon\to 0} \mathsf{K}_*^T(\varepsilon)\boldsymbol{e}_0 = \fint_Q \mathsf{L}_*^T(\boldsymbol{z})(\boldsymbol{E}_*(\boldsymbol{z}) + \boldsymbol{e}_0)d\boldsymbol{z} = \mathsf{K}_*^T\boldsymbol{e}_0.$$

□

We are now ready to prove theorem 3.8.

Proof of theorem 3.8. Part (i). Let us fix $\boldsymbol{x}_0 \in \Omega$—a Lebesgue point of $\mathsf{L}_*(\boldsymbol{x})$, $\varepsilon > 0$, and cube Q_r of side r centered at the point $\boldsymbol{x}_0$. Next, let us make a periodic composite with period cell Q_r and local tensor $\mathsf{L}_\varepsilon|Q_r$. Let $\mathsf{K}_*(r, \varepsilon)$ be its effective tensor. Clearly, $\mathsf{K}_*(r, \varepsilon) \in G(U)$. By part (c) of theorem 3.31 $\mathsf{L}_\varepsilon|Q_r \overset{H}{\rightharpoonup} \mathsf{L}_*|Q_r$. Therefore, by the iterated homogenization lemma 3.37

$$\lim_{\varepsilon\to 0} \mathsf{K}_*(r, \varepsilon) = \mathsf{K}_*(r),$$

where $\mathsf{K}_*(r)$ is the effective tensor of a periodic composite with period cell Q_r and the local tensor $\mathsf{L}_*(\boldsymbol{x})$, $\boldsymbol{x} \in Q_r$. Rescaling the cube Q_r to the unit cube Q we see that $\mathsf{K}_*(r)$ is also the effective tensor of a Q-periodic composite with the local tensor $\mathsf{L}_*(\boldsymbol{x}_0 + r\boldsymbol{z})$, $\boldsymbol{z} \in Q$. We observe that $\mathsf{L}_*(\boldsymbol{x}_0 + r\boldsymbol{z}) \to \mathsf{L}_*(\boldsymbol{x}_0)$ in $L^1(Q)$, as $r \to 0$, since $\boldsymbol{x}_0$ is a Lebesgue point of $\mathsf{L}_*(\boldsymbol{x})$. Hence, by lemma 3.2 $\mathsf{L}_*(\boldsymbol{x}_0 + r\boldsymbol{z}) \xrightarrow{H} \mathsf{L}_*(\boldsymbol{x}_0)$. Applying the iterated homogenization lemma 3.37 again we conclude that

$$\lim_{r \to 0} \mathsf{K}_*(r) = \mathsf{L}_*(\boldsymbol{x}_0).$$

Thus, $\mathsf{L}_*(\boldsymbol{x}_0) \in G(U)$.

Part (ii). We can use the geometric structure of G-closed sets corollary 3.21 to represent each function $\mathsf{L} \in G(U)^\Omega$ as $\mathsf{L} = \mathsf{A}(\mathsf{F})$, where $\mathsf{F} \in \mathscr{C}^\Omega$, A is a fixed analytic diffeomorphism and $\mathscr{C}$ is a closed convex subset of $\mathrm{End}(\mathscr{T})$. Let $\mathsf{L}_0 \in G(U)$ and define

$$\mathsf{F}_0(\boldsymbol{x}) = \begin{cases} \mathsf{F}(\boldsymbol{x}), & \boldsymbol{x} \in \Omega, \\ \mathsf{A}^{-1}(\mathsf{L}_0), & \boldsymbol{x} \in \Omega^c. \end{cases}$$

Thus, F_0: $\mathbb{R}^d \to \mathscr{C}$. Let ρ: $\mathbb{R}^d \to [0, +\infty)$ be a standard convolution kernel, i.e., $\rho(\boldsymbol{x})$ is nonnegative, supported on the unit ball B_1, and

$$\int_{B_1} \rho(\boldsymbol{x}) d\boldsymbol{x} = 1.$$

Then the function

$$\mathsf{F}_\varepsilon(\boldsymbol{x}) = \int_{B_1} \mathsf{F}_0(\boldsymbol{x} + \varepsilon \boldsymbol{z}) \rho(\boldsymbol{z}) d\boldsymbol{z}$$

is continuous and takes values in $\mathscr{C}$, since $\mathscr{C}$ is a closed convex set. Additionally, $\mathsf{F}_\varepsilon \to \mathsf{F}$, as $\varepsilon \to 0$ in measure in Ω. Therefore, $\mathsf{L}_\varepsilon = \mathsf{A}(\mathsf{F}_\varepsilon)$ converges to L, as $\varepsilon \to 0$ in measure in Ω. By lemma 3.2 it follows that $\mathsf{L}_\varepsilon \xrightarrow{H} \mathsf{L}$. We have constructed a family L_ε so that $\mathsf{L}_\varepsilon \in G(U)^\Omega$ and is continuous.

Now let us fix a continuous $\mathsf{L} \in G(U)^\Omega$. Let us cover Ω by a lattice of cubes $Q_{\boldsymbol{k}}(2r) = (-r, r)^d + 2r\boldsymbol{k}$ of side $2r > 0$, $\boldsymbol{k} \in \mathbb{Z}^d$:

$$\Omega \subset \bigcup_{\boldsymbol{k} \in I_r} \overline{Q_{\boldsymbol{k}}(2r)}, \qquad I_r = \{\boldsymbol{k} \in \mathbb{Z}^d: Q_{\boldsymbol{k}}(2r) \bigcap \Omega \neq \varnothing\}.$$

For each $\boldsymbol{k} \in I_r$ we choose a point $\boldsymbol{\xi}_{\boldsymbol{k}}^r \in Q_{\boldsymbol{k}}(2r) \cap \Omega$. The function

$$\mathsf{L}_r(\boldsymbol{x}) = \mathsf{L}(\boldsymbol{\xi}_{\boldsymbol{k}}^r), \quad \boldsymbol{x} \in Q_{\boldsymbol{k}}(2r)$$

is piecewise constant on the cubes of the lattice and converges to $\mathsf{L}(\boldsymbol{x})$, as $r \to 0$ pointwise (regardless of the specific choice of points $\boldsymbol{\xi}_{\boldsymbol{k}}^r$), since $\mathsf{L}(\boldsymbol{x})$ is continuous. Therefore, we have constructed a family of functions $\mathsf{L}_r \in G(U)^\Omega$ that are piecewise constant on the cubes of the lattice and $\mathsf{L}_r \xrightarrow{H} \mathsf{L}$, as $r \to 0$.

Now let us fix an arbitrary lattice of cubes and a function $\mathsf{L} \in G(U)^\Omega$, constant on each cube of the lattice. Then, in each cube we place a periodic composite with values in U, H-convergent to the constant value of $\mathsf{L}(\boldsymbol{x})$ in that cube. By part (d) of theorem 3.31 we have constructed a sequence of functions in U^{Ω} that H-converges to $\mathsf{L}(\boldsymbol{x})$. It remains to recall that the set $\mathscr{M}(\alpha, \beta)^{\Omega}$ is metrizable in the H-topology. Thus, U^{Ω} is H-dense in $G(U)^{\Omega}$. □

Corollary 3.38. *The set $G(U)$ is G-closed.*

Proof. Let L_* be the effective tensor of a periodic composite with local tensor $\mathsf{L} \in G(U)^{Q}$. By part (ii) of theorem 3.8 there is $\mathsf{L}_\varepsilon \in U^{Q}$ such that $\mathsf{L}_\varepsilon \overset{H}{\rightharpoonup} \mathsf{L}$. Let $\mathsf{K}_*(\varepsilon)$ be the effective tensor of the periodic composite with local tensor $\mathsf{L}_\varepsilon(\boldsymbol{x})$. Then $\mathsf{K}_*(\varepsilon) \in G(U)$. By the iterated homogenization lemma 3.37 $\mathsf{K}_*(\varepsilon) \to \mathsf{L}_*$, as $\varepsilon \to 0$. Therefore, $\mathsf{L}_* \in G(U)$. □

3.3.5 On convergence of fluxes

In this section we explain the role of convergence of fluxes in the definition of H-convergence. Whether or not this requirement is essential depends on the presence of nonzero bilinear null-Lagrangians. In the early work of Spagnolo [22] the convergence of fluxes was not necessary because in the context of conductivity there are no nonzero *quadratic* null-Lagrangians. In general, the assumption of symmetry of material tensors eliminates the need for requiring convergence of fluxes only if the space of quadratic null-Lagrangians is $\{0\}$. In particular, it is easy to check that for three-dimensional conductivity, elasticity and piezoelectricity there are no nonzero quadratic null-Lagrangians. However, the full definition 3.1 of H-convergence must be used in the context of thermoelectricity (both in 2D and 3D) as well as in the 2D and 3D conductivity in the presence of the magnetic field (Hall effect).

Definition 3.39. *A bilinear form $B(\boldsymbol{u}, \boldsymbol{v})$ on $\mathscr{T}$ is called a null-Lagrangian if*

$$\int_{\mathbb{R}^d} B(\boldsymbol{E}_1(\boldsymbol{x}), \boldsymbol{E}_2(\boldsymbol{x}))d\boldsymbol{x} = 0 \tag{3.56}$$

for all $\{\boldsymbol{E}_1, \boldsymbol{E}_2\} \subset \mathscr{E}$.

The null-Lagrangians can be characterized geometrically by means of the Plancherel identity. It is easy to see that B is a null-Lagrangian if and only if

$$B(\boldsymbol{e}_1, \boldsymbol{e}_2) = 0 \text{ for all } \{\boldsymbol{e}_1, \boldsymbol{e}_2\} \in \mathscr{E}_{\eta}, \text{ and all } |\eta| = 1. \tag{3.57}$$

In our theory null-Lagrangians appear in a somewhat hidden form. The following lemma is our link to the null-Lagrangians.

Lemma 3.40. *Assume that the symbol $\Gamma(\eta)$ is such that $C_0^\infty(\mathbb{R}^d; \mathscr{T}) \cap \mathscr{E}$ is dense in $\mathscr{E}$. Let Ω be an open and bounded subset of $\mathbb{R}^d$ and $\mathsf{N} \in L^\infty(\Omega; \mathrm{End}(\mathscr{T}))$ be such that*

$$\int_{\Omega} (\mathsf{N}(\boldsymbol{x})\boldsymbol{E}_1(\boldsymbol{x}), \boldsymbol{E}_2(\boldsymbol{x}))d\boldsymbol{x} = 0 \tag{3.58}$$

for all $\{\boldsymbol{E}_1, \boldsymbol{E}_2\} \subset \mathscr{E}_0(\Omega)$. *Then the bilinear forms* $B_x(\boldsymbol{u}, \boldsymbol{v}) = (\mathsf{N}(\boldsymbol{x})\boldsymbol{u}, \boldsymbol{v})$ *are null-Lagrangians for a.e.* $\boldsymbol{x} \in \Omega$.

Proof. We split the proof into two steps. In the first step we show that

$$\int_{\Omega} (\mathsf{N}(\boldsymbol{x}_0)\boldsymbol{E}_1(\boldsymbol{x}), \boldsymbol{E}_2(\boldsymbol{x}))d\boldsymbol{x} = 0 \tag{3.59}$$

for a.e. $\boldsymbol{x}_0 \in \Omega$ and all $\{\boldsymbol{E}_1, \boldsymbol{E}_2\} \subset \mathscr{E}_0(\Omega)$. In the second step we show that if $\mathsf{N} \in \mathrm{End}(\mathscr{T})$ satisfies (3.59) then $B(\boldsymbol{u}, \boldsymbol{v}) = (\mathsf{N}\boldsymbol{u}, \boldsymbol{v})$ must be a null-Lagrangian.

Step 1. We start with the observation that by lemma 2.2, the relation (3.58) is equivalent to $\mathsf{N}\boldsymbol{E} \in \overline{\mathscr{J}(\Omega)}$ for any $\boldsymbol{E} \in \mathscr{E}_0(\Omega)$. Let us fix a Lebesgue point $\boldsymbol{x}_0 \in \Omega$ of $\mathsf{N}(\boldsymbol{x})$ and choose $\delta_0 > 0$ so small that $\boldsymbol{E}_\delta(\boldsymbol{x}) = (X_0(\Omega)\boldsymbol{E})((\boldsymbol{x} - \boldsymbol{x}_0)/\delta) \in \mathscr{E}_0^0(\Omega)$ for all $\delta \in (0, \delta_0)$. But then $\mathsf{N}\boldsymbol{E}_\delta|\Omega \in \overline{\mathscr{J}(\Omega)}$. Hence,

$$\mathsf{N}(\boldsymbol{x}_0 + \delta\boldsymbol{z})\boldsymbol{E}(\boldsymbol{z}) \in \overline{\mathscr{J}(\Omega)}.$$

Observe that

$$\mathsf{N}(\boldsymbol{x}_0 + \delta\boldsymbol{z})\boldsymbol{E}(\boldsymbol{z}) \to \mathsf{N}(\boldsymbol{x}_0)\boldsymbol{E}(\boldsymbol{z}), \qquad \delta \to 0$$

in $L^2(\Omega; \mathscr{T})$, since $\boldsymbol{x}_0$ is a Lebesgue point of $\mathsf{N}(\boldsymbol{x})$. Thus, $\mathsf{N}(\boldsymbol{x}_0)\boldsymbol{E} \in \overline{\mathscr{J}(\Omega)}$ for a.e. $\boldsymbol{x}_0 \in \Omega$ and all $\boldsymbol{E} \in \mathscr{E}_0(\Omega)$. Hence, the constant tensor $\mathsf{N}(\boldsymbol{x}_0)$ satisfies (3.59).

Step 2. In this step $\mathsf{N} \in \mathrm{End}(\mathscr{T})$ is a constant tensor satisfying (3.59) and $B(\boldsymbol{u}, \boldsymbol{v}) = (\mathsf{N}\boldsymbol{u}, \boldsymbol{v})$. Let D be an arbitrary open and bounded subset of $\mathbb{R}^d$, and let $\boldsymbol{x}_0 \in D$ and $\boldsymbol{x}_1 \in \Omega$ be arbitrary but fixed. Then there exists $\delta > 0$, so that $\tilde{\boldsymbol{E}}(\boldsymbol{x}) = \boldsymbol{E}(\boldsymbol{x}_0 + (\boldsymbol{x} - \boldsymbol{x}_1)/\delta) \in \mathscr{E}_0(\Omega)$ for any $\boldsymbol{E} \in \mathscr{E}_0(D)$. Hence, changing variables $\boldsymbol{z} = \boldsymbol{x}_0 + (\boldsymbol{x} - \boldsymbol{x}_1)/\delta$, we have for any $\{\boldsymbol{E}_1, \boldsymbol{E}_2\} \subset \mathscr{E}_0(D)$

$$0 = \int_{\Omega} B(\tilde{\boldsymbol{E}}_1, \tilde{\boldsymbol{E}}_2)d\boldsymbol{x} = \delta^d \int_{\mathbb{R}^d} B(\boldsymbol{E}_1, \boldsymbol{E}_2)d\boldsymbol{z} = \delta^d \int_{D} B(\boldsymbol{E}_1, \boldsymbol{E}_2)d\boldsymbol{z}.$$

Thus, B satisfies (3.59) with Ω replaced by D. In particular, B satisfies (3.59) with Ω replaced by B_n for all $n \geqslant 1$. Now, let us fix $\{\boldsymbol{E}_1, \boldsymbol{E}_2\} \subset \mathscr{E}$. By assumption there exists a sequence $\boldsymbol{E}_n^{(j)} \in \mathscr{E}_0(B_n)$, $j = 1, 2$, such that take $X_0(B_n)\boldsymbol{E}_n^{(j)} \to \boldsymbol{E}_j$ in $L^2(\mathbb{R}^d; \mathscr{T})$, as $n \to \infty$. Thus,

$$\begin{aligned}\int_{\mathbb{R}^d} B(\boldsymbol{E}_1, \boldsymbol{E}_2)d\boldsymbol{x} &= \lim_{n\to\infty} \int_{\mathbb{R}^d} B\left(X_0(B_n)\boldsymbol{E}_n^{(1)}, X_0(B_n)\boldsymbol{E}_n^{(2)}\right)d\boldsymbol{x} \\ &= \lim_{n\to\infty} \int_{B_n} B(\boldsymbol{E}_n^{(1)}, \boldsymbol{E}_n^{(2)})d\boldsymbol{x} = 0.\end{aligned}$$

Thus, $B(\boldsymbol{u}, \boldsymbol{v}) = (\mathsf{N}\boldsymbol{u}, \boldsymbol{v})$ is a null-Lagrangian and the lemma is proved. □

We remark that the property (3.58) is equivalent to

$$\Gamma_\Omega(\mathsf{N}(\boldsymbol{x})\boldsymbol{E}) = 0 \text{ for all } \boldsymbol{E} \in \mathscr{E}_0(\Omega). \tag{3.60}$$

The set $\mathfrak{N}$ of null-Lagrangians is a subspace of $\mathrm{End}(\mathscr{T})$, which is closed with respect to transposition: $\mathsf{N}^T \in \mathfrak{N}$, provided $\mathsf{N} \in \mathfrak{N}$. Therefore, both the symmetric and the antisymmetric parts of a null-Lagrangian are null-Lagrangians. Thus,

$$\mathfrak{N} = \mathfrak{N}_s \oplus \mathfrak{N}_a, \qquad \mathfrak{N}_s = \mathfrak{N} \bigcap \mathrm{Sym}(\mathscr{T}), \qquad \mathfrak{N}_a = \mathfrak{N} \bigcap \mathrm{Skew}(\mathscr{T}),$$

where $\mathrm{Skew}(\mathscr{T})$ is the set of all antisymmetric operators on $\mathscr{T}$.

Now, we are ready to examine the necessity of convergence of fluxes for H-convergence. For this purpose we assume that

(A1) $\{\mathsf{L}_\varepsilon, \mathsf{L}_*\} \subset \mathcal{M}(\alpha, \beta)^\Omega$

(A2) For every $\boldsymbol{f} \in \mathcal{E}_0(\Omega)$ $\boldsymbol{E}_\varepsilon \rightharpoonup \boldsymbol{E}_*$, as $\varepsilon \to 0$ in $L^2(\Omega; \mathcal{T})$, where $\boldsymbol{E}_\varepsilon \in \mathcal{E}_0(\Omega)$ is the unique solution of $\Gamma_\Omega(\mathsf{L}_\varepsilon \boldsymbol{E}_\varepsilon) = \boldsymbol{f}$ and $\boldsymbol{E}_* \in \mathcal{E}_0(\Omega)$ is the unique solution of $\Gamma_\Omega(\mathsf{L}_* \boldsymbol{E}_*) = \boldsymbol{f}$.

Our goal is either to prove that $\mathsf{L}_\varepsilon \boldsymbol{E}_\varepsilon \rightharpoonup \mathsf{L}_* \boldsymbol{E}_*$, establishing H-convergence $\mathsf{L}_\varepsilon \overset{H}{\rightharpoonup} \mathsf{L}_*$, or find an example, where the weak limit of $\mathsf{L}_\varepsilon \boldsymbol{E}_\varepsilon$ differs from $\mathsf{L}_* \boldsymbol{E}_*$. The latter is easy to construct, provided $\mathfrak{N} \neq \{0\}$. Let $\mathsf{N} \in \mathfrak{N} \setminus \{0\}$. Let $\delta \neq 0$ be so small in absolute value that $\mathsf{K}_* = \mathsf{L}_* + \delta \mathsf{N} \in \mathcal{M}(\alpha, \beta)^\Omega$. Then K_* satisfies (A1) and (A2).

Let $\boldsymbol{E}_* \in \mathcal{E}_0(\Omega)$ be such that $\mathsf{N}\boldsymbol{E}_* \neq 0$. This can always be arranged if $V_{\mathcal{E}} = \mathcal{T}$. Let $\boldsymbol{E}_\varepsilon$ be the unique solution of $\Gamma_\Omega(\mathsf{L}_\varepsilon \boldsymbol{E}_\varepsilon) = \Gamma_\Omega(\mathsf{L}_* \boldsymbol{E}_*) = \Gamma_\Omega(\mathsf{K}_* \boldsymbol{E}_*)$. Then the weak limit of $\mathsf{L}_\varepsilon \boldsymbol{E}_\varepsilon$ (or a weakly convergent subsequence thereof) cannot be equal to both $\mathsf{L}_* \boldsymbol{E}_*$ and $(\mathsf{L}_* + \delta \mathsf{N})\boldsymbol{E}_*$.

We now prove the converse, namely that if $\mathfrak{N} = \{0\}$ then conditions (A1) and (A2) imply that $\mathsf{L}_\varepsilon \overset{H}{\rightharpoonup} \mathsf{L}_*$. The following simple lemma gives the reason.

Lemma 3.41. *Suppose* L_ε *and* L_* *satisfy assumptions (A1) and (A2) above. Suppose* $\mathsf{L}_\varepsilon \overset{H}{\rightharpoonup} \mathsf{K}_*$, *as* $\varepsilon \to 0$. *Then* $\mathsf{N}(\boldsymbol{x}) = \mathsf{K}_*(\boldsymbol{x}) - \mathsf{L}_*(\boldsymbol{x})$ *satisfies (3.58).*

Proof. Let $\boldsymbol{E} \in \mathcal{E}_0(\Omega)$ be arbitrary. Let $\boldsymbol{E}_\varepsilon$ be the unique solution of $\Gamma_\Omega(\mathsf{L}_\varepsilon \boldsymbol{E}_\varepsilon) = \Gamma_\Omega(\mathsf{L}_* \boldsymbol{E})$. Then $\boldsymbol{E}_\varepsilon \rightharpoonup \boldsymbol{E}$, as $\varepsilon \to 0$ in $L^2(\Omega; \mathcal{T})$ by assumption (A1). But, by definition of H-convergence we conclude that $\boldsymbol{E}$ also solves $\Gamma_\Omega(\mathsf{K}_* \boldsymbol{E}) = \Gamma_\Omega(\mathsf{L}_* \boldsymbol{E})$. Thus, $\mathsf{N}(\boldsymbol{x})$ satisfies (3.60), which is equivalent to (3.58). □

Combining this lemma with compactness theorem (theorem 3.3) and lemma 3.40 gives the desired result, which we state formally as a theorem.

Theorem 3.42. *Assume that the symbol* $\Gamma(\eta)$ *is* $\mathcal{E}$*-nondegenerate and satisfies the compensated compactness property. Suppose* $\mathsf{L}_\varepsilon(\boldsymbol{x})$ *and* $\mathsf{L}_*(\boldsymbol{x})$ *have the properties (A1), (A2) and that* $\mathfrak{N} = \{0\}$. *Then* $\mathsf{L}_\varepsilon \overset{H}{\rightharpoonup} \mathsf{L}_*$.

Proof. By the compactness theorem, there exists an H-convergent subsequence $\mathsf{L}_{\varepsilon_k} \overset{H}{\rightharpoonup} \mathsf{K}_*$. Then, by lemma 3.41 the difference $\mathsf{N}(\boldsymbol{x}) = \mathsf{L}_*(\boldsymbol{x}) - \mathsf{K}_*(\boldsymbol{x})$ satisfies (3.58). Applying lemma 3.40 we conclude that $\mathsf{N}(\boldsymbol{x}) \in \mathfrak{N} = \{0\}$ for a.e. $\boldsymbol{x} \in \Omega$. Thus, all H-limits of H-convergent subsequences of L_ε are equal to L_*. The theorem follows. □

In most physical contexts the tensors of material properties are symmetric. In that case we have the following theorem

Theorem 3.43. *Assume that the symbol* $\Gamma(\eta)$ *is* $\mathcal{E}$*-nondegenerate and satisfies the compensated compactness property. Suppose* $\mathsf{L}_\varepsilon(\boldsymbol{x})$ *and* $\mathsf{L}_*(\boldsymbol{x})$ *have the properties (A1), (A2) and that* $\mathsf{L}_\varepsilon^T = \mathsf{L}_\varepsilon$.

(i) *If* $\mathfrak{N}_a = \{0\}$, *then* $\mathsf{L}_*^T = \mathsf{L}_*$.
(ii) *If* $\mathsf{L}_*^T = \mathsf{L}_*$ *and* $\mathfrak{N}_s = \{0\}$, *then* $\mathsf{L}_\varepsilon \overset{H}{\rightharpoonup} \mathsf{L}_*$.

Proof. Part (i). Let $\{\boldsymbol{E}_1, \boldsymbol{E}_2\} \subset \mathscr{E}_0(\Omega)$ be arbitrary. Let $\boldsymbol{E}_\varepsilon^{(j)} \in \mathscr{E}_0(\Omega)$ solve

$$\Gamma_\Omega(\mathsf{L}_\varepsilon \boldsymbol{E}_\varepsilon^{(j)}) = \Gamma_\Omega(\mathsf{L}_* \boldsymbol{E}_j), \quad j = 1, 2.$$

Then $\boldsymbol{E}_\varepsilon^{(j)} \rightharpoonup \boldsymbol{E}_j$, $j = 1, 2$ in $L^2(\Omega; \mathscr{T})$ by assumption. Then

$$\begin{aligned}(\boldsymbol{E}_\varepsilon^{(1)}, \mathsf{L}_*\boldsymbol{E}_2)_{L^2(\Omega)} &= (\boldsymbol{E}_\varepsilon^{(1)}, \Gamma_\Omega(\mathsf{L}_*\boldsymbol{E}_2))_{L^2(\Omega)} = (\boldsymbol{E}_\varepsilon^{(1)}, \Gamma_\Omega(\mathsf{L}_\varepsilon\boldsymbol{E}_\varepsilon^{(2)}))_{L^2(\Omega)} \\ &= (\Gamma_\Omega(\mathsf{L}_\varepsilon\boldsymbol{E}_\varepsilon^{(1)}), \boldsymbol{E}_\varepsilon^{(2)})_{L^2(\Omega)} = (\mathsf{L}_*\boldsymbol{E}_1, \boldsymbol{E}_\varepsilon^{(2)})_{L^2(\Omega)}.\end{aligned}$$

Passing to the limit as $\varepsilon \to 0$ we obtain that

$$((\mathsf{L}_* - \mathsf{L}_*^T)\boldsymbol{E}_1, \boldsymbol{E}_2)_{L^2(\Omega)} = 0$$

for all $\{\boldsymbol{E}_1, \boldsymbol{E}_2\} \subset \mathscr{E}_0(\Omega)$. Therefore, according to lemma 3.40, $\mathsf{N}(\boldsymbol{x}) = \mathsf{L}_*(\boldsymbol{x}) - \mathsf{L}_*^T(\boldsymbol{x}) \in \mathfrak{N}_s = \{0\}$ for a.e. $\boldsymbol{x} \in \Omega$.

Part (ii). The proof repeats the proof of theorem 3.42. We extract an H-convergent subsequence $\mathsf{L}_{\varepsilon_k} \overset{H}{\rightharpoonup} \mathsf{K}_*$. By part (a) of theorem 3.31 we conclude that $\mathsf{K}_*^T = \mathsf{K}_*$. Therefore, the difference $\mathsf{N}(\boldsymbol{x}) = \mathsf{L}_*(\boldsymbol{x}) - \mathsf{K}_*(\boldsymbol{x}) \in \mathfrak{N}_s = \{0\}$ for a.e. $\boldsymbol{x} \in \Omega$. Thus, all H-limits of H-convergent subsequences of L_ε are equal to L_*. Part (ii) is proved. □

References

[1] Allaire G 2002 *Shape Optimization by the Homogenization Method vol 146 of Applied Mathematical Sciences* (New York: Springer)

[2] Babadjian J-F and Barchiesi M 2009 A variational approach to the local character of g-closure: the convex case *Ann. Inst. Henri Poincare (C) Non Linear Anal.* **26** 351–73

[3] Backus G E 1962 Long-wave elastic anisotropy produced by horizontal layering *J. Geophys. Res.* **67** 4427–40

[4] Cherkaev A and Zhang Y 2011 Optimal anisotropic three-phase conducting composites: plane problem *Int. J. Solids Struct.* **48** 2800–13

[5] De Giorgi E and Spagnolo S 1973 Sulla convergenza degli integrali dell'energia per operatori ellittici del secondo ordine *Boll. Un. Mat. Ital. (4)* **8** 391–411

[6] Eyre D J and Milton G W 1999 A fast numerical scheme for computing the response of composites using grid refinement *Eur. Phys. J. Appl. Phys.* **6** 41–7

[7] Francfort G A and Milton G W 1987 Optimal bounds for conduction in two-dimensional, multiphase, polycrystalline media *J. Stat. Phys.* **46** 161–77

[8] Francfort G A and Milton G W 1994 Sets of conductivity and elasticity tensors stable under lamination *Commun. Pure Appl. Math.* **47** 257–79

[9] Grabovsky Y 1993 The G-closure of two well-ordered anisotropic conductors *Proc. R. Soc. Edinburgh* **123A** 423–32

[10] Grabovsky Y and Milton G W 1998 Rank one plus a null-Lagrangian is an inherited property of two-dimensional compliance tensors under homogenization *Proc. R. Soc. Edinburgh A* **128** 283–99

[11] Hollister S J and Kikuchi N 1992 A comparison of homogenization and standard mechanics analyses for periodic porous composites *Comput. Mech.* **10** 73–95

[12] Lurie K A and Cherkaev A V 1981 *G*-closure of a set of anisotropic conducting media in the case of two dimensions *Dokl. Akad. Nauk SSSR* **259** 328–31 In Russian

[13] Lurie K A and Cherkaev A V 1984 Exact estimates of conductivity of composites formed by two isotropically conducting media taken in prescribed proportion *Proc. R. Soc. Edinburgh* **99A** 71–87

[14] Lurie K A and Cherkaev A V 1986 Exact estimates of a binary mixture of isotropic components *Proc. R. Soc. Edinburgh* **104A** 21–38

[15] Michel J, Moulinec H and Suquet P 2001 A computational scheme for linear and non-linear composites with arbitrary phase contrast *Int. J. Numer. Methods Eng.* **52** 139–60

[16] Milton G W 1990 On characterizing the set of possible effective tensors of composites: the variational method and the translation method *Commun. Pure Appl. Math.* **43** 63–125

[17] Milton G W 2002 *The Theory of Composites Cambridge Monographs on Applied and Computational Mathematics* (Cambridge: Cambridge University Press)

[18] Murat F 1981 Compacité par compensation: condition nécessaire et suffisante de continuité faible sous une hypothèse de rang constant *Ann. Scuola Norm. Sup. Pisa Cl. Sci. (4)* **8** 69–102

[19] Murat F and Tartar L 1997 H-convergence *Topics in the Mathematical Modelling of Composite Materials* ed A Cherkaev and R V Kohn *Progress in Non-linear Differential equations and Their Applications* vol 31 (Boston, MA: Birkhäuser) ch 3, pp 21–43

[20] Raitums U 2001 On the local representation of G-closure *Arch. Ration. Mech. Anal.* **158** 213–34

[21] Sánchez-Palencia E 1980 *Nonhomogeneous Media and Vibration Theory* (Berlin: Springer)

[22] Spagnolo S 1968 Sulla convergenza di soluzioni di equazioni paraboliche ed ellittiche *Ann. Scuola Norm. Sup. Pisa (3) 22 571–97; errata, ibid. (3)* **32** 673

[23] Tartar L 1979 Compensated compactness and applications to partial differential equations *Nonlinear Analysis and Mechanics: Heriot-Watt Symp., vol IV Research Notes in Mathematics* vol 39 (Boston, MA: Pitman) pp 136–212

[24] Tartar L 1979 Estimation de coefficients homogénéisés *Computing Methods in Applied Sciences and Engineering (Proc. 3rd Int. Sympos., Versailles, 1977), I vol 704 of Lecture Notes in Mathematics* (Berlin: Springer) pp 364–73

[25] Tartar L 1990 *H*-measures, a new approach for studying homogenisation, oscillations and concentration effects in partial differential equations *Proc. R. Soc. Edinburgh* A **115** 193–230

[26] Tartar L 2000 An introduction to the homogenization method in optimal design *Optimal Shape Design (Tróia, 1998) Lecture Notes in Mathematics* vol 1740 (Berlin: Springer) pp 47–156

[27] Zhikov V V 1991 Estimates for the averaged matrix and the averaged tensor *Russ. Math Surv.* **46** 65–136

Part II

General theory of exact relations and links

IOP Publishing

Composite Materials (Second Edition)

Mathematical theory and exact relations

Yury Grabovsky

Chapter 4

Exact relations

4.1 Introduction

Formulas for effective behavior of composite materials have been sought out throughout the entire history of mathematical analysis of composite media. Clausius-Mossotti [4, 31] (a.k.a. Maxwell-Garnett [11, 25]) formulas for effective conductivity of composites[1] and Einstein's formula for effective viscosity of dilute suspensions [10] are probably the most well-known examples in physics. It is understood that these 'formulas' are actually approximations valid asymptotically in the small volume fraction limit. The actual effective behavior of a composite depends on its microstructure. Nevertheless, exact algebraic formulas relating bulk properties of composites and their constituents do exist, and have been attracting steady attention of physicists and engineers. Not only are such formulas surprising, they possess a peculiar aesthetic beauty. This part of the book will explain how *all* such formulas can be obtained systematically for every material property of interest.

As an illustration, let us describe two earliest and most well-known examples of such formulas. The first, due to Hill [16, 17], says that any composite made with isotropic elastic materials with the same shear modulus μ_0 must always be isotropic (even if geometrically its microstructure is highly anisotropic!) and have the shear modulus μ_0. Moreover, it bulk modulus κ_* can be expressed in terms of the volume average of the local elastic tensor:

$$(3\kappa_* + 4\mu_0)^{-1} = \langle(3\kappa(z) + 4\mu_0)^{-1}\rangle. \tag{4.1}$$

The second example refers to 2D conductivity and was first observed by Keller [20] in a periodic array of circular cylindrical inclusions (see [8, 27, 32] for its subsequent development). If the local conductivity tensor $\boldsymbol{\sigma}(z)$ has the property $\det \boldsymbol{\sigma}(z) = d_0$ for all z, then the effective conductivity tensor $\boldsymbol{\sigma}_*$ must have the same property, $\det \boldsymbol{\sigma}_* = d_0$, *regardless of microstructure*.

[1] For a history of this discovery see e.g. [21].

Examples of such exact algebraic formulas for coupled field material properties, such as thermoelasticity and piezoelectricity abound. In this part of the book we are going to describe a theory that can generate *complete lists* of all such formulas. The complete lists that have been computed are presented in part III of the book, which is intended to be used as a reference. To begin, we need to give a formal definition of an exact relation. We will use the general Hilbert space formalism developed in the first part of this book in order to obtain a unified theory of exact relations equally applicable to all material properties to which the homogenization theory of part I applies.

Definition 4.1. *An exact relation is a G-closed smooth submanifold* $\mathbb{M} \subset \mathrm{End}^+(\mathscr{T})$.

According to this definition, the algebraic formulas relating effective behavior of composites to that of their constituents are now viewed geometrically, as describing a smooth submanifold (surface) $\mathbb{M}$ in the space $\mathrm{End}^+(\mathscr{T})$ of tensors of material properties. For a submanifold $\mathbb{M}$ to be an exact relation it has to be G-closed, or *stable under homogenization*. In other words, the effective tensor of every composite made of any number of materials taken from $\mathbb{M}$ must again belong to $\mathbb{M}$.

Our first example of Hill's exact relation describes a whole family of such manifolds $\mathbb{M}_{\mu_0}$ parameterized by the constant shear modulus μ_0. Geometrically, each manifold $\mathbb{M}_{\mu_0}$ looks like a (part of a) one-dimensional affine line in the 21-dimensional space of elasticity tensors given by

$$\mathbb{M}_{\mu_0} = \{\mathsf{C} \in \mathbb{E}^3_+ \colon \mathsf{C}\varepsilon = \lambda(\mathrm{Tr}\,\varepsilon)\boldsymbol{I} + 2\mu_0\varepsilon,\ \varepsilon \in \mathrm{Sym}(\mathbb{R}^3),\ \lambda \in \mathbb{R}\}. \tag{4.2}$$

where $\mathbb{E}^3_+ = \mathrm{Sym}^+(\mathrm{Sym}(\mathbb{R}^3))$ denotes the set of all positive definite three-dimensional elasticity tensors.

Our second example of Keller-Dykhne exact relation also describes a whole family of such manifolds $\mathbb{M}_{d_0}$, parameterized by the constant determinant d_0 of 2D conductivity tensors. Geometrically, it is a (part of a) sheet of a two-sheeted hyperboloid in the three-dimensional space of 2D conductivity tensors, given by

$$\mathbb{M}_{d_0} = \{\boldsymbol{\sigma} \in \mathrm{Sym}^+(\mathbb{R}^2) \colon \det \boldsymbol{\sigma} = d_0\}. \tag{4.3}$$

Observe that formula (4.1) does not seem to fit into the concept of an exact relation that we have just defined. At the first glance, at least. Nevertheless, the theory of exact relations developed in this chapter is perfectly capable of capturing all such formulas as well. In chapter 5 the abstract nature of our approach will become an advantage, permitting us to obtain volume fraction relations, like (4.1), as well as links between effective behavior of a pair of composites sharing the same microstructure, but differing in the properties of their constituents.

There are several classes of exact relations that are physically natural, or obvious. One is $\mathbb{M} = \mathrm{Sym}^+(\mathscr{T}) \subset \mathrm{End}^+(\mathscr{T})$. Onsager's reciprocity relations [33, 34] show that symmetry of material tensors has a deep statistical mechanics foundation. Our theory in part I of this book shows that the effective tensor of a composite made of reciprocal constituents must be reciprocal.

Another natural class of exact relations was briefly mentioned in part I of the book in remark 2.7.

$$\mathbb{M} = \mathrm{End}^+_{\mathbb{C}}(\mathscr{T}_{\mathbb{C}}) \subset \mathrm{End}^+_{\mathbb{R}}(\mathscr{T}_{\mathbb{R}})$$

and

$$\mathbb{M} = \mathfrak{H}^+(\mathscr{T}_{\mathbb{C}}) \subset \mathrm{Sym}^+_{\mathbb{R}}(\mathscr{T}_{\mathbb{R}}),$$

where $\mathscr{T}_{\mathbb{C}}$ is a complex Hermitian space and $\mathscr{T}_{\mathbb{R}}$ is the same set $\mathscr{T}_{\mathbb{C}}$ regarded as a real vector space. Complex linear operators on $\mathscr{T}_{\mathbb{C}}$ form a proper subset of real linear operators on $\mathscr{T}_{\mathbb{R}}$, and complex Hermitian operators $\mathfrak{H}(\mathscr{T}_{\mathbb{C}})$ on $\mathscr{T}_{\mathbb{C}}$ form a proper subset of real symmetric operators on $\mathscr{T}_{\mathbb{R}}$.

Finally, an important class of exact relations are the *uniform field relations* (UFR). Let us consider composites made with a given set of r materials $\mathsf{L}_1, \ldots, \mathsf{L}_r$. Suppose that there are field values $\{e_1, \ldots, e_k\} \subset \mathscr{T}$, such that

$$\mathsf{L}_1 e_j = \mathsf{L}_2 e_j = \ldots = \mathsf{L}_r e_j = \tau_j, \quad j = 1, \ldots, k.$$

Then, the effective tensor L_* of any composite made with r such materials will also satisfy $\mathsf{L}_* e_j = \tau_j$, $j = 1, \ldots, k$. Physically, this is clear, since all r of the constituents behave as a single material in response to the applied fields $e_1, \ldots, e_k$. Mathematically, this is explained by observing that the cell problem (3.10) admits uniform field solutions $\boldsymbol{E}(z) = e_j$, $j = 1, \ldots, k$. For this reason such exact relations are called the uniform field relations (UFR) [6, 7, 22]. Hill's exact relation is an example of UFR, where the fields $\{e_1, \ldots, e_k\}$ can be chosen arbitrarily from the subspace of symmetric, trace-free matrices, as long as they span the subspace. Curiously, formula (4.1) cannot be derived by means of a uniform field argument, and our general theory explains why. As examples (4.1) and (4.3) show, the above natural classes of exact relations, do not exhaust all possible ones in all but a few simplest cases, such as 3D conductivity. And even in these cases the general theory provides a guarantee that no as yet undiscovered exact relations remain.

The theory emerges from the observation that if a submanifold $\mathbb{M}$ is an exact relation, then it must contain the effective tensor of any simple laminate whose constituents are in $\mathbb{M}$. The general lamination formula (3.36) implies that the submanifold $\mathbb{K}_n = W_n(\mathbb{M})$ must be convex and, hence, flat. Choosing the reference medium $\mathsf{L}_0 \in \mathbb{M}$, and recalling that $W_n(\mathsf{L}_0) = 0$ we can ensure that the convex sets $\mathbb{K}_n$ belong to subspaces Π_n of the same dimension as $\mathbb{M}$. These simple observations can be expressed as a family of algebraic constraints on $\mathbb{M}$, parameterized by unit vectors $\boldsymbol{n}$. The analysis of these algebraic constraints leads us to conclude that the subspaces Π_n do not depend on $\boldsymbol{n}$ and satisfy a set of beautiful algebraic conditions, which are necessary for the submanifold $\mathbb{M}$ to describe an exact relation. The physical meaning of these conditions is that the submanifold $\mathbb{M}$ is closed under lamination, i.e., $\mathbb{M}$ must contain the effective tensor of every laminate made of materials taken from $\mathbb{M}$.

The passage from laminar to general microstructures occurs via formula (3.26), valid for all composites. This formula can be expanded into the Neumann series (3.35), when when $\mathsf{L}(z)$ is sufficiently close to L_0. Each term of the Neumann series

(3.35) represents a sum of products[2] of operators (matrices) in $\Pi = \Pi_{n}$ interlaced with operators $\Lambda_{\Gamma_0(n)}(k)$, defined in (3.24). We can conclude that each term is in Π if Π is a multialgebra (i.e., a subspace of operators on $\mathscr{T}$ closed with respect to a set of interlacing matrix products). This explanation, while a little vague, shows how and why the theory works.

This chapter is organized as follows. In section 4.2 we examine the algebraic constraints on $\mathbb{M}$ (or rather the subspace Π) that come from the requirement that $\mathbb{M}$ is closed under lamination. In section 4.3 we analyze the Neumann series (3.35) and prove that each term belongs to Π, provided Π satisfies some additional, but easily verifiable algebraic conditions. In section 4.4 we define UFRs formally and introduce *polycrystalline exact relations*, which are the actual focus of this part of the book. The polycrystalline exact relations are physically more natural, allowing any given anisotropic constituent to enter the composite in arbitrary orientations. At the same time they are both mathematically beautiful and feasibly computable in every physical context. Section 4.5 concludes the chapter with rigorous proofs of statements made in section 4.3.

4.2 L-relations

The goal of this section is to characterize all submanifolds $\mathbb{M} \subset \mathrm{End}^{+}(\mathscr{T})$ of non-zero co-dimension, that are stable under lamination [12].

Definition 4.2. *A submanifold* $\mathbb{M} \subset \mathrm{End}^{+}(\mathscr{T})$ *is called an L-relation if the effective tensor of any laminate made with any two materials, whose tensors lie in* $\mathbb{M}$, *belongs to* $\mathbb{M}$.

Being an L-relation places rather strong constraints on $\mathbb{M}$, because for each pair of materials from $\mathbb{M}$ both the volume fraction and the direction of lamination can be varied arbitrarily. Figure 4.1 shows lamination paths parameterized by the volume fraction for four different choices of the lamination direction on the exact relation manifold (4.3).

By corollary 3.21, if $\mathbb{M}$ is an L-relation, then the sets $W_{n}(\mathbb{M})$ must be convex submanifolds of $\mathrm{End}(\mathscr{T})$ of the same dimension as $\mathbb{M}$ for all unit vectors $\boldsymbol{n}$, since all maps W_{n}, defined in (3.37), (3.21), are analytic and injective on $\mathrm{End}^{+}(\mathscr{T})$ by lemma 3.23. Thus, for each unit vector $\boldsymbol{n}$ the set $W_{n}(\mathbb{M})$ must be a convex, relatively open subset of an affine subspace $\Pi_{n} \subset \mathrm{End}(\mathscr{T})$. Observing that $W_{n}(\mathsf{L}_0) = 0$, which is obvious from formula (3.21), we may assume, without loss of generality, that all Π_{n} are linear subspaces of $\mathrm{End}(\mathscr{T})$, provided $\mathsf{L}_0 \in \mathbb{M}$. Let us fix a unit vector $\boldsymbol{n}_0$ and denote $\Pi = \Pi_{n_0}$. If $\mathbb{M}$ is an L-relation we must have

$$\mathbb{M} = \{\mathsf{L} \in \mathrm{End}^{+}(\mathscr{T}) \colon W_{n_0}(\mathsf{L}) \in \Pi\}. \tag{4.4}$$

[2] The fact that some of the operators are Fourier multipliers does not change validity of our reasoning, since the Fourier transform of a $\Pi_{\mathbb{C}}$-valued function is also $\Pi_{\mathbb{C}}$-valued, where $\Pi_{\mathbb{C}} = \Pi \otimes \mathbb{C}$ is the complexification of Π.

Figure 4.1. Lamination paths for four different lamination directions on the Keller-Dykhne exact relation manifold (4.3).

Now, let us vary the direction of the unit vector $\boldsymbol{n}$. The composition map $\Lambda_n = W_n \circ W_{n_0}^{-1}$ must map a sufficiently small neighborhood of 0 in $\Pi = \Pi_{n_0}$ to a neighborhood of 0 in Π_n. We easily compute, using formulas (3.21), (3.37) for W_n,

$$(W_n \circ W_{n_0}^{-1})(\mathsf{K}) = \Lambda_n(\mathsf{K}) = [\mathsf{I} - \mathsf{K}\mathsf{A}(\boldsymbol{n})]^{-1}\mathsf{K}, \quad \mathsf{A}(\boldsymbol{n}) = \Gamma_0(\boldsymbol{n}_0) - \Gamma_0(\boldsymbol{n}), \tag{4.5}$$

since for K in a sufficiently small neighborhood of 0 the operator $\mathsf{I} - \mathsf{K}\mathsf{A}(\boldsymbol{n})$ is always invertible[3]. Expanding $\Lambda_n(\mathsf{K})$ in power series around $\mathsf{K} = 0$

$$\Lambda_n(\mathsf{K}) = \sum_{n=0}^{\infty} \mathsf{K}(\mathsf{A}(\boldsymbol{n})\mathsf{K})^n = \mathsf{K} + \mathsf{K}\mathsf{A}(\boldsymbol{n})\mathsf{K} + \ldots + \mathsf{K}(\mathsf{A}(\boldsymbol{n})\mathsf{K})^n + \ldots \tag{4.6}$$

we conclude that each term of the expansion (4.6) must belong to Π_n. Looking at the first term we see that $\mathsf{K} \in \Pi_n$ for every K in a small neighborhood of $0 \in \Pi$. Thus $\Pi \subset \Pi_n$. But, since all the subspaces Π_n are diffeomorphic images of $\mathbb{M}$ they must have the same dimension as $\mathbb{M}$, and we conclude that $\Pi_n = \Pi$ for all unit vectors $\boldsymbol{n}$.

Looking at the second term in (4.6) we see that the subspace Π must satisfy

$$\mathsf{K}\mathsf{A}(\boldsymbol{n})\mathsf{K} \subset \Pi \tag{4.7}$$

for all $\mathsf{K} \in \Pi$ and all unit vectors $\boldsymbol{n}$. We can reformulate (4.7) as follows. If we set $\mathsf{K} = \mathsf{K}_1 + \mathsf{K}_2$ in (4.7) and expand, then we get that for any $\{\mathsf{K}_1, \mathsf{K}_2\} \subset \Pi$ and any direction $\boldsymbol{n}$

$$\mathsf{K}_1\mathsf{A}(\boldsymbol{n})\mathsf{K}_2 + \mathsf{K}_2\mathsf{A}(\boldsymbol{n})\mathsf{K}_1 \in \Pi. \tag{4.8}$$

Since Π is a subspace we can make all possible linear combinations of expressions in (4.8) with the same K_1 and K_2 and different $\boldsymbol{n}$. Therefore,

$$\mathsf{K}_1\mathsf{A}\mathsf{K}_2 + \mathsf{K}_2\mathsf{A}\mathsf{K}_1 \in \Pi, \quad \forall \{\mathsf{K}_1, \mathsf{K}_2\} \subset \Pi, \mathsf{A} \in \mathscr{A}, \tag{4.9}$$

[3] In fact the operator $\mathsf{I} - \mathsf{K}\mathsf{A}(\boldsymbol{n})$ is invertible for *any* $\mathsf{K} = W_{n_0}(\mathsf{L})$ by corollary 3.25.

where

$$\mathscr{A} = \mathrm{Span}\{\boldsymbol{\Gamma}_0(\boldsymbol{n}) - \boldsymbol{\Gamma}_0(\boldsymbol{n}_0): \boldsymbol{n} \in \mathbb{R}^d, |\boldsymbol{n}| = 1\}. \tag{4.10}$$

It is easy to see that the subspace $\mathscr{A}$ is independent of $\boldsymbol{n}_0$, because

$$\boldsymbol{\Gamma}_0(\boldsymbol{n}) - \boldsymbol{\Gamma}_0(\boldsymbol{n}_1) = \boldsymbol{\Gamma}_0(\boldsymbol{n}) - \boldsymbol{\Gamma}_0(\boldsymbol{n}_0) - (\boldsymbol{\Gamma}_0(\boldsymbol{n}_1) - \boldsymbol{\Gamma}_0(\boldsymbol{n}_0)).$$

A more natural way to see it is to define $\mathscr{A}$ without reference to $\boldsymbol{n}_0$. Indeed, $\mathscr{A}$ is the vector space parallel to the affine hull of the set $\{\boldsymbol{\Gamma}_0(\boldsymbol{n}): |\boldsymbol{n}| = 1\}$. In summary, in order for a submanifold $\mathbb{M}$ given by (4.4) to be an exact relation it is necessary for $W_{\boldsymbol{n}_0}(\mathbb{M})$ to be a relatively open subset of a vector space $\Pi \subset \mathrm{End}(\mathscr{T})$ satisfying (4.9). We observe that no further information about the subspace Π can be obtained from the expansion (4.6), since all other terms in (4.6) will necessarily be in Π, if Π satisfies (4.9). This is proved by induction in n, since

$$\mathsf{K}(\mathsf{A}(\boldsymbol{n})\mathsf{K})^n = \frac{1}{2}\{\mathsf{KA}(\boldsymbol{n})[\mathsf{K}(\mathsf{A}(\boldsymbol{n})\mathsf{K})^{n-1}] + [\mathsf{K}(\mathsf{A}(\boldsymbol{n})\mathsf{K})^{n-1}]\mathsf{A}(\boldsymbol{n})\mathsf{K}\}.$$

The property (4.9) of a subspace $\Pi \subset \mathrm{End}(\mathscr{T})$ can be interpreted algebraically. It means that Π is closed with respect to a family of commutative, but non-associative multiplications

$$\mathsf{K}_1 *_{\mathsf{A}} \mathsf{K}_2 = \frac{1}{2}(\mathsf{K}_1\mathsf{A}\mathsf{K}_2 + \mathsf{K}_2\mathsf{A}\mathsf{K}_1), \quad \mathsf{A} \in \mathscr{A}. \tag{4.11}$$

Each of the products in the family (4.11) is an example of a Jordan product [18, 19], making the subspace Π into a Jordan subalgebra of $\mathrm{End}(\mathscr{T})$. Hence, we introduce the following terminology:

Definition 4.3. *Let $\mathscr{A}$ be a subspace in* $\mathrm{End}(\mathscr{T})$. *The subspace* $\Pi \subset \mathrm{End}(\mathscr{T})$ *is called a Jordan $\mathscr{A}$-multialgebra if it is closed with respect to the family of Jordan multiplications (4.11).*

The next theorem shows that algebraic property (4.9) is also sufficient for stability under lamination.

Theorem 4.4. *Suppose that the subspace $\Pi \subset \mathrm{End}(\mathscr{T})$ is a Jordan $\mathscr{A}$-multialgebra, where $\mathscr{A}$ is given by (4.10). Then for any $\mathsf{L}_0 \in \mathrm{End}^+(\mathscr{T})$ and any unit vector $\boldsymbol{n}_0$ the submanifold $\mathbb{M} \subset \mathrm{End}^+(\mathscr{T})$, given by (4.4), is an L-relation and does not depend on the choice of $\boldsymbol{n}_0$ in its definition.*

Proof. Let us take $\{\mathsf{L}_1, \mathsf{L}_2\} \subset \mathbb{M}$ and make a laminate with normal $\boldsymbol{n}$ and volume fractions $\theta, 1 - \theta$, respectively. By theorem 3.20, its effective tensor L_* must satisfy $W_{\boldsymbol{n}}(\mathsf{L}_*) = \theta W_{\boldsymbol{n}}(\mathsf{L}_1) + (1 - \theta)W_{\boldsymbol{n}}(\mathsf{L}_2)$. We note that corollary 3.35 guarantees that $\mathsf{L}_* \in \mathrm{End}^+(\mathscr{T})$, since $\{\mathsf{L}_1, \mathsf{L}_2\} \subset \mathrm{End}^+(\mathscr{T})$, by definition of $\mathbb{M}$. Our first step is to show that $\{W_{\boldsymbol{n}}(\mathsf{L}_1), W_{\boldsymbol{n}}(\mathsf{L}_2)\} \subset \Pi$.

Lemma 4.5. *Let* Π *be a Jordan* $\mathscr{A}$*-multialgebra. If* $\mathsf{K}_0 \in \Pi$, $\mathsf{A}_0 \in \mathscr{A}$, *and* $\mathsf{I} - \mathsf{K}_0\mathsf{A}_0$ *is an invertible operator on* $\mathscr{T}$, *then* $(\mathsf{I} - \mathsf{K}_0\mathsf{A}_0)^{-1}\mathsf{K}_0 \in \Pi$.

Proof. We first observe that the polynomial $p(t) = \det(\mathsf{I} - t\mathsf{K}_0\mathsf{A}_0)$ is such that $p(0) = 1$ and $p(1) \neq 0$. Therefore, there exists a curve $\gamma \subset \mathbb{C}$, joining 0 and 1, and its open neighborhood $U_\gamma \subset \mathbb{C}$, such that there are no zeros of $p(t)$ in U_γ. Then $1/p(t)$ is analytic on U_γ. Let $\mathsf{N} \in \mathrm{End}(\mathscr{T})$ be in the orthogonal complement of Π. Then the function

$$\phi(t) = ((\mathsf{I} - t\mathsf{K}_0\mathsf{A}_0)^{-1}\mathsf{K}_0, \mathsf{N})_{\mathrm{End}(\mathscr{T})}, \quad t \in \mathbb{C}$$

is also analytic on U_γ, since $p(t)\phi(t)$ is a polynomial in t. However, expanding $(\mathsf{I} - t\mathsf{K}_0\mathsf{A}_0)^{-1}$ into the Neumann series, we see that $\phi(t) = 0$ on a sufficiently small neighborhood of 0 in $\mathbb{C}$. The analyticity of $\phi(t)$ implies that $\phi(t) = 0$ for all $t \in U_\gamma$. In particular, $\phi(1) = 0$. The lemma follows. □

Corollary 4.6. *Let* $\mathbb{M}$ *be given by (4.4), where* Π *satisfies (4.9). Then* $W_{\boldsymbol{n}}(\mathbb{M}) \subset \Pi$ *for any unit vector* $\boldsymbol{n}$.

Proof. Let $\mathsf{L} \in \mathbb{M}$. Then $\mathsf{K} = W_{\boldsymbol{n}_0}(\mathsf{L}) \in \Pi$. For any unit vector $\boldsymbol{n}$ we compute

$$W_{\boldsymbol{n}}(\mathsf{L}) = W_{\boldsymbol{n}}(W_{\boldsymbol{n}_0}^{-1}(\mathsf{K})) = \boldsymbol{\Lambda}_{\boldsymbol{n}}(\mathsf{K}),$$

where the map $\boldsymbol{\Lambda}_{\boldsymbol{n}}$ is defined in (4.5). The operator $\mathsf{I} - \mathsf{K}\mathsf{A}(\boldsymbol{n})$ is invertible, by corollary 3.25. Then, by lemma 4.5 we conclude that $W_{\boldsymbol{n}}(\mathsf{L}) = \boldsymbol{\Lambda}_{\boldsymbol{n}}(\mathsf{K}) \in \Pi$. □

Corollary 4.6 implies that $\{W_{\boldsymbol{n}}(\mathsf{L}_1), W_{\boldsymbol{n}}(\mathsf{L}_2)\} \subset \Pi$ and hence

$$W_{\boldsymbol{n}}(\mathsf{L}_*) = \theta W_{\boldsymbol{n}}(\mathsf{L}_1) + (1 - \theta) W_{\boldsymbol{n}}(\mathsf{L}_2) \in \Pi.$$

Thus, $\mathsf{L}_* \in \mathbb{M}_{\boldsymbol{n}}$, where we (temporarily) define $\mathbb{M}_{\boldsymbol{n}}$ by (4.4) with $\boldsymbol{n}$ playing the role of $\boldsymbol{n}_0$. Now, corollary 4.6 implies that $\mathbb{M}_{\boldsymbol{n}_0} \subset \mathbb{M}_{\boldsymbol{n}}$ for any unit vector $\boldsymbol{n}$. Hence $\mathbb{M}_{\boldsymbol{n}} = \mathbb{M}_{\boldsymbol{n}_0}$, since the unit vector $\boldsymbol{n}_0$ was chosen arbitrarily. Theorem 4.4 is proved now. □

4.3 Sufficient conditions for stability under homogenization

In this section we formulate simple algebraic sufficient conditions for an L-relation submanifold $\mathbb{M}$ to be an exact relation, i.e., to be G-closed.

Definition 4.7. *We say that a subspace* $\Pi \subset \mathrm{End}(\mathscr{T})$ *has the j-chain property,* $j \geqslant 2$, *if for every* $\{\mathsf{K}_1, \ldots, \mathsf{K}_j\} \subset \Pi$ *and every* $\{\mathsf{A}_1, \ldots, \mathsf{A}_{j-1}\} \subset \mathscr{A}$ *we have*

$$\mathsf{K}_1\mathsf{A}_1\mathsf{K}_2\mathsf{A}_2 \ldots \mathsf{K}_{j-1}\mathsf{A}_{j-1}\mathsf{K}_j + \mathsf{K}_j\mathsf{A}_{j-1}\mathsf{K}_{j-1} \ldots \mathsf{A}_2\mathsf{K}_2\mathsf{A}_1\mathsf{K}_1 \in \Pi. \tag{4.12}$$

We remark that the 2-chain property is the definition of Jordan $\mathscr{A}$-multialgebras. We can now state our sufficient conditions.

Theorem 4.8. *Suppose that the subspace* $\Pi \subset \text{End}(\mathscr{T})$ *has 2, 3 and 4-chain properties, then the submanifold* $\mathbb{M}$ *given by (4.4) is an exact relation.*

The proof can be found in section 4.5. When tensors of material properties are symmetric, as in most of our case studies in part III, we can give a neater algebraic interpretation of the sufficient conditions in theorem 4.8. In particular, if the reference medium tensor L_0 is symmetric, then so is $\mathscr{A} \subset \text{Sym}(\mathscr{T})$, due to lemma 3.12.

Definition 4.9. *A subspace* $\Pi' \subset \text{End}(\mathscr{T})$ *is called an associative* $\mathscr{A}$*-multialgebra, if* $\mathsf{K}_1\mathsf{A}\mathsf{K}_2 \in \Pi'$ *for every* $\mathsf{A} \in \mathscr{A}$ *and every* $\{\mathsf{K}_1, \mathsf{K}_2\} \subset \Pi'$.

We observe that for every associative $\mathscr{A}$-multialgebra Π'

$$\Pi'_{\text{sym}} = \Pi' \bigcap \text{Sym}(\mathscr{T}) \tag{4.13}$$

is a Jordan $\mathscr{A}$-multialgebra that has the j-chain property for any $j \geqslant 2$, since $\mathscr{A} \subset \text{Sym}(\mathscr{T})$.

Definition 4.10. *Assume* $\mathscr{A} \subset \text{Sym}(\mathscr{T})$. *Let* Π' *be the smallest associative* $\mathscr{A}$*-multialgebra, containing a Jordan* $\mathscr{A}$*-multialgebra* $\Pi \subset \text{Sym}(\mathscr{T})$. *If* $\Pi = \Pi'_{\text{sym}}$, *then we say that the Jordan* $\mathscr{A}$*-multialgebra* Π *is reflexive*[4].

The following theorem characterizes reflexive Jordan $\mathscr{A}$-multialgebras.

Theorem 4.11. *Suppose* $\mathscr{A} \subset \text{Sym}(\mathscr{T})$. *Then a Jordan* $\mathscr{A}$*-multialgebra* $\Pi \subset \text{Sym}(\mathscr{T})$ *is reflexive if and only if it has the 3 and 4-chain properties.*

The proof can be found in section 4.5.

Corollary 4.12. *If a Jordan* $\mathscr{A}$*-multialgebra* Π *is reflexive, then* Π *corresponds to an exact relation.*

It is often easier to exhibit explicitly an associative $\mathscr{A}$-multialgebra Π', whose symmetric operators comprise the given Jordan $\mathscr{A}$-multialgebra Π than to check the 3 and 4-chain properties directly. Remarkably, *all* Jordan multialgebras

[4] Our notion of reflexivity is very similar, but different from the standard terminology, where reflexivity refers to the same property of the Jordan algebra with respect to its natural embedding into the space of symmetric elements of the associative enveloping algebra with natural involution [18].

corresponding to the results in part III of this book are reflexive. Thus, the 2-chain property is a necessary condition for an exact relation, while 2, 3 and 4-chain properties (equivalent to reflexivity) are sufficient.

Whether the converse to theorem 4.8 or corollary 4.12 is true is not known. In other words we cannot rule out the existence of an exact relation that fails either 3 or 4-chain property. However, remarkably, in two space dimensions the 3-chain property is actually necessary.

Theorem 4.13. *In two space dimensions the 3-chain property is necessary for a corresponding L-relation to be exact.*

The proof, first published in [13], is placed in section 4.5. Using this theorem we have constructed in [14] an example of an L-relation that is not exact in the context of multifield response materials [28–30] coupling 4 curl-free fields to 4 divergence-free fields in two space dimensions. The Jordan $\mathscr{A}$-multialgebra that fails the 3-chain condition in two space dimensions and the corresponding L-relation are given for the sake of completeness in chapter 16. In [14] this example was used to construct a function with a lot of symmetries that is rank-one convex, but not quasiconvex.

We remark that in the classical theory of Jordan algebras [18, 26] (i.e., when $\dim \mathscr{A} = 1$) the 3-chain property is a consequence of the 2-chain property. There are also many examples of special Jordan algebras for which the 4-chain property is not satisfied. In this light, the fact that each and every Jordan $\mathscr{A}$-multialgebra in physical examples of part III is reflexive looks surprising. In order to explain this 'happy coincidence' we have examined reflexivity of all Jordan subalgebras in $\mathrm{Sym}(\mathbb{R}^n)$ in chapter 15. As indicated in [26], section 1.5 only N-dimensional subalgebras containing so called 'spin factors' that live in the space of $2^N \times 2^N$ matrices could be non-reflexive. The smallest non-reflexive subalgebra of $\mathrm{Sym}(\mathbb{R}^n)$ occurs when $n = 8$, which is the dimension of $\mathscr{T}$ in our example of an L-relation that is not exact in section 16. The explanation of the universal reflexivity in part III of the book is that, from the algebraic standpoint, all of the examples from physics of two-dimensional composites are too low dimensional to contain any non-reflexive Jordan $\mathscr{A}$-multialgebras. The analysis of three-dimensional multifield composites in section 6.1, where all SO(3)-invariant Jordan $\mathscr{A}$-multialgebras are reflexive, suggests that the somewhat ugly rules of multiplication of irreducible representations of SO (3), described in chapter 17, would result in the universal reflexivity for all three-dimensional polycrystalline composites. In other words, we conjecture that all three-dimensional polycrystalline L-relations must be exact relations in the sense of definition 4.1.

4.4 Special types of exact relations

4.4.1 Polycrystalline exact relations

The definition of an exact relation and more generally, of a G-closed set, treats different orientations of the same anisotropic material as distinct materials.

However, from the practical point of view it is not natural to demand that an anisotropic constituent be used only in one particular orientation. Composites in which any orientation of an anisotropic constituent is allowed, are called polycrystalline. When we deal with polycrystalline composites we assume that tensors of material properties are symmetric. This is done for physical reasons, since conditions that break symmetry of material tensors usually also break their rotational equivariance, as in the Hall effect [1, 2] (see chapter 8).

Definition 4.14. *A G-closed submanifold* $\mathbb{M} \subset \mathrm{Sym}^+(\mathscr{T})$ *is called a polycrystalline exact relation if* $\boldsymbol{R} \cdot \mathsf{L} \in \mathbb{M}$ *for all* $\boldsymbol{R} \in \mathrm{SO}(d)$ *(*$d = 2$ *or 3), where* $\boldsymbol{R} \cdot \mathsf{L}$ *denotes the tensor* L *in a rotated coordinate system.*

Since the presence of an anisotropic material tensor in $\mathbb{M}$ causes $\mathbb{M}$ to contain its entire $\mathrm{SO}(d)$-orbit, there are far fewer polycrystalline exact relations, than general ones. Partly for this reason, it is much easier to characterize all polycrystalline exact relations. The rotational equivariance postulate 2.5, that we did not needed thus far, will become essential in much of part II, which focuses in large part on polycrystalline exact relations. We did not want to make postulate 2.5 into a standing assumption in this book because we also address situations, where the isotropy of space is broken by an external magnetic field, for example. In the absence of special symmetry-breaking circumstances the tensor of material properties can usually be represented by symmetric matrices (see, e.g., [3]).

It is intuitively clear that a polycrystal with statistically isotropic texture must be isotropic, i.e., have an effective tensor L, such that

$$\boldsymbol{R} \cdot \mathsf{L} = \mathsf{L}, \qquad \boldsymbol{R} \in \mathrm{SO}(d),\, d = 2,\, 3. \tag{4.14}$$

It means that any polycrystalline exact relation submanifold $\mathbb{M}$ must contain an isotropic tensor. In the theory of polycrystalline exact relations we will always use an isotropic reference medium $\mathsf{L}_0 \in \mathbb{M}$. Then, the rotational equivariance postulate 2.5 together with formula (3.14) for $\boldsymbol{\Gamma}_0(\boldsymbol{n})$ imply

$$\boldsymbol{R} \cdot \boldsymbol{\Gamma}_0(\boldsymbol{n}) = \boldsymbol{\Gamma}_0(\boldsymbol{R}\boldsymbol{n}) \tag{4.15}$$

for every $\boldsymbol{R} \in \mathrm{SO}(d)$ and every unit vector $\boldsymbol{n}$. Hence, we also have

$$\boldsymbol{R} \cdot W_{\boldsymbol{n}}(\mathsf{L}) = W_{\boldsymbol{R}\boldsymbol{n}}(\boldsymbol{R} \cdot \mathsf{L}), \qquad \boldsymbol{R} \in \mathrm{SO}(d). \tag{4.16}$$

Recall that for every $\mathsf{K} \in \Pi$, that is sufficiently close to zero and any unit vector $\boldsymbol{n}$, there exists $\mathsf{L} \in \mathbb{M}$ (depending on both K and $\boldsymbol{n}$), such that $\mathsf{K} = W_{\boldsymbol{n}}(\mathsf{L})$. Thus, for any rotation $\boldsymbol{R} \in \mathrm{SO}(d)$ we have $\boldsymbol{R} \cdot \mathsf{K} = W_{\boldsymbol{R}\boldsymbol{n}}(\boldsymbol{R} \cdot \mathsf{L}) \in \Pi$, since $\boldsymbol{R} \cdot \mathsf{L} \in \mathbb{M}$. It follows that the Jordan multialgebra Π is a rotationally-invariant subspace of $\mathrm{Sym}(\mathscr{T})$. It is easy to see that the subspace $\mathscr{A}$, given by (4.10) must also be rotationally-invariant:

$$\begin{aligned}\boldsymbol{R} \cdot (\boldsymbol{\Gamma}_0(\boldsymbol{n}) - \boldsymbol{\Gamma}_0(\boldsymbol{n}_0)) &= \boldsymbol{\Gamma}_0(\boldsymbol{R}\boldsymbol{n}) - \boldsymbol{\Gamma}_0(\boldsymbol{R}\boldsymbol{n}_0) \\ &= (\boldsymbol{\Gamma}_0(\boldsymbol{R}\boldsymbol{n}) - \boldsymbol{\Gamma}_0(\boldsymbol{n}_0)) - (\boldsymbol{\Gamma}_0(\boldsymbol{R}\boldsymbol{n}_0) - \boldsymbol{\Gamma}_0(\boldsymbol{n}_0)) \in \mathscr{A}.\end{aligned}$$

In fact, there is a simpler alternative description of $\mathscr{A}$ that will permit us to compute the subspace $\mathscr{A}$ in a fairly straightforward way.

Lemma 4.15. *Suppose that the rotational equivariance postulate 2.5 holds and the reference medium* L_0 *is isotropic. Then there exists a single tensor* $\tilde{\Gamma}\in\mathrm{Sym}(\mathscr{T})$, *such that*

$$\mathscr{A} = \mathrm{Span}\{\boldsymbol{R}\cdot\tilde{\Gamma}:\boldsymbol{R}\in\mathrm{SO}(d)\},\quad d=2,3. \tag{4.17}$$

Proof. To prove the lemma we will need the existence of the Haar measure μ on SO(d), which is the unique probability measure that is both right and left-invariant, [15]. We define

$$\bar{\Gamma} = \int_{\mathrm{SO}(d)} \boldsymbol{Q}\cdot\Gamma_0(\boldsymbol{n}_0)d\mu(\boldsymbol{Q}). \tag{4.18}$$

The left invariance of the Haar measure implies that for any $\boldsymbol{R}\in\mathrm{SO}(d)$

$$\begin{aligned}\boldsymbol{R}\cdot\bar{\Gamma} &= \int_{\mathrm{SO}(d)}(\boldsymbol{RQ})\cdot\Gamma_0(\boldsymbol{n}_0)d\mu(\boldsymbol{Q}) = \int_{\mathrm{SO}(d)}\boldsymbol{Q}'\cdot\Gamma_0(\boldsymbol{n}_0)d\mu(\boldsymbol{R}^{-1}\boldsymbol{Q}')\\ &= \int_{\mathrm{SO}(d)}\boldsymbol{Q}'\cdot\Gamma_0(\boldsymbol{n}_0)d\mu(\boldsymbol{Q}') = \bar{\Gamma}.\end{aligned}$$

Thus, $\bar{\Gamma}$ is isotropic. The right invariance of the Haar measure together with (4.15) implies that $\bar{\Gamma}$ is independent of the choice of the unit vector $\boldsymbol{n}_0$. We can think of $\bar{\Gamma}$ as the isotropic part of $\Gamma(\boldsymbol{n})$ for any unit vector $\boldsymbol{n}$. In fact, it is the orthogonal projection of $\Gamma_0(\boldsymbol{n})$ onto $\mathrm{Sym}_{\mathrm{SO}(d)}(\mathscr{T})$—the subspace of all isotropic tensors in $\mathrm{Sym}(\mathscr{T})$.

Let us fix an arbitrary unit vector $\boldsymbol{n}_0$ and define $\tilde{\Gamma} = \Gamma_0(\boldsymbol{n}_0) - \bar{\Gamma}$. Let us temporarily denote the right-hand side of (4.17) by $\tilde{\mathscr{A}}$. Then, for any other unit vector $\boldsymbol{n}$ we can find a rotation $\boldsymbol{R}\in\mathrm{SO}(d)$, such that $\boldsymbol{R}\boldsymbol{n}_0=\boldsymbol{n}$. Then, using (4.15), we have

$$\Gamma_0(\boldsymbol{n}) - \Gamma_0(\boldsymbol{n}_0) = \boldsymbol{R}\cdot\Gamma_0(\boldsymbol{n}_0) - \Gamma_0(\boldsymbol{n}_0) = \boldsymbol{R}\cdot\tilde{\Gamma}-\tilde{\Gamma}\in\tilde{\mathscr{A}}. \tag{4.19}$$

It follows that $\mathscr{A}\subset\tilde{\mathscr{A}}$. To prove the reverse inclusion we observe that

$$\int_{\mathrm{SO}(d)}\boldsymbol{R}\cdot\tilde{\Gamma}d\mu(\boldsymbol{R}) = 0. \tag{4.20}$$

Integrating (4.19) and applying (4.20) we conclude that

$$-\tilde{\Gamma} = \int_{\mathrm{SO}(d)}(\boldsymbol{R}\cdot\tilde{\Gamma}-\tilde{\Gamma})d\mu(\boldsymbol{R}) = \int_{\mathrm{SO}(d)}(\Gamma(\boldsymbol{R}\boldsymbol{n}_0) - \Gamma(\boldsymbol{n}_0))d\mu(\boldsymbol{R})\in\mathscr{A},$$

and the reverse inclusion follows. □

We remark that formula

$$\tilde{\Gamma} = \Gamma_0(\boldsymbol{n}_0) - \bar{\Gamma} \tag{4.21}$$

shows that all tensors $\tilde{\Gamma}$ generating $\mathscr{A}$ are related to one another by a rotation.

There are a lot fewer rotationally invariant subspaces than arbitrary ones in $\mathrm{Sym}(\mathscr{T})$. Moreover, the tools of representation theory of SO(2) and SO(3), originally developed for the study of angular momentum in quantum mechanics (see e.g. [9]), give a relatively straightforward method for describing them. One immediate simplification is a corollary of lemma 4.15.

Corollary 4.16. *Suppose $\tilde{\Gamma}$ is the generating vector for $\mathscr{A}$ in the sense of (4.17). Then an SO(d)-invariant subspace $\Pi \subset \mathrm{Sym}(\mathscr{T})$ is a Jordan $\mathscr{A}$-multialgebra, if and only if it is a Jordan algebra with multiplication $\mathsf{K}_1 *_{\tilde{\Gamma}} \mathsf{K}_2$, defined in (4.11).*

Proof. If Π is a Jordan $\mathscr{A}$-multialgebra, then it is obviously closed with respect to the multiplication $\mathsf{K}_1 *_{\tilde{\Gamma}} \mathsf{K}_2$, since $\tilde{\Gamma} \in \mathscr{A}$. Conversely, suppose that Π is an SO(d)-invariant subspace $\Pi \subset \mathrm{Sym}(\mathscr{T})$, closed with respect to the multiplication $\mathsf{K}_1 *_{\tilde{\Gamma}} \mathsf{K}_2$. Then, for any $\mathsf{K} \in \Pi$ and $\boldsymbol{R} \in \mathrm{SO}(d)$

$$\mathsf{K}(\boldsymbol{R} \cdot \tilde{\Gamma})\mathsf{K} = \boldsymbol{R} \cdot ((\boldsymbol{R}^{-1} \cdot \mathsf{K})\tilde{\Gamma}(\boldsymbol{R}^{-1} \cdot \mathsf{K})) \in \Pi,$$

because Π is an SO(d)-invariant subspace. But then $\mathsf{KAK} \in \Pi$ for every $\mathsf{A} \in \mathrm{Span}\{\boldsymbol{R} \cdot \tilde{\Gamma} : \boldsymbol{R} \in \mathrm{SO}(d)\} = \mathscr{A}$. □

We will discuss the explicit characterization of all SO(d)-invariant subspaces of $\mathrm{Sym}(\mathscr{T})$ in section 6.1.2. However, in many cases one can identify all rotationally invariant subspaces 'by inspection'. For example, in the case of conductivity, there are only two rotationally invariant proper subspaces in $\mathrm{Sym}(\mathbb{R}^d)$, for $d = 2$ or 3. They are $\mathbb{R}\boldsymbol{I}_d = \{\alpha \boldsymbol{I}_d : \alpha \in \mathbb{R}\}$ and

$$\mathrm{Sym}_0(\mathbb{R}^d) = \{\boldsymbol{X} \in \mathrm{End}(\mathbb{R}^d) : \boldsymbol{X}^T = \boldsymbol{X}, \mathrm{Tr}\, \boldsymbol{X} = 0\}. \tag{4.22}$$

It is easy to check that for conductivity $\boldsymbol{\Gamma}_0(\boldsymbol{n}) = \sigma_0^{-1} \boldsymbol{n} \otimes \boldsymbol{n}$, where $\sigma_0 \boldsymbol{I}_d$ is an isotropic reference medium. Hence, for conductivity $\mathscr{A} = \mathrm{Sym}_0(\mathbb{R}^d)$. It is also very easy to verify which of the two rotationally invariant subspaces, if any, are Jordan $\mathscr{A}$-multialgebras. When $d = 3$ we verify that neither of the two are. However, when $d = 2$ we discover that $\Pi = \mathrm{Sym}_0(\mathbb{R}^2)$ is in fact a Jordan $\mathscr{A}$-multialgebra. This proves that any polycrystalline G-closed set in $\mathrm{Sym}^+(\mathbb{R}^3)$ containing more than one point must have a non-empty interior. For 2D conductivity, the hypersurfaces $\mathbb{M}_{d_0}$, given by (4.3) are G-closed. They correspond to the rotationally invariant Jordan $\mathscr{A}$-multialgebra $\Pi = \mathrm{Sym}_0(\mathbb{R}^2)$. This algebra is reflexive, because $\Pi' = \Pi$, as one can easily verify directly,

$$\boldsymbol{K}_1 \boldsymbol{A} \boldsymbol{K}_2 \in \mathrm{Sym}_0(\mathbb{R}^2), \quad \forall \{\boldsymbol{K}_1, \boldsymbol{K}_2\} \subset \Pi = \mathrm{Sym}_0(\mathbb{R}^2)\ \forall \boldsymbol{A} \in \mathscr{A} = \mathrm{Sym}_0(\mathbb{R}^2).$$

4.4.2 Uniform field relations

We have already mentioned the uniform field relations (UFR) in section 4.1. Here we will discuss them from the point of view of our general theory of exact relations.

There is a simple observation made by Lurie and Cherkaev [22, 23] (although in the context of thermal expansion it dates back to the work of Cribb [6]) that for any given uniform fields $\{\boldsymbol{u}, \boldsymbol{v}\} \subset \mathscr{T}$ the submanifolds of the form

$$\mathscr{U}(\boldsymbol{u}, v) = \{\mathsf{L} \in \mathrm{End}^+(\mathscr{T}) \colon \mathsf{L}u = v\} \tag{4.23}$$

are G-closed. Consequently, any intersections of them is also an exact relation, given by

$$\mathbb{M} = \{\mathsf{L} \in \mathrm{End}^+(\mathscr{T}) \colon \mathsf{L}\boldsymbol{u}_1 = v_1, \dots, \mathsf{L}\boldsymbol{u}_k = v_k\} \tag{4.24}$$

Since the uniform fields $\boldsymbol{u}_i$, v_i, $i = 1, \dots, k$ are non-uniquely determined by the UFR $\mathbb{M}$, it is more convenient to identify each UFR by a subspace $U \subset \mathscr{T}$ and a reference medium $\mathsf{L}_0 \in \mathbb{M}$:

$$\mathbb{M} = (\mathsf{L}_0 + \mathrm{Ann}(U)) \bigcap \mathrm{End}^+(\mathscr{T}), \tag{4.25}$$

where

$$\mathrm{Ann}(U) = \{\mathsf{K} \in \mathrm{End}(\mathscr{T}) \colon \mathsf{K}\boldsymbol{u} = 0 \text{ for all } \boldsymbol{u} \in U\}$$

is the 'annihilator' of U. In fact, the Jordan $\mathscr{A}$-multialgebra corresponding to $\mathbb{M}$ is exactly $\Pi = \mathrm{Ann}(U)$.

Theorem 4.17. *Suppose* $\mathbb{M}$ *is given by (4.25). Then* $W_{\boldsymbol{n}}(\mathbb{M})$ *with reference tensor* $\mathsf{L}_0 \in \mathbb{M}$ *is an open subset in* $\Pi = \mathrm{Ann}(U)$. *Moreover,* Π *is always a reflexive Jordan* $\mathscr{A}$*-multialgebra in the sense of definition 4.10.*

Proof. While the problem of computing $\mathbb{M}$ from Π is technically non-trivial (we devote chapter 6 to this issue), computing the subspace Π from $\mathbb{M}$ is straightforward. From elementary differential geometry it is known that the differential $dW_{\boldsymbol{n}}$ at L_0 is an isomorphism between the tangent space of $\mathbb{M}$ at L_0 and Π. It is easy to compute that $dW_{\boldsymbol{n}}(\mathsf{K}) = -\mathsf{K}$ for any $\mathsf{K} \in T_{\mathsf{L}_0}\mathbb{M}$ (tangent space of $\mathbb{M}$ at L_0). Thus, $\Pi = T_{\mathsf{L}_0}\mathbb{M}$. The statement of the theorem becomes obvious now. To prove reflexivity we only need to observe that Π is also an associative $\mathscr{A}$-multialgebra. □

It is a simple observation that a UFR is polycrystalline if and only if $U \subset \mathscr{T}$ in (4.25) is rotationally invariant. Since for polycrystals we have assumed that material tensors must be symmetric:

$$\mathbb{M} = (\mathsf{L}_0 + \mathrm{Ann}(U)) \bigcap \mathrm{Sym}^+(\mathscr{T}), \qquad \Pi = \mathrm{Ann}(U) \bigcap \mathrm{Sym}(\mathscr{T}). \tag{4.26}$$

Example 4.18. The list of all rotationally invariant proper subspaces of $\mathscr{T} = \mathrm{Sym}(\mathbb{R}^d)$ shows that there are exactly 2 polycrystalline uniform field relations in both two and three-dimensional elasticity. The UFR corresponding to $\mathbb{R}\boldsymbol{I}_d$ is Hill's exact relation (4.2) [16, 17]. The UFR corresponding to $\mathrm{Sym}_0(\mathbb{R}^d)$ says that the set of all Hooke's laws C such that $\mathsf{C}\boldsymbol{I}_d = d\kappa_0\boldsymbol{I}_d$, where κ_0 is a given constant (bulk modulus) is an exact relation. This exact relation is also due to Hill [17], as well as Lurie, Cherkaev and Fedorov [24].

4.5 Proofs of theorems 4.8, 4.11, 4.13†

4.5.1 Proof of theorem 4.8

Observe that $W_n(\mathsf{L}_0) = 0$. Therefore, W_n maps a sufficiently small neighborhood of L_0 in $\mathbb{M}$ onto a small neighborhood of 0 in Π. Let us show that without loss of generality we can assume that $\mathsf{L}(z) \in \mathbb{M}$ is sufficiently close to L_0, or, equivalently, that $W_n(\mathsf{L}(z))$ is sufficiently small (uniformly in $z \in Q$).

Lemma 4.19. *Suppose* $\mathbb{M}$ *is given by (4.4), where* Π *is a Jordan* $\mathscr{A}$*-multialgebra. Let us assume that there exists* $\epsilon > 0$*, such that* $\mathsf{L}_* \in \mathbb{M}$ *for every periodic composite* $\mathsf{L}(z) \in \mathbb{M}$*, satisfying* $|W_n(\mathsf{L}(z))| < \epsilon$*. Then* $\mathsf{L}_* \in \mathbb{M}$ *for any periodic composite with uniformly bounded local tensor* $\mathsf{L}(z) \in \mathbb{M}$*.*

Proof. Let $\mathsf{L}(z)$ be uniformly bounded and $\mathsf{L}(z) \in \mathbb{M}$ for all $z \in Q$. Then $\mathsf{K}(z) = W_n(\mathsf{L}(z))$ is a uniformly bounded function, satisfying $\mathsf{K}(z) \in \Pi$, by (4.4). Thus, there exists $\delta > 0$ so small that for every $0 < \theta < \delta$ we have, $\theta|\mathsf{K}(z)| < \epsilon$. Let $\mathsf{L}^\theta(z) = W_n^{-1}(\theta\mathsf{K}(z))$. Then, by the lamination formula (3.38), $\mathsf{L}^\theta(z)$ is the effective tensor of the laminate of $\mathsf{L}(z)$ and L_0 taken in volume fractions θ and $1 - \theta$, respectively, with lamination direction $\boldsymbol{n}$, for each fixed $z \in Q$. Theorem 4.4 then implies that $\mathsf{L}^\theta(z) \in \mathbb{M}$, for every $0 \leqslant \theta \leqslant 1$, and every $z \in Q$. By assumption, $\mathsf{L}_*^\theta \in \mathbb{M}$ for every $0 \leqslant \theta < \delta$. That means that $f(\theta) = (W_n(\mathsf{L}_*^\theta), \mathsf{P})_{\mathrm{End}(\mathscr{T})} = 0$ for every P in the orthogonal complement to Π in $\mathrm{End}(\mathscr{T})$ and every $0 \leqslant \theta < \delta$. By formula (3.26)

$$W_n(\mathsf{L}_*^\theta) = \langle \mathsf{K}(z)(\theta^{-1}\mathsf{I} - \Lambda_n\mathsf{K}(z))^{-1}\rangle \quad \forall\,\theta \in (0, 1],$$

where $\Lambda_n = \Lambda_{\mathsf{M}}$, given by (3.24) with $\mathsf{M} = \boldsymbol{\Gamma}_0(\boldsymbol{n})$. By corollary 3.16 $\mathsf{M} = \boldsymbol{\Gamma}_0(\boldsymbol{n})$ has the universal invertibility property and hence, by lemma 3.17 the operators $\theta^{-1}\mathsf{I} - \Lambda_n\mathsf{K}$ must be invertible for all $\theta \in (0, 1]$. Thus, θ^{-1} is not in the spectrum of $\Lambda_n\mathsf{K}$ for all $\theta \in (0, 1]$. But then the function $f(\theta)$ must be analytic on a neighborhood of $\theta \in [0, 1]$ in the complex plane [35]. It follows that $f(\theta) = 0$ for all $\theta \in [0, 1]$, since $f(\theta) = 0$ for all $0 \leqslant \theta < \delta$. This proves that $W_n(\mathsf{L}_*) \in \Pi$, which implies that $\mathsf{L}_* \in \mathbb{M}$. □

Next we combine lemma 4.19 with the Neumann expansion (3.35) for the effective tensor of a periodic composite in order to obtain a purely algebraic criterion for stability of a manifold $\mathbb{M}$ under homogenization.

Lemma 4.20. *Let* $\Pi_{\mathbb{C}} = \{\mathsf{K}_1 + i\mathsf{K}_2 | \{\mathsf{K}_1, \mathsf{K}_2\} \subset \Pi\}$ *be the complexification of* Π*. Then* $\mathbb{M}$*, given by (4.4), is G-closed if and only if*

$$\langle(\mathsf{K}\Lambda_n)^k\mathsf{K}(z)\rangle \in \Pi_{\mathbb{C}}, \quad k \geqslant 0, \tag{4.27}$$

for every $\mathsf{K} \in L^\infty(Q; \Pi_{\mathbb{C}})$*, which holds if and only if* $\langle(\mathsf{K}\Lambda_n)^k\mathsf{K}(z)\rangle \in \Pi$ *for every* $\mathsf{K} \in L^\infty(Q; \Pi)$.

Proof. If $\mathbb{M}$ is G-closed, and $\mathsf{K} \in L^\infty(Q; \Pi)$ then for sufficiently small (in absolute value) ϵ we have $\mathsf{L}^\epsilon(z) = W_n^{-1}(\epsilon\mathsf{K}(x)) \in \mathbb{M}$. Hence, $\mathsf{L}_*^\epsilon \in \mathbb{M}$, which, in turn, implies that $W_n(\mathsf{L}_*^\epsilon) \in \Pi$. By (3.35)

$$W_n(\mathsf{L}_*^{\epsilon}) = \sum_{k=0}^{\infty} \epsilon^{k+1} \langle (\mathsf{K}\Lambda_n)^k \mathsf{K}(z) \rangle. \tag{4.28}$$

Hence, (4.27) holds for every $\mathsf{K} \in L^\infty(Q; \Pi)$.

Conversely, if (4.27) holds for every $\mathsf{K} \in L^\infty(Q; \Pi)$, then formula (4.28) proves that $W_n(\mathsf{L}_*) \in \Pi$, provided $\mathsf{L}(z) \in \mathbb{M}$ and $\mathsf{K}(z) = W_n(\mathsf{L}(z))$ is sufficiently small. Lemma 4.19 now guarantees that $\mathbb{M}$ is G-closed. It is obvious that if (4.27) holds for every $\mathsf{K} \in L^\infty(Q; \Pi_{\mathbb{C}})$, then it also holds for every $\mathsf{K} \in L^\infty(Q; \Pi)$. To finish the proof of the lemma, we only need to show that if (4.27) holds for every $\mathsf{K} \in L^\infty(Q; \Pi)$, then it also holds for every $\mathsf{K} \in L^\infty(Q; \Pi_{\mathbb{C}})$. Indeed, for any $\{\mathsf{K}_1, \mathsf{K}_2\} \subset L^\infty(Q; \Pi)$ and $\lambda \in \mathbb{C}$ we define $\mathsf{K}_\lambda(z) = \mathsf{K}_1(z) + \lambda \mathsf{K}_2(z)$. Observe that $\mathsf{K}_\lambda \in L^\infty(Q; \Pi)$ for any $\lambda \in \mathbb{R}$, and therefore, for any $\mathsf{P} \in \Pi^\perp$, and any $k \geqslant 1$ we have

$$p(\lambda) = (\langle (\mathsf{K}_\lambda \Lambda_n) \rangle^k \mathsf{K}_\lambda(z), \mathsf{P})_{\mathrm{End}(\mathscr{T})} = 0, \qquad \lambda \in \mathbb{R}.$$

Observe that $p(\lambda)$ is a polynomial in λ. Therefore, if it vanishes on $\mathbb{R}$ it must also vanish on $\mathbb{C}$. Hence, $p(i) = 0$, and (4.27) is proved for any $\mathsf{K} \in L^\infty(Q; \Pi_{\mathbb{C}})$. □

Now we convert (4.27) into a rather messy, but entirely algebraic condition. For this purpose we will need extra notation, which will not be used outside of this section. Let S_k, $k \geqslant 2$, denote the group of all permutations of k elements, called the symmetric group. For $\boldsymbol{v} = (\boldsymbol{l}_1, \ldots, \boldsymbol{l}_k) \in (\mathbb{Z}^d)^k$ and for $\sigma \in S_k$ we define

$$\sigma(\boldsymbol{v}) = (\boldsymbol{l}_{\sigma(1)}, \ldots, \boldsymbol{l}_{\sigma(k)}) \in (\mathbb{Z}^d)^k. \tag{4.29}$$

Let $\mathscr{O}(\boldsymbol{v}) = \{\sigma(\boldsymbol{v}) \in (\mathbb{Z}^d)^k | \sigma \in S_k\}$ be the orbit of $\boldsymbol{v}$ under the action of the permutation group S_k. The number of elements in $\mathscr{O}(\boldsymbol{v})$ will depend not only on k, but also on how many vectors in the list $\boldsymbol{v}$ coincide. Let $\mathscr{F} \subset \mathbb{Z}^d$ contain at most k vectors. Let $\mathfrak{T}_k(\mathscr{F})$ denote the set of all k-tuples $\boldsymbol{v} = (\boldsymbol{l}_1, \ldots, \boldsymbol{l}_k) \in (\mathbb{Z}^d)^k$, such that $\{\boldsymbol{l}_1, \ldots, \boldsymbol{l}_k\} \subset \mathscr{F}$, and have the property

$$\sum_{i=1}^{k} \boldsymbol{l}_i = 0. \tag{4.30}$$

Obviously, if $\boldsymbol{v} \in \mathfrak{T}_k(\mathscr{F})$ then so is $\sigma(\boldsymbol{v})$ for any $\sigma \in S_k$. We denote by $\mathscr{Z}_k(\mathscr{F})$ the set of all S_k orbits in $\mathfrak{T}_k(\mathscr{F})$. The set $\mathfrak{T}_k(\mathscr{F})$ is a disjoint union of all orbits in $\mathscr{Z}_k(\mathscr{F})$. Of course, the sets $\mathfrak{T}_k(\mathscr{F})$ can empty, depending on $\mathscr{F}$, in which case the sets $\mathscr{Z}_k(\mathscr{F})$ would also be empty. Finally we denote by $\mathfrak{S}_k(\mathbb{Z}^d)$ the set of all subset $\mathscr{F} \subset \mathbb{Z}^d$, such $|\mathscr{F}| \leqslant k$. Using this notation we can now formulate an algebraic criterion for (4.27) to hold.

Lemma 4.21. *Condition (4.27) is satisfied if and only if for any $k \geqslant 2$, any $\mathscr{F} \in \mathfrak{S}_k(\mathbb{Z}^d)$, and for any function $\hat{\mathsf{K}}: \mathscr{F} \to \Pi_{\mathbb{C}}$ we have*

$$\sum_{\mathscr{O} \in \mathscr{Z}_k(\mathscr{F})} \sum_{(\boldsymbol{l}_1, \ldots \boldsymbol{l}_k) \in \mathscr{O}} \left(\prod_{s=1}^{k-1} \hat{\mathsf{K}}(\boldsymbol{l}_s) \mathsf{A}_n \left(\sum_{j=1}^{s} \boldsymbol{l}_j \right) \right) \hat{\mathsf{K}}(\boldsymbol{l}_k) \in \Pi_{\mathbb{C}}, \tag{4.31}$$

where

$$\mathsf{A}_n(\boldsymbol{k}) = \begin{cases} \Gamma_0\left(\dfrac{\boldsymbol{Q}^T\boldsymbol{k}}{|\boldsymbol{Q}^T\boldsymbol{k}|}\right) - \Gamma_0(\boldsymbol{n}), & \boldsymbol{k} \neq 0, \\ 0, & \boldsymbol{k} = 0. \end{cases} \tag{4.32}$$

Proof. For $\mathsf{K} \in L^\infty(Q; \Pi)$ we denote $\mathsf{T}_k(\boldsymbol{z}) = ((\mathsf{K}\Lambda_{\boldsymbol{n}})^k\mathsf{K})(\boldsymbol{z})$. Taking the Fourier transform of the recursion formula $\mathsf{T}_k(\boldsymbol{z}) = \mathsf{K}(\boldsymbol{z})(\Lambda_{\boldsymbol{n}}\mathsf{T}_{k-1})(\boldsymbol{z})$ and using induction in k, we can prove that

$$\hat{\mathsf{T}}_{k-1}(\boldsymbol{m}) = \sum_{l_1+\ldots+l_k=\boldsymbol{m}} \left(\prod_{s=1}^{k-1} \hat{\mathsf{K}}(\boldsymbol{l}_s)\mathsf{A}_{\boldsymbol{n}}\left(\boldsymbol{m} - \sum_{j=1}^{s}\boldsymbol{l}_j\right)\right)\hat{\mathsf{K}}(\boldsymbol{l}_k),$$

where $\boldsymbol{m}, \boldsymbol{l}_j \in \mathbb{Z}^d$. Thus, we get

$$\langle(\mathsf{K}\Lambda_{\boldsymbol{n}})^{k-1}\mathsf{K}(\boldsymbol{z})\rangle = \hat{\mathsf{T}}_{k-1}(0) = \sum_{l_1+\ldots+l_k=0} \left(\prod_{s=1}^{k-1} \hat{\mathsf{K}}(\boldsymbol{l}_s)\mathsf{A}_{\boldsymbol{n}}\left(\sum_{j=1}^{s}\boldsymbol{l}_j\right)\right)\hat{\mathsf{K}}(\boldsymbol{l}_k). \tag{4.33}$$

Observe that the sum in (4.33) can be split into parts. The summation in each part goes over all the distinct permutations of a list of vectors $\boldsymbol{l}_j$, satisfying (4.30). Then we can write

$$\langle(\mathsf{K}\Lambda_{\boldsymbol{n}})^{k-1}\mathsf{K}(\boldsymbol{z})\rangle = \sum_{\mathcal{O}\in\mathscr{Z}}\sum_{(l_1,\ldots,l_k)\in\mathcal{O}} \left(\prod_{s=1}^{k-1} \hat{\mathsf{K}}(\boldsymbol{l}_s)\mathsf{A}_{\boldsymbol{n}}\left(\sum_{j=1}^{s}\boldsymbol{l}_j\right)\right)\hat{\mathsf{K}}(\boldsymbol{l}_k), \tag{4.34}$$

where $\mathscr{Z}$ denote the set of all S_k orbits, whose elements satisfy (4.30). The set $\mathscr{Z}$ can be further split into disjoints parts labeled by the set of $\mathbb{Z}^d$ vectors contained in an orbit in $\mathscr{Z}$:

$$\mathscr{Z} = \bigcup_{\mathscr{F}\in\mathfrak{S}_k(\mathbb{Z}^d)} \mathscr{Z}_k(\mathscr{F}).$$

Thus we obtain the representation

$$\langle(\mathsf{K}\Lambda_{\boldsymbol{n}})^{k-1}\mathsf{K}(\boldsymbol{z})\rangle = \sum_{\mathscr{F}\in\mathfrak{S}_k(\mathbb{Z}^d)}\sum_{\mathcal{O}\in\mathscr{Z}_k(\mathscr{F})}\sum_{(l_1,\ldots,l_k)\in\mathcal{O}} \left(\prod_{s=1}^{k-1} \hat{\mathsf{K}}(\boldsymbol{l}_s)\mathsf{A}_{\boldsymbol{n}}\left(\sum_{j=1}^{s}\boldsymbol{l}_j\right)\right)\hat{\mathsf{K}}(\boldsymbol{l}_k). \tag{4.35}$$

If (4.31) holds, then every term of the outer sum in (4.35) is in $\Pi_{\mathbb{C}}$, since $\hat{\mathsf{K}}(\boldsymbol{p}) \in L^\infty(Q; \Pi_{\mathbb{C}})$ for any $\boldsymbol{p} \in \mathbb{Z}^d$ and any $\mathsf{K} \in L^\infty(Q; \Pi)$. Thus, property (4.27) is a consequence of (4.31).

Now suppose that (4.27) holds. Let us fix an arbitrary subset $\mathscr{F} = \{\boldsymbol{p}_1,\ldots, \boldsymbol{p}_s\} \subset \mathbb{Z}^d$, where $s \leqslant k$. Let $\tilde{\mathsf{K}}:\mathscr{F} \to \Pi_{\mathbb{C}}$ be an arbitrary function. Let

$$\mathsf{K}(\boldsymbol{z}) = \sum_{j=1}^{s}\tilde{\mathsf{K}}(\boldsymbol{p}_j)e^{2\pi i\boldsymbol{p}_j\cdot\boldsymbol{Q}\boldsymbol{z}}.$$

Then $\mathsf{K}(z) \in L^\infty(Q; \Pi_\mathbb{C})$ and its Fourier coefficients are

$$\hat{\mathsf{K}}(\boldsymbol{m}) = \begin{cases} \tilde{\mathsf{K}}(\boldsymbol{m}), & \text{if } \boldsymbol{m} \in \mathscr{F}, \\ 0, & \text{otherwise.} \end{cases}$$

According to (4.35)

$$\langle(\mathsf{K}\Lambda_{\boldsymbol{n}})^{k-1}\mathsf{K}(z)\rangle = \sum_{\mathscr{O}\in\mathscr{Z}_k(\mathscr{F})}\sum_{(\boldsymbol{l}_1,\dots,\boldsymbol{l}_k)\in\mathscr{O}}\left(\prod_{s=1}^{k-1}\hat{\mathsf{K}}(\boldsymbol{l}_s)\mathsf{A}_{\boldsymbol{n}}\left(\sum_{j=1}^{s}\boldsymbol{l}_j\right)\right)\hat{\mathsf{K}}(\boldsymbol{l}_k),$$

and (4.31) is proved. □

We now show that lemma 4.21 implies theorem 4.8. Let $\eta \in S_k$ be the permutation defined by $\eta(j) = k + 1 - j$. In other words it maps $1, 2, \dots, k$ into $k, k-1, \dots, 2, 1$. Therefore, η is an element of order 2 in the group S_k. It acts on $\mathscr{O}(v)$ and splits $\mathscr{O}(v)$ into a disjoint union of orbits of the group $\{1, \eta\}$. If an orbit contains two elements then their sum has the form (4.12). If the orbit contains a single element then this element is 1/2 of the sum of two copies of itself, which is again of the form (4.12). This shows that if Π has the j-chain property for all $j \geqslant 2$, then (4.30) is satisfied, and hence, by lemma 4.21 and 4.20, the submanifold $\mathbb{M}$ is G-closed. The proof of theorem 4.8 is completed by the following lemma.

Lemma 4.22. *Suppose* Π *has the j-chain property described in definition 4.3.1 for* $j = 2, 3$ *and 4, then* Π *satisfies the j-chain property for all* $j \geqslant 2$.

Proof. The proof is a straightforward modification of the proof of Cohn's theorem [5, 18]. For the sake of completeness we give a sketch of the proof. The lemma is proved by induction in j. If $j \leqslant 4$ the statement is true by assumption. Let $j > 4$, and $\{\mathsf{A}_1, \dots, \mathsf{A}_{j-1}\} \subset \mathsf{A}$ and $\{\mathsf{K}_1, \dots, \mathsf{K}_j\} \subset \Pi$. Let us rewrite the 2-chain property defining Jordan multialgebras as

$$\mathsf{K}_1\mathsf{A}\mathsf{K}_2 = -\mathsf{K}_2\mathsf{A}\mathsf{K}_1 + \mathsf{K}' \tag{4.36}$$

for some $\mathsf{K}' \in \Pi$ and any $\mathsf{A} \in \mathscr{A}$. This identity allows us to swap successively the positions of adjacent K's in any j-chain, leaving $(j-1)$-chains as remainder. Using 3-chain property we can write

$$\mathsf{K}_1\mathsf{A}_1\mathsf{K}_2\mathsf{A}_2\mathsf{K}_3 = -\mathsf{K}_3\mathsf{A}_2\mathsf{K}_2\mathsf{A}_1\mathsf{K}_1 + \mathsf{K}'',$$

where $\mathsf{K}'' \in \Pi$. Then, repeatedly swapping K s, using (4.36) we obtain

$$\mathsf{K}_3\mathsf{A}_2\mathsf{K}_2\mathsf{A}_1\mathsf{K}_1 = -\mathsf{K}_1\mathsf{A}_2\mathsf{K}_2\mathsf{A}_1\mathsf{K}_3 + \text{a sum of } 2 - \text{chains}.$$

Therefore,

$$\mathsf{K}_1\mathsf{A}_1\mathsf{K}_2\mathsf{A}_2\mathsf{K}_3 = \mathsf{K}_1\mathsf{A}_2\mathsf{K}_2\mathsf{A}_1\mathsf{K}_3 + \mathsf{K}'' + \text{a sum of } 2 - \text{chains}$$

for some $\mathsf{K}'' \in \Pi$. This allows us to swap successively the positions of adjacent A's in any j-chain, leaving $(j-1)$-chains and $(j-2)$-chains as remainder. By first reversing the order of the A's and then reversing the order of the K's we see that

$$\begin{aligned}\mathsf{K}_1\mathsf{A}_1\mathsf{K}_2\mathsf{A}_2\mathsf{K}_3\mathsf{A}_3 \dots \mathsf{K}_{j-1}\mathsf{A}_{j-1}\mathsf{K}_j &= (-1)^{j(j-1)/2}\mathsf{K}_j\mathsf{A}_{j-1}\mathsf{K}_{j-1} \dots \mathsf{A}_3\mathsf{K}_3\mathsf{A}_2\mathsf{K}_2\mathsf{A}_1\mathsf{K}_1 \\ &\quad + \text{a sum of } (j-1) - \text{chains and } (j-2) - \text{chains},\end{aligned} \tag{4.37}$$

where $j(j-1)/2$ is the number of swaps of adjacent K's needed to achieve this reordering. If the sign of $(-1)^{j(j-1)/2}$ is negative, then the inductive hypothesis establishes the j-chain property. However, if the sign is positive (as it is when $j=4$), then we use the 4-chain property once, replacing the chain header

$$\mathsf{K}_1\mathsf{A}_1\mathsf{K}_2\mathsf{A}_2\mathsf{K}_3\mathsf{A}_3\mathsf{K}_4 \quad \text{with} \quad -\ \mathsf{K}_4\mathsf{A}_3\mathsf{K}_3\mathsf{A}_2\mathsf{K}_2\mathsf{A}_1\mathsf{K}_1 + \mathsf{K}'$$

where $\mathsf{K}' \in \Pi$. Now, if we swap the A's by the 3-chain property and the K's by the 2-chain property, we obtain, according to (4.37),

$$\begin{aligned}\mathsf{K}_1\mathsf{A}_1\mathsf{K}_2\mathsf{A}_2\mathsf{K}_3\mathsf{A}_3\mathsf{K}_4 = &-\mathsf{K}_1\mathsf{A}_1\mathsf{K}_2\mathsf{A}_2\mathsf{K}_3\mathsf{A}_3\mathsf{K}_4 + \mathsf{K}' \\ &+ \text{a sum of } 3-\text{chains and } 2-\text{chains}.\end{aligned} \tag{4.38}$$

Now, if the sign of $(-1)^{j(j-1)/2}$ in (4.37) is positive, we first apply (4.38) to the j-chain header, followed by (4.37) obtaining the desired relation

$$\begin{aligned}\mathsf{K}_1\mathsf{A}_1\mathsf{K}_2\mathsf{A}_2\mathsf{K}_3\mathsf{A}_3 \ldots \mathsf{K}_{j-1}\mathsf{A}_{j-1}\mathsf{K}_j = &-\mathsf{K}_j\mathsf{A}_{j-1}\mathsf{K}_{j-1} \ldots \mathsf{A}_3\mathsf{K}_3\mathsf{A}_2\mathsf{K}_2\mathsf{A}_1\mathsf{K}_1 \\ &+ \text{a sum of } (j-1)-\text{chains and } (j-2)-\text{chains}\end{aligned}$$

that permits us to apply the inductive hypothesis to shorter chains and conclude the proof. □

4.5.2 Proof of theorem 4.11

Let us first assume that $\Pi = \Pi'_{\text{sym}}$, where Π' is the smallest associative $\mathscr{A}$-multialgebra containing Π. Then for every $\{\mathsf{K}_1,\ldots,\mathsf{K}_j\} \subset \Pi$ and every $\{\mathsf{A}_1,\ldots,\mathsf{A}_{j-1}\} \subset \mathscr{A}$ we have

$$\tilde{\mathsf{K}} = \mathsf{K}_1\mathsf{A}_1\mathsf{K}_2\mathsf{A}_2 \ldots \mathsf{K}_{j-1}\mathsf{A}_{j-1}\mathsf{K}_j \in \Pi'.$$

Therefore,

$$\mathsf{K}_1\mathsf{A}_1\mathsf{K}_2\mathsf{A}_2 \ldots \mathsf{K}_{j-1}\mathsf{A}_{j-1}\mathsf{K}_j + \mathsf{K}_j\mathsf{A}_{j-1}\mathsf{K}_{j-1} \ldots \mathsf{A}_2\mathsf{K}_2\mathsf{A}_1\mathsf{K}_1 = \tilde{\mathsf{K}}+\tilde{\mathsf{K}}^T \in \Pi'_{\text{sym}} = \Pi.$$

In particular, 3 and 4-chain properties will hold.

Conversely, if the 3 and 4-chain properties are satisfied, then by lemma 4.22, the j-chain properties hold for all j. To finish the proof it remains to observe that the smallest associative $\mathscr{A}$-multialgebra Π' containing Π has an explicit characterization:

$$\Pi' = \text{Span}\{\mathsf{K}_1\mathsf{A}_1 \ldots \mathsf{A}_{j-1}\mathsf{K}_j\colon \{\mathsf{K}_1,\ldots,\mathsf{K}_j\} \subset \Pi,\ \{\mathsf{A}_1,\ldots,\mathsf{A}_{j-1}\} \subset \mathscr{A}, j \geqslant 1\}$$

It is now evident that if Π has j-chain properties for all $j \geqslant 1$, then $\tilde{\mathsf{K}}+\tilde{\mathsf{K}}^T \in \Pi$ for any $\tilde{\mathsf{K}}\in\Pi'$. Hence, $\Pi'_{\text{sym}} = \Pi$.

4.5.3 Proof of theorem 4.13

Let us show that in 2D the necessary and sufficient condition (4.31) from lemma 4.21 for $k=3$ implies the 3-chain property. First we observe that any unit vector $\boldsymbol{n}$ can be approximated with any degree of accuracy by $\boldsymbol{Q}^T\boldsymbol{k}_0/|\boldsymbol{Q}^T\boldsymbol{k}_0|$ for some $\boldsymbol{k}_0 \in \mathbb{Z}^2$. Hence, without loss of generality, we can regard that $\boldsymbol{n} = \boldsymbol{Q}^T\boldsymbol{k}_0/|\boldsymbol{Q}^T\boldsymbol{k}_0|$. In that case

$\mathsf{A}_n(\boldsymbol{k}_0) = 0$, according to (4.32). Let $\boldsymbol{k}_1 \in \mathbb{Z}^2$ be an arbitrary vector linearly independent with $\boldsymbol{k}_0$. and define $\boldsymbol{l}_1 = m_1\boldsymbol{k}_1$, $\boldsymbol{l}_2 = -m_0\boldsymbol{k}_0 - m_1\boldsymbol{k}_1$, $\{m_0, m_1\} \subset \mathbb{Z}\setminus\{0\}$, so that $\boldsymbol{l}_3 = -\boldsymbol{l}_1 - \boldsymbol{l}_2 = m_0\boldsymbol{k}_0$. Observe that $\boldsymbol{l}_1$, $\boldsymbol{l}_2$ and $\boldsymbol{l}_3$ are non-zero and distinct. Therefore $\mathscr{Z}_3(\{\boldsymbol{l}_1, \boldsymbol{l}_2, \boldsymbol{l}_3\}) = \{\mathscr{O}(\boldsymbol{l}_1, \boldsymbol{l}_2, \boldsymbol{l}_3)\}$. Condition (4.31) for $k = 3$ and $\mathscr{F} = \{\boldsymbol{l}_1, \boldsymbol{l}_2, \boldsymbol{l}_3\}$ becomes

$$\mathsf{K}_2\mathsf{A}_n(m_0\boldsymbol{k}_0 + m_1\boldsymbol{k}_1)\mathsf{K}_3\mathsf{A}_n(\boldsymbol{k}_1)\mathsf{K}_1 + \mathsf{K}_1\mathsf{A}_n(\boldsymbol{k}_1)\mathsf{K}_3\mathsf{A}_n(m_0\boldsymbol{k}_0 + m_1\boldsymbol{k}_1)\mathsf{K}_2 \in \Pi$$

for all $\{\mathsf{K}_1, \mathsf{K}_2, \mathsf{K}_3\} \subset \Pi$. Now, if we vary $\boldsymbol{k}_1 \in \mathbb{Z}^2\setminus\mathbb{R}\boldsymbol{k}_0$ and $\{m_0, m_1\} \subset \mathbb{Z}\setminus\{0\}$, we observe, that the set of pairs of unit vectors

$$\mathscr{P} = \left\{\left(\frac{\boldsymbol{k}_1}{|\boldsymbol{k}_1|}, \frac{m_0\boldsymbol{k}_0 + m_1\boldsymbol{k}_1}{|m_0\boldsymbol{k}_0 + m_1\boldsymbol{k}_1|}\right): \boldsymbol{k}_1 \in \mathbb{Z}^2\setminus\mathbb{R}\boldsymbol{k}_0,\ \{m_0, m_1\} \subset \mathbb{Z}\setminus\{0\}\right\}$$

is a dense subset in $\mathbb{S}^1 \times \mathbb{S}^1$, where $\mathbb{S}^1$ denotes the unit circle. Indeed, for any unit vector $\boldsymbol{u}$ we can choose $\boldsymbol{k}_1 \in \mathbb{Z}^2$, such that $\boldsymbol{k}_1/|\boldsymbol{k}_1|$ approximates $\boldsymbol{u}$ with any degree of accuracy. Then

$$\frac{m_0\boldsymbol{k}_0 + m_1\boldsymbol{k}_1}{|m_0\boldsymbol{k}_0 + m_1\boldsymbol{k}_1|} = \frac{s\boldsymbol{N} + r\boldsymbol{u}}{|s\boldsymbol{N} + r\boldsymbol{u}|}, \quad \boldsymbol{N} = \frac{\boldsymbol{k}_0}{|\boldsymbol{k}_1|}, \quad r = \frac{m_1}{|m_0|}, \quad s = \mathrm{sign}(m_0).$$

By our construction the vectors $\boldsymbol{u}$ and $\boldsymbol{N}$ are linearly independent in $\mathbb{R}^2$. It is now clear that for any unit vector $\boldsymbol{v}$ we can choose the sign s and a non-zero rational number r, so that $(s\boldsymbol{N} + r\boldsymbol{u})/|s\boldsymbol{N} + r\boldsymbol{u}|$ approximates unit vector $\boldsymbol{v}$ with any degree of accuracy. Therefore, by continuity, we conclude that for any $\{\boldsymbol{q}_1, \boldsymbol{q}_2\} \subset \mathbb{S}^1$ we have

$$\mathsf{K}_2\mathsf{A}_n(\boldsymbol{q}_2)\mathsf{K}_3\mathsf{A}_n(\boldsymbol{q}_1)\mathsf{K}_1 + \mathsf{K}_1\mathsf{A}_n(\boldsymbol{q}_1)\mathsf{K}_3\mathsf{A}_n(\boldsymbol{q}_2)\mathsf{K}_2 \in \Pi$$

for all $\{\mathsf{K}_1, \mathsf{K}_2, \mathsf{K}_3\} \subset \Pi$. Fixing all elements in the above formula except either $\boldsymbol{q}_1$ or $\boldsymbol{q}_2$, and taking linear combinations we conclude that the 3-chain property holds. Theorem 4.13 is proved.

References

[1] Bergman D J 1982 Self-duality and the low field Hall effect in 2D and 3D metal-insulator composites. *Percolation Structures and Processes* vol 5 of Annals of the Israel Physical Society, ed G Deutscher, R Zallen and J Adler (Bristol: IOP Publishing) 297–321

[2] Briane M and Milton G 2009 Homogenization of the three-dimensional Hall effect and change of sign of the Hall coefficient *Arch. Ration. Mech. Anal.* **193** 715–36

[3] Callen H B 1960 *Thermodynamics* (New York: Wiley)

[4] Clausius R 1879 *Die mechanische Behandlung der Electricität* (Braunschweig: Vieweg)

[5] Cohn P M 1954 On homomorphic images of special Jordan algebras *Can. J. Math.* **6** 253–64

[6] Cribb J L 1968 Shrinkage and thermal expansion of a two-phase material *Nature Lond.* **220** 576–7

[7] Dvorak G J 1990 On uniform fields in heterogeneous media *Proc. R. Soc. Lond.* A **431** 89–110

[8] Dykhne A M 1971 Conductivity of a two-dimensional two-phase system *Sov. Phys. JETP* **32** 63–5 [Zh. Eksp. Teor. Fiz., 59, (1970) p.110–115.]

[9] Edmonds A R 1996 *Angular Momentum in Quantum Mechanics* (Princeton, NJ: Princeton University Press)

[10] Einstein A 1906 Eine neue Bestimmung der Moleküldimensionen *Ann. Phys., Lpz.* **324** 289–306

[11] Garnett J C M 1904 Colours in metal glasses and in metallic films *Phil. Trans. R. Soc.* A **203** 385–420

[12] Grabovsky Y 1998 Exact relations for effective tensors of polycrystals. I: necessary conditions *Arch. Ration. Mech. Anal.* **143** 309–30

[13] Grabovsky Y 2004 Algebra, geometry and computations of exact relations for effective moduli of composites *Advances in Multifield Theories of Continua with Substructure* Modeling and Simulation in Science, Engineering and Technology G Capriz and P M Mariano (Boston, MA: Birkhäuser) 167–97

[14] Grabovsky Y 2018 From microstructure-independent formulas for composite materials to rank-one convex, non-quasiconvex functions *Arch. Ration. Mech. Anal.* **227** 607–36

[15] Haar A 1933 Der Massbegriff in der Theorie der Kontinuierlichen Gruppen *Ann. Math.* **34** 147–69

[16] Hill R 1963 Elastic properties of reinforced solids: some theoretical principles *J. Mech. Phys. Solids* **11** 357–72

[17] Hill R 1964 Theory of mechanical properties of fibre-strengthened materials: I. Elastic behaviour *J. Mech. Phys. Solids* **12** 199–212

[18] Jacobson N 1968 *Structure and Representations of Jordan Algebras* American Mathematical Society Colloquium Publications, vol XXXIX (Providence, RI: American Mathematical Society)

[19] Jordan P, Neumann J V and Wigner E 1934 On an algebraic generalization of the quantum mechanical formalism *Ann. Math.* **35** 29–64

[20] Keller J B 1964 A theorem on the conductivity of a composite medium *J. Math. Phys.* **5** 548–9

[21] Landauer R 1978 Electrical conductivity in inhomogeneous media *AIP Conf. Proc.* **40** 2–45

[22] Lurie K A and Cherkaev A V 1986 Effective characteristics of composite materials and the optimal design of structural elements. (Russian) *Usp. Mekh. = Adv. Mech.* **9** 3–81 English translation in [23]

[23] Lurie K A and Cherkaev A V 1997 Effective characteristics of composite materials and the optimal design of structural elements *Topics in the Mathematical Modeling of Composite Materials* Progress in Nonlinear Differential equations and Their Applications vol 31 ed A Cherkaev and R Kohn (Boston, MA: Birkhäuser) 175–258

[24] Lurie K A, Cherkaev A V and Fedorov A V 1984 On the existence of solutions to some problems of optimal design for bars and plates *J. Optim. Theory Appl.* **42** 247–81

[25] Maxwell J C 1881 *A Treatise on Electricity and Magnetism* vol 1 2nd edn (Oxford: Clarendon)

[26] McCrimmon K 2004 *A Taste of Jordan Algebras* (Berlin: Springer)

[27] Mendelson K S 1975 A theorem on the conductivity of two-dimensional heterogeneous medium *J. Appl. Phys.* **46** 4740–1

[28] Milgrom M 1997 Some more exact results concerning multifield moduli of two-phase composites *J. Mech. Phys. Solids* **45** 399–404

[29] Milgrom M and Shtrikman S 1989 Linear response of polycrystals to coupled fields: exact relations among the coefficients *Phys. Rev. B (Solid State)* **40** 5991–4

[30] Milgrom M and Shtrikman S 1989 Linear response of two-phase composites with cross moduli: exact universal relations *Phys. Rev.* **40** 1568–75

[31] Mossotti O F 1850 Discussione analitica sull'influenza che l'azione di un mezzo dielettrico ha sulla distribuzione dell'elettricità alla superficie di più corpi elettrici disseminati in esso *Memorie di Matematica e di Fisica della Società Italiana delle Scienze Residente in Modena* **24** 49–74

[32] Nevard J and Keller J B 1985 Reciprocal relations for effective conductivities of anisotropic media *J. Math. Phys.* **26** 2761–5

[33] Onsager L 1931 Reciprocal relations in irreversible processes. I *Phys. Rev.* **37** 405–26

[34] Onsager L 1931 Reciprocal relations in irreversible processes. II *Phys. Rev.* **38** 2265–79

[35] Rudin W 1973 *Functional Analysis* McGraw-Hill Series in Higher Mathematics (New York: McGraw-Hill)

Chapter 5

Links

5.1 Links as exact relations

In addition to exact relations, the literature on composite media contains many microstructure-independent formulas that at first glance do not quite fit into the setting of exact relations. These formulas relate effective tensors of two composites that have the same microstructure, but differ only in terms of material properties of their constituents. Perhaps the earliest link is due to Mendelson [6]. It says that if a two-dimensional periodic composite with local conductivity tensor $\boldsymbol{\sigma}(z)$ has an effective conductivity $\boldsymbol{\sigma}_*$, then the periodic composite with the local tensor

$$\tilde{\boldsymbol{\sigma}}(z) = \frac{\boldsymbol{\sigma}(z)}{\det \boldsymbol{\sigma}(z)} \tag{5.1}$$

has the effective conductivity tensor $\tilde{\boldsymbol{\sigma}}_*$, given by

$$\tilde{\boldsymbol{\sigma}}_* = \frac{\boldsymbol{\sigma}_*}{\det \boldsymbol{\sigma}_*}. \tag{5.2}$$

For example, if we have a two-phase composite made with two isotropic materials with conductivities α and β, then $\tilde{\boldsymbol{\sigma}} = (1/\alpha)\boldsymbol{I}_2$, wherever $\boldsymbol{\sigma} = \alpha\boldsymbol{I}_2$ and $\tilde{\boldsymbol{\sigma}} = (1/\beta)\boldsymbol{I}_2$, wherever $\boldsymbol{\sigma} = \beta\boldsymbol{I}_2$. Observing that $\alpha\beta\tilde{\boldsymbol{\sigma}} = \alpha\boldsymbol{I}_2$, wherever $\boldsymbol{\sigma} = \beta\boldsymbol{I}_2$ and $\beta\boldsymbol{I}_2$, wherever $\boldsymbol{\sigma} = \alpha\boldsymbol{I}_2$, we obtain the 'phase-interchange' formula [4, 5]:

$$\boldsymbol{\sigma}_*(\beta, \alpha) = \frac{\alpha\beta\boldsymbol{\sigma}_*(\alpha, \beta)}{\det \boldsymbol{\sigma}_*(\alpha, \beta)}.$$

If the composite is isotropic, i.e., $\boldsymbol{\sigma}_*(\alpha, \beta) = \sigma_*(\alpha, \beta)\boldsymbol{I}_2$, then the composite with interchanged materials is also isotropic, satisfying

$$\sigma_*(\alpha, \beta)\sigma_*(\beta, \alpha) = \alpha\beta.$$

doi:10.1088/978-0-7503-6249-8ch5

If the microstructure is insensitive to phase interchange, such as the checkerboard geometry [1], then

$$\sigma_*(\alpha, \beta) = \sigma_*(\beta, \alpha) = \sqrt{\alpha\beta}.$$

If we apply Mendelson's formula (5.2) to a polycrystal $\boldsymbol{\sigma}(z) = \boldsymbol{R}(z)\sigma_0\boldsymbol{R}(z)^T$, where $\boldsymbol{R}(z) \in SO(3)$ describes the polycrystalline texture, we obtain

$$\tilde{\boldsymbol{\sigma}}(z) = \frac{\boldsymbol{\sigma}(z)}{\det \boldsymbol{\sigma}_0}.$$

Hence,

$$\tilde{\boldsymbol{\sigma}}_* = \frac{\boldsymbol{\sigma}_*}{\det \boldsymbol{\sigma}_0}.$$

However, according to (5.2) we also have

$$\tilde{\boldsymbol{\sigma}}_* = \frac{\boldsymbol{\sigma}_*}{\det \boldsymbol{\sigma}_*}.$$

The two expressions for $\tilde{\boldsymbol{\sigma}}_*$ agree, only if $\det \boldsymbol{\sigma}_* = \det \boldsymbol{\sigma}_0$, which is an exact relation for two-dimensional conducting composites, discovered first by Keller [4]. The aim of the foregoing discussion was to show that links are more general and deeper statements about the effective behavior of composites than exact relations. However, their theory is more complicated than the theory of exact relations, and delves deeper into the algebraic structure of Jordan multialgebras. Also, links do not have to be limited to a single physical context. For example, we may be interested in links between the effective elastic and conducting properties of one and the same composite. The idea is to regard the pair of local tensors $\mathsf{L}_1(z)$, $\mathsf{L}_2(z)$ of material properties at each point z in the period cell Q, as block-components of a single tensor

$$\hat{\mathsf{L}}(z) = \begin{bmatrix} \mathsf{L}_1(z) & 0 \\ 0 & \mathsf{L}_2(z) \end{bmatrix} \tag{5.3}$$

of an imaginary 'multi-physics' material, whose coupling coefficients are zero. To make more efficient use of page space, we will use a more compact notation $\hat{\mathsf{L}}(z) = [\mathsf{L}_1(z), \mathsf{L}_2(z)]$, instead of (5.3). In this formalism links become exact relations in $\mathbb{V} = \mathrm{End}(\mathscr{T}_1) \oplus \mathrm{End}(\mathscr{T}_2)$, and hence, can be described by theorem 4.4, as submanifolds

$$\hat{\mathbb{M}} = \{\hat{\mathsf{L}} \in \mathrm{End}^+(\mathscr{T}_1) \oplus \mathrm{End}^+(\mathscr{T}_2)\colon \hat{W}_{\boldsymbol{n}_0}(\hat{\mathsf{L}}) \in \hat{\Pi}\}, \tag{5.4}$$

independent of the choice of the unit vector $\boldsymbol{n}_0$. In this formula $\hat{\mathsf{L}}$ is given by (5.3) and

$$\hat{W}_{\boldsymbol{n}_0}(\hat{\mathsf{L}}) = [W^{(1)}_{\boldsymbol{n}_0}(\mathsf{L}_1), W^{(2)}_{\boldsymbol{n}_0}(\mathsf{L}_2)],$$

where

$$W^{(i)}_{\boldsymbol{n}}(\mathsf{L}) = \left(\mathsf{I} + (\mathsf{L} - \mathsf{L}^{(i)}_0)\Gamma^{(i)}_0(\boldsymbol{n})\right)^{-1}(\mathsf{L} - \mathsf{L}^{(i)}_0), \quad i = 1, 2.$$

The subspace $\hat{\Pi} \subset \mathbb{V}$ has the structure of Jordan $\hat{\mathscr{A}}$-multialgebra, where

$$\hat{\mathscr{A}} = \text{Span}\{[\boldsymbol{\Gamma}_0^{(1)}(\boldsymbol{n}) - \boldsymbol{\Gamma}_0^{(1)}(\boldsymbol{n}_0), \boldsymbol{\Gamma}_0^{(2)}(\boldsymbol{n}) - \boldsymbol{\Gamma}_0^{(2)}(\boldsymbol{n}_0)]: |\boldsymbol{n}| = 1\}. \tag{5.5}$$

5.2 Algebraic structure of links

On the most basic level, a precondition for existence of any non-trivial links is the non-trivial structure of the subspace $\hat{\mathscr{A}} \subset \mathscr{A}_1 \oplus \mathscr{A}_2$, where

$$\mathscr{A}_i = \text{Span}\{\boldsymbol{\Gamma}_0^{(i)}(\boldsymbol{n}) - \boldsymbol{\Gamma}_0^{(i)}(\boldsymbol{n}_0)\}, \, i = 1, 2. \tag{5.6}$$

If $\hat{\mathscr{A}} = \mathscr{A}_1 \oplus \mathscr{A}_2$, then there are no relations between the two components A_1 and A_2 of $\hat{\mathsf{A}} = [\mathsf{A}_1, \mathsf{A}_2] \in \hat{\mathscr{A}}$. It follows, as is easy to check, that there are no non-trivial links between $\mathscr{A}_1$ and $\mathscr{A}_2$-multialgebras. Therefore, the degree of triviality of the structure of $\hat{\mathscr{A}}$ can be quantified by the subspaces

$$\mathfrak{A}_1 = \{\mathsf{A}_1 \in \mathscr{A}_1: [\mathsf{A}_1, 0] \in \hat{\mathscr{A}}\}, \qquad \mathfrak{A}_2 = \{\mathsf{A}_2 \in \mathscr{A}_2: [0, \mathsf{A}_2] \in \hat{\mathscr{A}}\}. \tag{5.7}$$

The larger the subspaces $\mathfrak{A}_i$ are, the more trivial are the links one can expect in a given context. In general, the subspace $\hat{\mathscr{A}} \subset \mathscr{A}_1 \oplus \mathscr{A}_2$ can be described by subspaces $\mathfrak{A}_1, \mathfrak{A}_2$ and a linear isomorphism $\Xi: \mathscr{A}_1/\mathfrak{A}_1 \to \mathscr{A}_2/\mathfrak{A}_2$, defined by the rule $\bar{\mathsf{A}}_2 = \Xi(\bar{\mathsf{A}}_1)$, if $[\mathsf{A}_1, \mathsf{A}_2] \in \hat{\mathscr{A}}$, where $\bar{\mathsf{A}}_i$ denotes the equivalence class of A_i in $\mathscr{A}_i/\mathfrak{A}_i$, $i = 1, 2$:

$$\hat{\mathscr{A}} = \{[\mathsf{A}_1, \mathsf{A}_2] \in \mathscr{A}_1 \oplus \mathscr{A}_2: \bar{\mathsf{A}}_2 = \Xi(\bar{\mathsf{A}}_1)\}. \tag{5.8}$$

For example, it was shown in [3] that for conductivity $\hat{\mathscr{A}} = \mathscr{A}_1 \oplus \mathscr{A}_2$, if and only if the reference media $\boldsymbol{\sigma}_0^{(1)}, \boldsymbol{\sigma}_0^{(2)}$ are not scalar multiples of one another. At the same time, it is easy to see that if $\boldsymbol{\sigma}_0^{(1)} = \alpha\boldsymbol{\sigma}_0^{(2)}$, then $\mathscr{A}_1 = \mathscr{A}_2 = \mathscr{A}$ and $\hat{\mathscr{A}} = \{[\boldsymbol{A}, \alpha\boldsymbol{A}]: \boldsymbol{A} \in \mathscr{A}\}$. In particular,

$$\mathfrak{A}_1 = \mathfrak{A}_2 = \{0\}. \tag{5.9}$$

In this case the one-to-one correspondence between the first and the second component of $\hat{\mathscr{A}}$ leads to the presence of non-trivial links, which were computed in [2, 3].

We will show that once all the Jordan $\mathscr{A}$-multialgebras have been computed, one can then compute all Jordan $\hat{\mathscr{A}}$-multialgebras, by examining the algebraic structure of Jordan $\mathscr{A}$-multialgebras. Specifically, for a Jordan $\hat{\mathscr{A}}$-multialgebra $\hat{\Pi}$ in $\mathbb{V} = \text{End}(\mathscr{T}_1) \oplus \text{End}(\mathscr{T}_2)$ we define

$$\Pi_1 = \{\mathsf{K}_1 \in \text{End}(\mathscr{T}_1): [\mathsf{K}_1, \mathsf{K}_2] \in \hat{\Pi} \text{ for some } \mathsf{K}_2 \in \text{End}(\mathscr{T}_2)\}, \tag{5.10}$$

$$\Pi_2 = \{\mathsf{K}_2 \in \text{End}(\mathscr{T}_2): [\mathsf{K}_1, \mathsf{K}_2] \in \hat{\Pi} \text{ for some } \mathsf{K}_1 \in \text{End}(\mathscr{T}_1)\}, \tag{5.11}$$

$$\mathscr{I}_1 = \{\mathsf{K}_1 \in \text{End}(\mathscr{T}_1): [\mathsf{K}_1, 0] \in \hat{\Pi}\} \subset \Pi_1, \tag{5.12}$$

$$\mathscr{I}_2 = \{\mathsf{K}_2 \in \text{End}(\mathscr{T}_2): [0, \mathsf{K}_2] \in \hat{\Pi}\} \subset \Pi_2. \tag{5.13}$$

Theorem 5.1. *The subspace* $\hat{\Pi} \subset \mathbb{V}$ *is a Jordan* $\hat{\mathscr{A}}$*-multialgebra if and only if all three conditions below are satisfied.*

(i) *the subspaces* Π_j *of* $\mathrm{End}(\mathscr{T}_j)$ *are Jordan* $\mathscr{A}_j$*-multialgebras,* $j = 1, 2$;

(ii) *The subspaces* $\mathscr{I}_j \subset \Pi_j$, $j = 1, 2$, *are Jordan ideals, i.e.,*

$$\mathsf{J}*_{\mathsf{A}_j} \mathsf{K} \in \mathscr{I}_j \text{ for all } \mathsf{J} \in \mathscr{I}_j,\ \ \mathsf{K} \in \Pi_j,\ \mathsf{A}_j \in \mathscr{A}_j \quad j = 1, 2,$$

where the Jordan products $\mathsf{K}_1 *_{\mathsf{A}} \mathsf{K}_2$ *are defined in (4.11). Moreover, the ideals* $\mathscr{I}_j$ *must contain minimal ideals*

$$\mathfrak{J}_j = \mathrm{Span}\{\mathsf{KAK}\colon \mathsf{K} \in \Pi_j,\ \ \mathsf{A} \in \mathfrak{A}_j\}, \quad j = 1, 2.$$

(iii) *The linear isomorphism* $\Phi\colon \Pi_1/\mathscr{I}_1 \to \Pi_2/\mathscr{I}_2$, *defined by the rule* $\overline{\mathsf{K}_2} = \Phi(\overline{\mathsf{K}_1})$, *whenever* $[\mathsf{K}_1, \mathsf{K}_2] \in \hat{\Pi}$, *satisfies*

$$\Phi(\bar{\mathsf{K}}\bar{\mathsf{A}}\bar{\mathsf{K}}) = \Phi(\bar{\mathsf{K}})\Xi(\bar{\mathsf{A}})\Phi(\bar{\mathsf{K}}), \qquad \bar{\mathsf{K}} \in \Pi_1/\mathscr{I}_1,\ \bar{\mathsf{A}} \in \mathscr{A}_1/\mathfrak{A}_1, \tag{5.14}$$

where $\Xi\colon \mathscr{A}_1/\mathfrak{A}_1 \to \mathscr{A}_2/\mathfrak{A}_2$ *is a linear isomorphism that describes* $\hat{\mathscr{A}}$ via *(5.8).*

Conversely, given the algebraic data Π_j, $\mathscr{I}_j$, $j = 1, 2$ *and* Φ *satisfying properties (i)–(iii) above, the Jordan* $\hat{\mathscr{A}}$*-multialgebra* $\hat{\Pi}$ *can be reconstructed by the formula*

$$\hat{\Pi} = \{[\mathsf{K}_1, \mathsf{K}_2] \in \Pi_1 \oplus \Pi_2\colon \overline{\mathsf{K}_2} = \Phi(\overline{\mathsf{K}_1})\}. \tag{5.15}$$

Proof. **Sufficiency.** Suppose that the algebraic data Π_j, $\mathscr{I}_j$, $j = 1, 2$ and Φ satisfy properties (i)–(iii) in the theorem. Let us show that a subspace $\hat{\Pi} \subset \mathbb{V}$ given by (5.15) is a Jordan $\hat{\mathscr{A}}$-multialgebra. Let $\hat{\mathsf{K}} = [\mathsf{K}_1, \mathsf{K}_2] \subset \hat{\Pi}$ and $\hat{\mathsf{A}} - [\mathsf{A}_1, \mathsf{A}_2] \in \hat{\mathscr{A}}$. Then

$$\hat{\mathsf{K}}\hat{\mathsf{A}}\hat{\mathsf{K}} = [\mathsf{K}_1\mathsf{A}_1\mathsf{K}_1, \mathsf{K}_2\mathsf{A}_2\mathsf{K}_2] \in \Pi_1 \oplus \Pi_2,$$

since Π_1 and Π_2 are Jordan $\mathscr{A}_1$ and $\mathscr{A}_2$-multialgebras, respectively. We observe that for any $\mathsf{J} \in \mathscr{I}_1$ and $\mathsf{A}_0 \subset \mathfrak{A}_1$

$$(\mathsf{K}_1 + \mathsf{J})(\mathsf{A}_1 + \mathsf{A}_0)(\mathsf{K}_1 + \mathsf{J}) - \mathsf{K}_1\mathsf{A}_1\mathsf{K}_1 = 2\mathsf{J}*_{\mathsf{A}_1 + \mathsf{A}_0} \mathsf{K}_1 + \mathsf{K}_1\mathsf{A}_0\mathsf{K}_1.$$

The first term on the right-hand side is in $\mathscr{I}_1$, since $\mathscr{I}_1$ is a Jordan ideal, while the second term is in $\mathfrak{J}_1 \subset \mathscr{I}_1$. Thus, the equivalence class of $\mathsf{K}_1\mathsf{A}_1\mathsf{K}_1$ in $\Pi_1/\mathscr{I}_1$ is independent of the choice of representatives $\mathsf{K}_1 \in \overline{\mathsf{K}_1} \in \Pi_1/\mathscr{I}_1$ and $\mathsf{A}_1 \in \overline{\mathsf{A}_1} \in \mathscr{A}_1/\mathfrak{A}_1$. Similarly, the equivalence class of $\mathsf{K}_2\mathsf{A}_2\mathsf{K}_2$ in $\Pi_2/\mathscr{I}_2$ is independent of the choice of representatives $\mathsf{K}_2 \in \overline{\mathsf{K}_2} \in \Pi_2/\mathscr{I}_2$ and $\mathsf{A}_2 \in \overline{\mathsf{A}_2} \in \mathscr{A}_2/\mathfrak{A}_2$. In other words,

$$\overline{\mathsf{K}_j\mathsf{A}_j\mathsf{K}_j} = \overline{\mathsf{K}_j}\ \overline{\mathsf{A}_j}\ \overline{\mathsf{K}_j}, \quad j = 1, 2. \tag{5.16}$$

By the property (5.14) we get

$$\Phi(\overline{K_1A_1K_1}) = \Phi(\overline{K_1})\Xi(\overline{A_1})\Phi(\overline{K_1}).$$

By definition of the subspace $\hat{\Pi}$ the assumption that $[K_1, K_2] \in \hat{\Pi}$ implies that $\overline{K_2} = \Phi(\overline{K_1})$. Similarly, the assumption that $[A_1, A_2] \in \hat{\mathscr{A}}$ implies that $\overline{A_2} = \Xi(\overline{A_1})$. Thus,

$$\Phi(\overline{K_1A_1K_1}) = \Phi(\overline{K_1})\Xi(\overline{A_1})\Phi(\overline{K_1}) = \overline{K_2}\ \overline{A_2}\ \overline{K_2} = \overline{K_2A_2K_2}.$$

Hence, $[K_1A_1K_1, K_2A_2K_2] \in \hat{\Pi}$ by definition of $\hat{\Pi}$. The sufficiency is proved now.

Necessity. Part (i). Suppose $K_1 \in \Pi_1$ and $A_1 \in \mathscr{A}_1$. Then, by definitions (5.10) Π_1 and (5.5) of $\hat{\mathscr{A}}$ there exist $K_2 \in \Pi_2$ and $A_2 \in \mathscr{A}_2$, such that $\hat{K} = [K_1, K_2] \in \hat{\Pi}$ and $\hat{A} = [A_1, A_2] \in \hat{\mathscr{A}}$. By assumption, $\hat{\Pi}$ is an $\hat{\mathscr{A}}$-multialgebra, and therefore, $\hat{K}\hat{A}\hat{K} \in \hat{\Pi}$, which implies that $K_1A_1K_1 \in \Pi_1$. The assertion about Π_2 is proved in the same way.

Part (ii). Let $J \in \mathscr{I}_1$, $K_1 \in \Pi_1$ and $A_1 \in \mathscr{A}_1$. Let $K_2 \in \Pi_2$ and $A_2 \in \mathscr{A}_2$ be as in the proof of part (i). Then $[J, 0] \in \hat{\Pi}$, by definition of $\mathscr{I}_1$. Thus,

$$[J, 0][A_1, A_2][K_1, K_2] + [K_1, K_2][A_1, A_2][J, 0] \in \hat{\Pi}.$$

It follows that $[JA_1K_1 + K_1A_1J, 0] \in \hat{\Pi}$. We conclude that, $J *_{A_1} K_1 \in \mathscr{I}_1$ for any $J \in \mathscr{I}_1$, $A_1 \in \mathscr{A}_1$ and $K_1 \in \Pi_1$, proving that $\mathscr{I}_1$ is an ideal. Switching the order of indices in the above proof, we also conclude that $\mathscr{I}_2$ is an ideal. To finish the proof of part (ii) we take any $K_1 \in \Pi_1$ and $A \in \mathfrak{A}_1$. Then, $[A, 0] \in \hat{\mathscr{A}}$ and there exist $K_2 \in \Pi_2$, such that $[K_1, K_2] \in \hat{\Pi}$. Then

$$[K_1AK_1, 0] = [K_1, K_2][A, 0][K_1, K_2] \in \hat{\Pi}.$$

This proves that $\mathfrak{J}_1 \subset \mathscr{I}_1$, and similarly, $\mathfrak{J}_2 \subset \mathscr{I}_2$.

Part (iii). The representation (5.15) follows from definitions of Π_j and $\mathscr{I}_j$, $j = 1, 2$ and Φ. Indeed, by the vector space structure of $\hat{\Pi}$ and $\hat{\mathscr{A}}$ we have $[K_1, K_2] \in \hat{\Pi}$ if and only if $\overline{K_2} = \Phi(\overline{K_1})$, and $[A_1, A_2] \in \hat{\mathscr{A}}$ if and only if $\overline{A_2} = \Xi(\overline{A_1})$. Thus,

$$[K_1A_1K_1, K_2A_2K_2] \in \hat{\Pi} \iff \overline{K_2A_2K_2} = \Phi(\overline{K_1A_1K_1}).$$

Formula (5.14) follows from (5.16), which is a consequence of part (ii), as we have shown in the proof of sufficiency. □

In order to ensure validity of a link corresponding to $\hat{\Pi} \subset \mathbb{V}$ for all microstructures, we apply theorem 4.8 to Jordan multialgebras in $\mathbb{V}$. If we rewrite the corresponding 3 and 4-chain properties for $\hat{\Pi}$ in terms of the algebraic data Π_j, $\mathscr{I}_j$, $j = 1, 2$, and Φ, we will obtain the following theorem.

Theorem 5.2. *The Jordan $\hat{\mathscr{A}}$-multialgebra $\hat{\Pi}$ has 3 and 4-chain properties if and only if all three conditions below are satisfied.*

(i) *Jordan multialgebras Π_j, $j = 1, 2$, have 3 and 4-chain properties*

(ii) *The ideals* $\mathscr{I}_j$, $j = 1, 2$ *have the ideal 3 and 4-chain properties:*

$$\mathsf{JA_1K_1A_2K_2} + \mathsf{K_2A_2K_1A_1J} \in \mathscr{I}_j, \tag{5.17}$$

$$\mathsf{JA_1K_1A_2K_2A_3K_3} + \mathsf{K_3A_3K_2A_2K_1A_1J} \in \mathscr{I}_j \tag{5.18}$$

for all $\mathsf{J} \in \mathscr{I}_j$, $\{\mathsf{K_1, K_2, K_3}\} \subset \Pi_j$, $\{\mathsf{A_1, A_2, A_3}\} \subset \mathscr{A}_j$, $j = 1, 2$

(iii) *The Jordan factor-multialgebra isomorphism* Φ *has the isomorphism 3 and 4-chain properties:*

$$\begin{aligned}&\Phi(\overline{\mathsf{K_0A_1K_1A'_1K_2} + \mathsf{K_2A'_1K_1A_1K_0}}) =\\ &\Phi(\overline{\mathsf{K_0}})\mathsf{A_2}\Phi(\overline{\mathsf{K_1}})\mathsf{A'_2}\,\Phi(\overline{\mathsf{K_2}}) + \Phi(\overline{\mathsf{K_2}})\mathsf{A'_2}\,\Phi(\overline{\mathsf{K_1}})\mathsf{A_2}\Phi(\overline{\mathsf{K_0}})\end{aligned} \tag{5.19}$$

$$\begin{aligned}&\Phi(\overline{\mathsf{K_0A_1K_1A'_1K_2A''_1K_3} + \mathsf{K_3A''_1K_2A'_1K_1A_1K_0}}) =\\ &\Phi(\overline{\mathsf{K_0}})\mathsf{A_2}\Phi(\overline{\mathsf{K_1}})\mathsf{A'_2}\,\Phi(\overline{\mathsf{K_2}})\mathsf{A''_2}\,\Phi(\overline{\mathsf{K_3}}) + \Phi(\overline{\mathsf{K_3}})\mathsf{A''_2}\,\Phi(\overline{\mathsf{K_2}})\mathsf{A'_2}\,\Phi(\overline{\mathsf{K_1}})\mathsf{A_2}\Phi(\overline{\mathsf{K_0}})\end{aligned} \tag{5.20}$$

for all $\{\mathsf{K_0, K_1, K_2, K_3}\} \subset \Pi_1$, *and all* $\{[\mathsf{A_1, A_2}], [\mathsf{A'_1, A'_2}], [\mathsf{A''_1, A''_2}]\} \subset \hat{\mathscr{A}}$.

Proof. **Necessity. Part (i).** Let π_i: $\mathbb{V} \to \mathrm{End}(\mathscr{T}_i)$, $i = 1, 2$, be the projection onto the ith component:

$$\pi_1[\mathsf{K_1, K_2}] = \mathsf{K_1}, \qquad \pi_2[\mathsf{K_1, K_2}] = \mathsf{K_2}.$$

By construction of the subspace $\hat{\mathscr{A}}$ it is clear that $\pi_i(\hat{\mathscr{A}}) = \mathscr{A}_i$, $i = 1, 2$. Thus, for any $\{\mathsf{A_1, A'_1}\} \subset \mathscr{A}_1$ and any $\{\mathsf{K_1, K'_1, K''_1}\} \subset \Pi_1$ there exist $\{\mathsf{A_2, A'_2}\} \subset \mathscr{A}_2$ and $\{\mathsf{K_2, K'_2, K''_2}\} \subset \Pi_2$, such that

$$\{[\mathsf{A_1, A_2}], [\mathsf{A'_1, A'_2}]\} \subset \hat{\mathscr{A}}, \qquad \{[\mathsf{K_1, K_2}], [\mathsf{K'_1, K'_2}], [\mathsf{K''_1, K''_2}]\} \subset \hat{\Pi}.$$

Then, by the assumed 3-chain property of $\hat{\Pi}$, we conclude that Π_1 must also have the 3-chain property. The 3-chain property for Π_2 and the 4-chain property are proved in the same way.

Part (ii). Suppose now that $\mathsf{J} \in \mathscr{I}_1$. Then $[\mathsf{J}, 0] \in \hat{\Pi}$. Using the 3 and 4-chain properties of $\hat{\Pi}$ with $[\mathsf{K_1, K_2}]$ replaced by $[\mathsf{J}, 0]$ we obtain the statement of part (ii) for $i = 1$. The statement for $i = 2$ is proved in the same way.

Part (iii). This is an immediate consequence of the 3 and 4-chain properties of $\hat{\Pi}$.

Sufficiency. We will prove the 3-chain property of $\hat{\Pi}$. The 4-chain property is proved in the same way. Let us take arbitrary

$$\{[\mathsf{A_1, A_2}], [\mathsf{A'_1, A'_2}]\} \subset \hat{\mathscr{A}}, \qquad \{[\mathsf{K_1, K_2}], [\mathsf{K'_1, K'_2}], [\mathsf{K''_1, K''_2}]\} \subset \hat{\Pi}.$$

We observe that

$$\mathsf{K_1A_1JA'_1K'_1} = 2(\mathsf{K_1} *_{\mathsf{A_1}} \mathsf{J})\mathsf{A'_1K'_1} - \mathsf{JA_1K_1A'_1K'_1}.$$

Therefore,

$$\mathsf{K_1A_1JA'_1K'_1} + \mathsf{K'_1A'_1JA_1K_1} = 4(\mathsf{K_1} *_{\mathsf{A_1}} \mathsf{J}) *_{\mathsf{A'_1}} \mathsf{K'_1} - \mathsf{JA_1K_1A'_1K'_1} - \mathsf{K'_1A'_1K_1A_1J} \in \mathscr{I}_1.$$

The same property is true for 4-chains as well, since they can be expressed in terms of 3-chains with one of the factors in $\mathscr{I}_1$ and a 4-chain (5.18). It follows that the equivalence classes in $\Pi_1/\mathscr{I}_1$ of 3 and 4-chains in (5.19) and (5.20) do not depend on the choices of representatives K_j of equivalence classes in $\Pi_1/\mathscr{I}_1$. It is now clear that part (iii) in the theorem implies validity of 3 and 4-chain properties of $\hat{\Pi}$. □

In the special case of polycrystalline links, i.e., when postulate 2.5 holds, all reference media are isotropic and all composites are polycrystalline, all the spaces in the algebraic structure of links are rotationally invariant and all the functions are rotationally equivariant.

Theorem 5.3. *Suppose that the reference medium* $\hat{\mathsf{L}}_0 = [\mathsf{L}_0^{(1)}, \mathsf{L}_0^{(2)}]$ *is isotropic (i.e.,* $\boldsymbol{R} \cdot \mathsf{L}_0^{(i)} = \mathsf{L}_0^{(i)}$ *for every* $\boldsymbol{R} \in \mathrm{SO}(d)$, $i = 1, 2$*). Suppose that* $\hat{\Pi}$ *is a rotationally invariant Jordan* $\hat{\mathscr{A}}$*-multialgebra. Then*

(i) *the subspaces* $\mathscr{A}_j$, Π_j, $\mathscr{I}_j$, $j = 1, 2$ *are rotationally invariant*

(ii) *The maps* Ξ *and* Φ*, defined in (5.8) and (5.14), respectively, are rotationally equivariant, i.e.,*

$$\Xi(\boldsymbol{R} \cdot \bar{\mathsf{A}}) = \boldsymbol{R} \cdot \Xi(\bar{\mathsf{A}}), \quad \Phi(\boldsymbol{R} \cdot \bar{\mathsf{K}}) = \boldsymbol{R} \cdot \Phi(\bar{\mathsf{K}}) \qquad \boldsymbol{R} \in \mathrm{SO}(d), \mathsf{A} \in \mathscr{A}_1, \mathsf{K} \in \Pi_1,$$

where the action of rotations on elements of the factor-space is defined in a natural way:

$$\boldsymbol{R} \cdot \bar{\mathsf{K}} = \overline{\boldsymbol{R} \cdot \mathsf{K}}, \quad \boldsymbol{R} \cdot \bar{\mathsf{A}} = \overline{\boldsymbol{R} \cdot \mathsf{A}}, \quad \boldsymbol{R} \in \mathrm{SO}(d), \mathsf{K} \in \Pi_1, \mathsf{A} \in \mathscr{A}_1.$$

The proof consists of straightforward verification of the required properties. In section 6.4 we will prove that property (5.9) holds for polycrystalline composites under a very mild (and probably unnecessary) assumption. In that case an important class of links, called the *global links*, where $\Pi_1 = \Pi_2 = \mathrm{Sym}(\mathscr{T})$, could be shown (see theorem 6.1 below) to come from Migram-Shtrikman covariance transformations (6.2).

5.3 Volume fraction formulas as links

In this section our goal is to develop the theory of formulas of the form

$$f(\mathsf{L}_*) = \langle f(\mathsf{L}(z)) \rangle, \tag{5.21}$$

for some f: $\mathrm{End}^+(\mathscr{T}) \to \mathbb{R}^m$, generalizing Hill's formula (4.1). An important feature of Hill's formula is that it serves as a 'companion' to Hill's exact relation (4.2). The idea is to observe that the effective tensor L_* of a periodic composite with local tensor $\mathsf{L}(z)$ is the volume average, if $\mathscr{E}_{\boldsymbol{n}} = \{0\} \subset \mathscr{T}$ for all unit vectors $\boldsymbol{n}$. This suggests that we should be able to obtain microstructure-independent formulas involving volume averages as links between our original composites and fictitious composites with tensor space of fields $\tilde{\mathscr{T}} = \mathbb{R}^N$, where $N = \dim \mathscr{T}$ and $\tilde{\mathscr{E}}_{\boldsymbol{n}} = \{0\}$. Thus,

$$\hat{\mathscr{T}} = \mathscr{T} \oplus \mathbb{R}^N, \qquad \hat{\mathscr{E}}_{\boldsymbol{n}} = \mathscr{E}_{\boldsymbol{n}} \oplus \{0\} \subset \mathscr{T} \oplus \mathbb{R}^N. \tag{5.22}$$

Hence,

$$\hat{\boldsymbol{\Gamma}}_0(\boldsymbol{n}) = [\boldsymbol{\Gamma}_0(\boldsymbol{n}), 0], \qquad \hat{\mathscr{A}} = \{[\mathsf{A}, 0]: \mathsf{A} \in \mathscr{A}\}. \tag{5.23}$$

Let $\hat{\Pi} \subset \mathrm{End}(\mathscr{T}) \oplus \mathrm{End}(\mathbb{R}^N)$ be a Jordan $\hat{\mathscr{A}}$-multialgebra. According to theorem 5.1 $\hat{\Pi}$ is completely determined by the algebraic data

$$\Pi = \{\mathsf{K} \in \mathrm{End}(\mathscr{T}): [\mathsf{K}, \tilde{\mathsf{K}}] \in \hat{\Pi} \text{ for some } \tilde{\mathsf{K}} \in \mathrm{End}(\mathbb{R}^N)\}, \tag{5.24}$$

$$\tilde{\Pi} = \{\tilde{\mathsf{K}} \in \mathrm{End}(\mathbb{R}^N): [\mathsf{K}, \tilde{\mathsf{K}}] \in \hat{\Pi} \text{ for some } \mathsf{K} \in \mathrm{End}(\mathscr{T})\}, \tag{5.25}$$

$$\mathscr{I} = \{\mathsf{K} \in \mathrm{End}(\mathscr{T}): [\mathsf{K}, 0] \in \hat{\Pi}\}, \tag{5.26}$$

$$\tilde{\mathscr{I}} = \{\tilde{\mathsf{K}} \in \mathrm{End}(\mathbb{R}^N): [0, \tilde{\mathsf{K}}] \in \hat{\Pi}\}. \tag{5.27}$$

Using special form (5.23) of the subspace $\hat{\mathscr{A}}$ we obtain that $[\mathsf{K}, \tilde{\mathsf{K}}] \in \hat{\Pi}$ if and only if $[\mathsf{KAK}, 0] \in \hat{\Pi}$, for any $\mathsf{K} \in \Pi$ and $\mathsf{A} \in \mathscr{A}$. Thus, $\Pi^2 \subset \mathscr{I}$, where

$$\Pi^2 = \mathrm{Span}\{\mathsf{K} *_{\mathsf{A}} \mathsf{K}': \{\mathsf{K}, \mathsf{K}'\} \subset \Pi, \mathsf{A} \in \mathscr{A}\}. \tag{5.28}$$

Obviously, Π^2 is an ideal in Π, and the factor algebra Π/Π^2 has a trivial multiplicative structure, since all Jordan products of elements in Π are in Π^2, which is zero in Π/Π^2. In fact, it is clear that the pair (Π, Π^2) contains all the information about Jordan $\hat{\mathscr{A}}$-multialgebras with $\Pi_1 = \Pi$, where Π_1 is defined by (5.10). Thus, all formulas involving volume averages are determined by Jordan $\hat{\mathscr{A}}$-multialgebras of the form

$$\hat{\Pi} = \{[\mathsf{K}, \mathscr{P}_{(\Pi^2)^c}\mathsf{K}]: \mathsf{K} \in \Pi\}, \tag{5.29}$$

where $\mathscr{P}_{(\Pi^2)^c}$ is a projection onto any complement of Π^2 in Π along Π^2.

Examining sufficient conditions for stability under homogenization, given in theorem 5.2, we obtain the corresponding sufficient conditions for the volume fraction formula corresponding to $\hat{\Pi}$ to be microstructure-independent.

Theorem 5.4. *A Jordan $\hat{\mathscr{A}}$-multialgebra $\hat{\Pi}$ with $\hat{\mathscr{A}}$, given by (5.23) possesses 3 and 4-chain properties if and only if*

$$\mathsf{K}_1\mathsf{A}_1\mathsf{K}_2\mathsf{A}_2\mathsf{K}_3 + \mathsf{K}_3\mathsf{A}_2\mathsf{K}_2\mathsf{A}_1\mathsf{K}_1 \in \Pi^2 \tag{5.30}$$

and

$$\mathsf{K}_1\mathsf{A}_1\mathsf{K}_2\mathsf{A}_2\mathsf{K}_3\mathsf{A}_3\mathsf{K}_4 + \mathsf{K}_4\mathsf{A}_3\mathsf{K}_3\mathsf{A}_2\mathsf{K}_2\mathsf{A}_1\mathsf{K}_1 \in \Pi^2 \tag{5.31}$$

for any $\{\mathsf{K}_1, \mathsf{K}_2, \mathsf{K}_3, \mathsf{K}_4\} \subset \Pi, \quad \{\mathsf{A}_1, \mathsf{A}_2, \mathsf{A}_3\} \subset \mathscr{A}$.

The link submanifold $\hat{\mathbb{M}}$, passing through the reference medium $[\mathsf{L}_0, \tilde{\mathsf{L}}_0]$ and corresponding to the Jordan $\hat{\mathscr{A}}$-multialgebra (5.29) can be written, according to (5.4), as follows

$$\hat{\mathbb{M}} = \{[\mathsf{L}, \tilde{\mathsf{L}}] \in \mathrm{End}^+(\mathscr{T}) \times \mathrm{End}^+(\mathbb{R}^N): \tilde{\mathsf{L}}_0 - \tilde{\mathsf{L}} = \mathscr{P}_{(\Pi^2)^c} W_{\boldsymbol{n}}(\mathsf{L})\}. \tag{5.32}$$

Thus, if $\mathsf{L}(z)$ is a local tensor of a periodic composite and L_* is its effective tensor, then the effective tensor of $\tilde{\mathsf{L}}(z) = \tilde{\mathsf{L}}_0 - \mathscr{P}_{(\Pi^2)^c} W_n(\mathsf{L}(z))$ is

$$\tilde{\mathsf{L}}_* = \tilde{\mathsf{L}}_0 - \mathscr{P}_{(\Pi^2)^c} W_n(\mathsf{L}_*).$$

However, our choice of $\tilde{\mathscr{E}}_n = \{0\}$ implies that

$$\tilde{\mathsf{L}}_* = \langle \tilde{\mathsf{L}}(z) \rangle = \tilde{\mathsf{L}}_0 - \mathscr{P}_{(\Pi^2)^c} \langle W_n(\mathsf{L}(z)) \rangle.$$

Hence, we obtain the generalization of Hills formula (4.1):

$$\mathscr{P}_{(\Pi^2)^c} W_n(\mathsf{L}_*) = \mathscr{P}_{(\Pi^2)^c} \langle W_n(\mathsf{L}) \rangle. \tag{5.33}$$

For future reference we summarize our results in the form of a theorem.

Theorem 5.5. *Suppose* Π *is a Jordan* $\mathscr{A}$*-multialgebra, corresponding to exact relation submanifold* $\mathbb{M}$ via *(4.4). This exact relation admits further formulas involving volume fractions if and only if* $\Pi^2 \neq \Pi$, *in which case* $\dim \Pi - \dim \Pi^2$ *additional formulas can be written in the form (5.33). These formulas are valid for all microstructures, provided the Jordan* $\mathscr{A}$*-multialgebra* Π *satisfies (5.30) and (5.31).*

References

[1] Dykhne A M 1971 Conductivity of a two-dimensional two-phase system *Sov. Phys. JETP* **32** 63–5 [Zh. Eksp. Teor. Fiz., 59 (1.70) 110–5.]

[2] Grabovsky Y 2004 Algebra, geometry and computations of exact relations for effective moduli of composites *Advances in Multifield Theories of Continua with Substructure* Modeling and Simulation in Science, Engineering and Technology G Capriz and P M Mariano (Boston, MA: Birkhäuser) 167–97

[3] Grabovsky Y 2009 Exact relations for effective conductivity of fiber-reinforced conducting composites with the Hall effect via a general theory *SIAM J. Math Anal.* **41** 973–1024

[4] Keller J B 1964 A theorem on the conductivity of a composite medium *J. Math. Phys.* **5** 548–9

[5] Mendelson K S 1975 Effective conductivities of two-phase material with cylindrical phase boundaries *J. Appl. Phys.* **46** 917–8

[6] Mendelson K S 1975 A theorem on the conductivity of two-dimensional heterogeneous medium *J. Appl. Phys.* **46** 4740–1

IOP Publishing

Composite Materials (Second Edition)
Mathematical theory and exact relations
Yury Grabovsky

Chapter 6

Computing exact relations and links

The general theory of exact relations and links developed in the foregoing chapters opens up a possibility of computing complete lists of all exact relations and links for most kinds of material properties, such as conductivity, elasticity, piezoelectricity, thermoelectricity and thermoelasticity. However, the practical difficulty of performing necessary calculations depends very much on the dimensionality of spaces $\mathscr{T}$ and $\mathscr{A}$. In fact, brute force calculations can become prohibitively difficult for anything more complicated than conducting composites. In this chapter we develop tools one can use to derive beautiful formulas like (4.1), (4.3) and (5.1) from the general theory in a systematic manner. Part III of this book lists the results of application of these methods and years of work of both the author and a large number of graduate and undergraduate students. The distinguishing feature of results in part III is that they are complete in the sense that all microstructure-independent formulas for effective behavior of composites are purely logical consequences of the listed facts.

The process of obtaining all exact relations and links consists of four major steps: finding all Jordan $\mathscr{A}$-multialgebras and their squares, finding all Jordan $\hat{\mathscr{A}}$-multialgebras, computing exact relations, including relations containing volume averages, and computing links. We will discuss methods for completing each of these steps and illustrate them by carrying out all calculations for two and three-dimensional conducting and elastic polycrystals.

6.1 Finding Jordan $\mathscr{A}$-multialgebras

6.1.1 Simplifying the subspace $\mathscr{A}$

The first step in computing all exact relations in any physical context is characterizing all Jordan $\mathscr{A}$-multialgebras. It is evident from formula (4.10) that the subspace $\mathscr{A}$ depends on L_0, which complicates matters, or makes them more interesting, depending on one's mood. The remedy comes from the idea of using symmetries of the problem (4.9). They play a dual role here. On the one hand they simplify the computation of Jordan $\mathscr{A}$-multialgebras, and on the other they can often be

doi:10.1088/978-0-7503-6249-8ch6

interpreted as global links. Specifically, we are looking for two invertible linear transformations

$$\Phi\colon \mathrm{End}(\mathscr{T}) \to \mathrm{End}(\mathscr{T}), \quad \Xi\colon \mathscr{A} \to \mathscr{A}_0,$$

such that

$$\Phi(\mathsf{K}_1 *_{\mathsf{A}} \mathsf{K}_2) = \Phi(\mathsf{K}_1) *_{\Xi(\mathsf{A})} \Phi(\mathsf{K}_2), \tag{6.1}$$

where the subspace $\mathscr{A}_0$ is 'simpler' than the subspace $\mathscr{A}$. Indeed, if $\Pi_0 \subset \mathrm{End}(\mathscr{T})$ satisfies

$$\mathsf{K}_1 *_{\mathsf{A}} \mathsf{K}_2 \in \Pi_0 \quad \forall \{\mathsf{K}_1, \mathsf{K}_2\} \subset \Pi_0 \ \forall \mathsf{A} \in \mathscr{A}_0,$$

then $\Pi = \Phi^{-1}(\Pi_0)$ is a Jordan $\mathscr{A}$-multialgebra.

A relatively large class of transformations (Φ, Ξ) satisfying (6.1) comes from Milgrom and Shtrikman's idea of 'covariance transformations' [13, 14]. In our context it amounts to the observations that for any $\{\mathsf{C}_1, \mathsf{C}_2\} \subset \mathrm{GL}(\mathscr{T})$ the matrix algebra identity

$$\mathsf{C}_1(\mathsf{KAK})\mathsf{C}_2 = (\mathsf{C}_1\mathsf{KC}_2)(\mathsf{C}_2^{-1}\mathsf{AC}_1^{-1})(\mathsf{C}_1\mathsf{KC}_2)$$

implies that

$$\Phi(\mathsf{K}) = \mathsf{C}_1\mathsf{KC}_2, \quad \Xi(\mathsf{A}) = \mathsf{C}_2^{-1}\mathsf{AC}_1^{-1} \tag{6.2}$$

satisfy (6.1). The goal is therefore, for each reference medium L_0 used in the construction of the subspace $\mathscr{A}$, to find $\{\mathsf{C}_1, \mathsf{C}_2\} \subset \mathrm{GL}(\mathscr{T})$, such that

$$\mathscr{A}_0 = \mathsf{C}_2^{-1}\mathscr{A}\mathsf{C}_1^{-1} \tag{6.3}$$

is independent of L_0.

For example, for conducting composites with not necessarily symmetric conductivity tensors (due to the Hall effect in the presence of the external magnetic field) we compute, [6, 7], for a reference medium $\boldsymbol{L}_0 \in \mathrm{End}^+(\mathbb{R}^d)$,

$$\Gamma_0(\boldsymbol{n}) = \frac{\boldsymbol{n} \otimes \boldsymbol{n}}{\boldsymbol{L}_0\boldsymbol{n} \cdot \boldsymbol{n}}, \qquad \mathscr{A} = \{\boldsymbol{A} \in \mathrm{Sym}(\mathbb{R}^d)\colon \mathrm{Tr}\,(\boldsymbol{A}\boldsymbol{L}_0) = 0\}. \tag{6.4}$$

It is easy to find the covariance transformation simplifying $\mathscr{A}$. Indeed, let $\boldsymbol{\sigma}_0$ be the symmetric part of $\boldsymbol{L}_0$. Then the desired covariance transformation is (6.2) with $\mathsf{C}_1 = \mathsf{C}_2 = \boldsymbol{\sigma}_0^{-1/2}$, so that

$$\mathscr{A}_0 = \boldsymbol{\sigma}_0^{1/2}\mathscr{A}\boldsymbol{\sigma}_0^{1/2} = \mathrm{Sym}_0(\mathbb{R}^d).$$

The set of all exact relations and links in the context of two-dimensional Hall effect conductivity has been computed in [6] and listed in section 8.1. The set of all exact relations and links in the context of three-dimensional Hall effect conductivity was computed in [16] and listed in section 8.2. We will see now that the covariance transformations (6.2), (6.3) are the only transformations satisfying (6.1) for polycrystalline composites.

The main observation here is corollary 4.16, reducing the search for rotationally invariant Jordan $\mathscr{A}$-multialgebras to the search for rotationally invariant Jordan algebras (i.e., Jordan algebras with a single Jordan multiplication). Typically, we want to relate polycrystalline exact relations passing through a given isotropic reference tensor L_0 to the exact relations passing through the identity $\mathsf{L}_0 = \mathsf{I}$, or at least a simpler reference medium than the original. Let $\tilde{\Gamma}$ and $\tilde{\Gamma}_0$ be the generating (in the sense of (4.17)) tensor for the subspaces $\mathscr{A}$ and $\mathscr{A}_0$, corresponding to isotropic reference media L_0 and I (or a simplified one), respectively. We are then looking for SO(d) equivariant isomorphisms Φ: Sym($\mathscr{T}$) $\to$ Sym($\mathscr{T}$), such that

$$\Phi(\mathsf{K}\tilde{\Gamma}_0\mathsf{K}) = \Phi(\mathsf{K})\tilde{\Gamma}\Phi(\mathsf{K}). \tag{6.5}$$

In that case, if Π is an SO(d)-invariant subspace closed with respect to the Jordan multiplication $*_{\tilde{\Gamma}_0}$, then $\Phi(\Pi)$ is also an SO(d)-invariant subspace closed with respect to the Jordan multiplication $*_{\tilde{\Gamma}}$, and therefore SO(d)-invariant Jordan $\mathscr{A}$-multialgebra.

Theorem 6.1. *If there exists a linear isomorphism* Φ: Sym($\mathscr{T}$) $\to$ Sym($\mathscr{T}$), *satisfying (6.5), then either tensors* $\tilde{\Gamma}$ *and* $\tilde{\Gamma}_0$ *or* $\tilde{\Gamma}$ *and* $-\tilde{\Gamma}_0$ *have the same signature (number of positive, negative and zero eigenvalues), in which case at least one of the sets*

$$\mathfrak{C}_{\pm}(\tilde{\Gamma}_0, \tilde{\Gamma}) = \{\mathsf{C} \in GL(\mathscr{T}): \tilde{\Gamma}_0 = \pm\mathsf{C}\tilde{\Gamma}\mathsf{C}^T\}$$

is nonempty. Moreover, the set of all invertible linear maps Φ: Sym($\mathscr{T}$) $\to$ Sym($\mathscr{T}$) *satisfying (6.5) is either* $\{\Phi(\mathsf{K}) = \mathsf{C}^T\mathsf{K}\mathsf{C}: \mathsf{C} \in \mathfrak{C}_+(\tilde{\Gamma}_0, \tilde{\Gamma})\}$ *or* $\{\Phi(\mathsf{K}) = -\mathsf{C}^T\mathsf{K}\mathsf{C}: \mathsf{C} \in \mathfrak{C}_-(\tilde{\Gamma}_0, \tilde{\Gamma})\}$, *or both.*

This theorem has been proved in [9, lemma 4.11] under the assumption that the operators $\tilde{\Gamma}_0$ and $\tilde{\Gamma}$ are invertible. The proof for the general case can be found in chapter 14. According to our discussion of (6.5), the relevant Jordan algebra isomorphisms must be rotationally equivariant, which means that they correspond to isotropic tensors in $\mathfrak{C}_{\pm}(\tilde{\Gamma}_0, \tilde{\Gamma})$. Another remark is that in the special case, when $\tilde{\Gamma}$ and $\tilde{\Gamma}_0$ have equal number of positive and negative eigenvalues (as for 2D thermoelectricity), then the set of all rotationally invariant Jordan isomorphisms correspond to isotropic tensors C from *both* $\mathfrak{C}_+(\tilde{\Gamma}_0, \tilde{\Gamma})$ and $\mathfrak{C}_-(\tilde{\Gamma}_0, \tilde{\Gamma})$.

6.1.2 Rotationally invariant subspaces

From now on we will focus exclusively on polycrystalline exact relations (see definition 4.14), which, according to our discussion in section 4.4.1, correspond to rotationally invariant Jordan $\mathscr{A}$-multialgebras. Rotational invariance leads to tremendous reduction in complexity. For example, for 3D elasticity $\mathscr{T} = \text{Sym}(\mathbb{R}^3)$ and Sym($\mathscr{T}$) is a 21-dimensional vector space of elastic tensors. If we ignore rotational invariance, then for each $0 < k < 21$ we must find all k-dimensional subspaces Π that satisfy (4.9). The set of all k-dimensional subspaces in a

21-dimensional space forms a $k(21-k)$-parametric family. If $k=10$ or 11, the search for exact relations would correspond to solving a large system of polynomial equations, corresponding to equation (4.9), in 110 variables! When we look for rotationally invariant Jordan multialgebras, our task would be reduced to examining only 16 nontrivial cases, in which the search manifolds would have dimensions 0 in 6, 1 in 8, and 2 in 2 of the 16 cases. Hence, the first step in our search for all solutions of (4.9) is to identify all rotationally invariant subspaces in $\mathrm{Sym}(\mathscr{T})$. This problem has already arisen in quantum mechanics, see, e.g., [2, 5]. Its solution is now a part of a well-developed *representation theory* of compact Lie groups [17].

We recall that for d-dimensional elasticity, $\mathscr{T}=\mathrm{Sym}(\mathbb{R}^d)$, $d=2, 3$, and the rotation of a 'vector' $\varepsilon\in\mathscr{T}$ by $\boldsymbol{R}\in SO(d)$ is given by

$$\boldsymbol{R}\cdot\varepsilon=\boldsymbol{R}\varepsilon\boldsymbol{R}^T. \tag{6.6}$$

This action of $SO(d)$ on $\mathscr{T}$ 'lifts' to $\mathrm{Sym}(\mathscr{T})$ in a *canonical way*, according to the 'distributive law'

$$\boldsymbol{R}\cdot(\mathsf{C}\varepsilon)=(\boldsymbol{R}\cdot\mathsf{C})(\boldsymbol{R}\cdot\varepsilon),\qquad \varepsilon\in\mathscr{T},\quad \boldsymbol{R}\in SO(d),\quad \mathsf{C}\in\mathrm{Sym}(\mathscr{T}), \tag{6.7}$$

which for elasticity translates into the explicit formula

$$(\boldsymbol{R}\cdot\mathsf{C})\varepsilon=\boldsymbol{R}(\mathsf{C}(\boldsymbol{R}^T\varepsilon\boldsymbol{R}))\boldsymbol{R}^T,\quad \varepsilon\in\mathrm{Sym}(\mathbb{R}^d),\quad \mathsf{C}\in\mathrm{Sym}(\mathrm{Sym}(\mathbb{R}^d)). \tag{6.8}$$

Formula (6.8) shows that the action of the rotation group on elasticity tensors is sufficiently complicated to rule out naive attempts to characterize all rotationally invariant subspaces of $\mathrm{Sym}(\mathrm{Sym}(\mathbb{R}^d))$, starting with definition (6.8). Therefore, the use of representation theory of compact Lie groups (see, e.g., [3]), and the groups of rotations $SO(2)$ and $SO(3)$, in particular, is unavoidable.

The main consequence of compactness is the *complete reducibility* property of all finite dimensional representations. This property becomes apparent if we construct a group-invariant inner product, making group elements act by orthogonal transformations. To construct such an inner product for elasticity we recall that the Frobenius inner product (2.5) on the strain space $\mathrm{Sym}(\mathbb{R}^d)$ is $SO(d)$-invariant:

$$\mathrm{Tr}\,((\boldsymbol{R}\cdot\boldsymbol{A})(\boldsymbol{R}\cdot\boldsymbol{B}))=\mathrm{Tr}\,(\boldsymbol{R}\boldsymbol{A}\boldsymbol{R}^T\boldsymbol{R}\boldsymbol{B}\boldsymbol{R}^T)=\mathrm{Tr}\,(\boldsymbol{A}\boldsymbol{B}).$$

Elasticity tensors are symmetric operators on $\mathrm{Sym}(\mathbb{R}^d)$ with respect to this inner product, and the corresponding Frobenius inner product

$$\langle\mathsf{C}_1,\mathsf{C}_2\rangle=\mathrm{Tr}\,(\mathsf{C}_1\mathsf{C}_2) \tag{6.9}$$

on $\mathrm{Sym}(\mathrm{Sym}(\mathbb{R}^d))$ is therefore also $SO(d)$-invariant. The $SO(d)$-invariance of the inner product implies that each rotation is an orthogonal transformation on the Euclidean vector space $\mathrm{Sym}(\mathrm{Sym}(\mathbb{R}^d))$ with the inner product (6.9). As a consequence, if a subspace $U\subset\mathrm{Sym}(\mathrm{Sym}(\mathbb{R}^d))$ is $SO(d)$-invariant, then so is its orthogonal complement. Thus, $\mathrm{Sym}(\mathrm{Sym}(\mathbb{R}^d))$ can be written as an orthogonal sum of $SO(d)$-invariant subspaces, each of which does not have any proper $SO(d)$-invariant subspaces. Such subspaces are called *irreducible representations* of $SO(d)$, or *irreps*, for short. The tools of representation theory of rotation groups $SO(2)$ and $SO(3)$

permit an easy identification of all irreps in $\mathrm{Sym}(\mathrm{Sym}(\mathbb{R}^d))$, $d = 2, 3$, by their isomorphism class. While this information is insufficient for our purposes, it is very useful for the eventual explicit characterization of all rotationally invariant subspaces in $\mathrm{Sym}(\mathscr{T})$.

We say that two irreps U and V are isomorphic if there exists a linear bijective map $f: U \to V$, such that $\boldsymbol{R} \cdot f(\boldsymbol{u}) = f(\boldsymbol{R} \cdot \boldsymbol{u})$ for every $\boldsymbol{u} \in U$ and $\boldsymbol{R} \in SO(d)$. The isomorphism classes of all irreps of rotation groups SO(2) and SO(3) are well-known. In both cases, the isomorphism classes of irreps are characterized by non-negative integers $m = 0, 1, 2, \ldots$, called *weights*. The irrep W_0 is one-dimensional, on which rotations act trivially: $\boldsymbol{R} \cdot \boldsymbol{w} = \boldsymbol{w}$ for all $\boldsymbol{w} \in W_0$. For this reason W_0 is sometimes called the *trivial irrep*. All SO(2) irreps W_m, $m \geqslant 1$ are two-dimensional and the action of SO(2) on W_m has a very simple geometric interpretation, since every rotation in $\boldsymbol{R}_\theta \in SO(2)$ can be uniquely identified by the angle $\theta \in \mathbb{R}/(2\pi\mathbb{Z})$ through which it rotates every vector in a plane. The W_m irrep possess a special basis $\{\boldsymbol{e}_1, \boldsymbol{e}_2\}$, such that any vector $\boldsymbol{u} \in W_m$ with coordinates (x, y) in this basis gets rotated by $\boldsymbol{R}_\theta$ to the vector $\boldsymbol{R}_\theta \cdot \boldsymbol{u}$ with coordinates (x', y'), representing the rotation of $(x, y) \in \mathbb{R}^2$ through the angle $m\theta \in \mathbb{R}/(2\pi\mathbb{Z})$.

The weight m of an SO(3) irrep W_m inherits its value from a copy of $SO(2) \subset SO(3)$, which represents all rotations around a fixed axis[1]. Such a copy of SO(2) is called a 'maximal torus' in representation theory. By the complete reducibility property we can write an SO(3) irrep as a direct sum of SO(2) irreps. According to the representation theory (see section 17.1), an irrep W_m of SO(3) is a direct sum of $m + 1$ distinct SO(2) irreps of weights $0, 1, 2, \ldots, m$, respectively, showing, in particular, that $\dim W_m = 2m + 1$. Hence, the weight of an SO(3) irrep W_m can be defined to be the highest weight of an SO(2) irrep in W_m. This structure of W_m also indicates that there is a unique one-dimensional weight zero (trivial) representation of SO(2) in W_m. A vector spanning this representation is called the zero weight vector in W_m. It is unique up to a scalar multiple. In other words, a zero weight vector is the unique (up to a scalar multiple) vector in W_m, which is invariant under the action of the maximal torus $SO(2) \subset SO(3)$. Recall that under the rotational equivariance postulate 2.5 we have $\boldsymbol{R} \cdot \boldsymbol{\Gamma}(\boldsymbol{e}_1) = \boldsymbol{\Gamma}(\boldsymbol{R}\boldsymbol{e}_1)$. Therefore, the vector $\boldsymbol{\Gamma}(\boldsymbol{e}_1)$ is a zero weight vector, since it is invariant under a copy of $SO(2) \subset SO(3)$ representing rotations in $\mathbb{R}^3$ around $\boldsymbol{e}_1 \in \mathbb{R}^3$.

We have already seen an example of a decomposition of $\mathrm{Sym}(\mathscr{T})$ into a direct sum of $SO(d)$ irreps in the context of conductivity, where $\mathscr{T} = \mathbb{R}^d$, $d = 2, 3$.

$$\mathrm{Sym}(\mathbb{R}^d) = \mathbb{R}\boldsymbol{I}_d \oplus \mathrm{Sym}_0(\mathbb{R}^d), \tag{6.10}$$

where $\mathrm{Sym}_0(\mathbb{R}^d)$ is defined in (4.22). In this example, the one-dimensional irrep $\mathbb{R}\boldsymbol{I}_d$ belongs to the W_0 isomorphism class, while $\mathrm{Sym}_0(\mathbb{R}^d)$ is an irrep of weight 2. This

[1] Even though there are infinitely many copies of $SO(2) \subset SO(3)$, all of them are conjugate to one another, and hence, each copy assigns one and the same weight to an irrep. For calculations one usually fixes a specific copy of SO(2), much in the same way as one chooses a coordinate system to solve a geometric problem.

information will also be useful for the decomposition of the space of elastic tensors $\mathbb{E}^d = \mathrm{Sym}(\mathrm{Sym}(\mathbb{R}^d))$ into a direct sum of $SO(d)$ irreps, $d = 2, 3$.

Starting with the decomposition (6.10) we obtain

$$\mathbb{E}^d = \mathrm{Sym}(\mathbb{R}\boldsymbol{I}_d) \oplus \mathrm{Sym}(\mathrm{Sym}_0(\mathbb{R}^d)) \oplus \mathrm{Sym}(\mathbb{R}\boldsymbol{I}_d,\, \mathrm{Sym}_0(\mathbb{R}^d)),$$

where $\mathrm{Sym}(U, V)$ denotes the set of all symmetric operators $T \in \mathrm{Sym}(U \oplus V)$ such that $T\colon U \to V$ (and consequently $T\colon V \to U$). The space $\mathrm{Sym}(\mathbb{R}\boldsymbol{I}_d) = \mathrm{End}(\mathbb{R}\boldsymbol{I}_d)$ of all linear maps on multiples of the identity is 1-dimensional and its elements are rotationally invariant. Hence, $\mathrm{Sym}(\mathbb{R}\boldsymbol{I}_d) \cong W_0$. It is also fairly obvious[2] that $\mathrm{Sym}(\mathbb{R}\boldsymbol{I}_d, \mathrm{Sym}_0(\mathbb{R}^d))$ is isomorphic to $\mathrm{Sym}_0(\mathbb{R}^d) \cong W_2$. Hence, we denote

$$\mathrm{Sym}(\mathbb{R}\boldsymbol{I}_d) = K_0', \qquad \mathrm{Sym}(\mathbb{R}\boldsymbol{I}_d,\, \mathrm{Sym}_0(\mathbb{R}^d)) = K_2', \tag{6.11}$$

where subscripts indicate the weight of the corresponding irrep. The primes in K_0' and K_2' are to distinguish these irreps from other irreps of the same weight coming from $\mathrm{Sym}(\mathrm{Sym}_0(\mathbb{R}^d))$. It will be useful to fix explicit isomorphisms between irreps of weight 0 and 2 in $\mathbb{E}^d$ and the standard irreps: $\mathbb{R}$, of weight 0, and $\mathrm{Sym}_0(\mathbb{R}^d)$, of weight 2, respectively. These isomorphisms will be conveniently denoted by the same letters as the corresponding subrepresentations of $\mathbb{E}^d$:

$$K_0'(\kappa)\varepsilon = \kappa(\mathrm{Tr}\,\varepsilon)\boldsymbol{I}_d, \qquad \varepsilon \in \mathrm{Sym}(\mathbb{R}^d),\ \kappa \in \mathbb{R}; \tag{6.12}$$

$$K_2'(\boldsymbol{A})\varepsilon = \mathrm{Tr}\,(\varepsilon)\boldsymbol{A} + \mathrm{Tr}\,(\varepsilon\boldsymbol{A})\boldsymbol{I}_d, \qquad \varepsilon \in \mathrm{Sym}(\mathbb{R}^d),\ \ \boldsymbol{A} \in \mathrm{Sym}_0(\mathbb{R}^d). \tag{6.13}$$

It remains to decompose the subrepresentation $\mathrm{Sym}(\mathrm{Sym}_0(\mathbb{R}^d))$ into irreducibles. Two of them (one in 2D) we can simply guess by constructing rotationally equivariant maps from $\mathbb{R}$ and $\mathrm{Sym}_0(\mathbb{R}^d)$ to $\mathrm{Sym}(\mathrm{Sym}_0(\mathbb{R}^d))$:

$$K_0(\mu)\varepsilon = 2\mu\varepsilon, \quad \mu \in \mathbb{R},\ \varepsilon \in \mathrm{Sym}_0(\mathbb{R}^d), \tag{6.14}$$

$$K_2(\boldsymbol{A})\varepsilon = \mathrm{dev}(\varepsilon\boldsymbol{A} + \boldsymbol{A}\varepsilon), \qquad \{\varepsilon, \boldsymbol{A}\} \subset \mathrm{Sym}_0(\mathbb{R}^d), \tag{6.15}$$

where

$$\mathrm{dev}(\varepsilon) = \varepsilon - \frac{1}{d}(\mathrm{Tr}\,\varepsilon)\boldsymbol{I}_d$$

is the deviatoric part of a symmetric $d \times d$ matrix. We note that $K_2(\boldsymbol{A}) = \{0\}$, when $d = 2$, due to the identity

$$\varepsilon\boldsymbol{A} + \boldsymbol{A}\varepsilon = \mathrm{Tr}\,(\varepsilon\boldsymbol{A})\boldsymbol{I}_2, \qquad \{\varepsilon, \boldsymbol{A}\} \subset \mathrm{Sym}_0(\mathbb{R}^2).$$

In order to view $K_0(\mu)$ and $K_2(\boldsymbol{A})$ as elements of $\mathbb{E}^d$ we write

$$K_0(\mu)\varepsilon = 2\mu\,\mathrm{dev}(\varepsilon), \qquad \varepsilon \in \mathrm{Sym}(\mathbb{R}^d),\ \mu \in \mathbb{R}, \tag{6.16}$$

[2] To state the obvious explicitly, we identify an operator $T \in \mathrm{Sym}(\mathbb{R}\boldsymbol{I}_d, \mathrm{Sym}_0(\mathbb{R}^d))$ with $T\boldsymbol{I}_d \in \mathrm{Sym}_0(\mathbb{R}^d)$ and then show that this correspondence is an $SO(d)$ isomorphism.

$$K_2(A)\varepsilon = \mathrm{dev}(\mathrm{dev}(\varepsilon)A + A\mathrm{dev}(\varepsilon)), \quad \varepsilon \in \mathrm{Sym}(\mathbb{R}^3),\ A \in \mathrm{Sym}_0(\mathbb{R}^3). \tag{6.17}$$

Hence,

$$K_0 = \{K_0(\mu)\colon \mu \in \mathbb{R}\}, \tag{6.18}$$

$$K_2 = \{K_2(A)\colon A \in \mathrm{Sym}_0(\mathbb{R}^3)\}. \tag{6.19}$$

It turns out that the orthogonal complement of $K_0 \oplus K_2$ (K_0 in 2D) in $\mathrm{Sym}(\mathrm{Sym}_0(\mathbb{R}^d))$ is isomorphic to the irrep W_4 of $SO(d)$. We denote this subspace $K_4 \subset \mathbb{E}^d$. For our purposes we need a more direct and explicit characterization of K_4. This has been accomplished in [1]. Let us summarize the results.

Let $\mathsf{C} \in \mathbb{E}^d$ and $\boldsymbol{n} \in \mathbb{R}^d$ be a unit vector. Then there is a unique symmetric $d \times d$ matrix $A_{\mathsf{C}}(\boldsymbol{n})$, such that for every $\boldsymbol{a} \in \mathbb{R}^d$

$$A_{\mathsf{C}}(\boldsymbol{n})\boldsymbol{a} \cdot \boldsymbol{a} = (\mathsf{C}(\boldsymbol{a} \otimes \boldsymbol{n}))\boldsymbol{n} \cdot \boldsymbol{a}.$$

The matrix $A_{\mathsf{C}}(\boldsymbol{n})$ is called the *acoustic tensor* of C. The name comes from the fact that the eigenvalues of $A_{\mathsf{C}}(\boldsymbol{n})$ are squares of the speeds of d different sound waves (a compression wave and $d - 1$ shear waves) propagating through the anisotropic elastic material in the direction $\boldsymbol{n}$. Observe that $\mathrm{Tr}\, A_{\mathsf{C}}(\boldsymbol{n})$ is quadratic in $\boldsymbol{n} \in \mathbb{R}^d$. Thus, there is a unique symmetric $d \times d$ matrix $\boldsymbol{M}_{\mathsf{C}}$, such that

$$\boldsymbol{M}_{\mathsf{C}}\boldsymbol{n} \cdot \boldsymbol{n} = \mathrm{Tr}\, A_{\mathsf{C}}(\boldsymbol{n}).$$

The subspace $K_4 \subset \mathbb{E}^d$ can now be described as follows.

$$K_4 = \{\mathsf{C} \in \mathbb{E}^d\colon \boldsymbol{M}_{\mathsf{C}} = 0,\ \mathsf{C}\boldsymbol{I}_d = 0\}. \tag{6.20}$$

To summarize,

$$\mathbb{E}^d = \mathrm{Sym}(\mathrm{Sym}(\mathbb{R}^d)) = \begin{cases} K_0' \oplus K_2' \oplus K_0 \oplus K_4, & d = 2, \\ K_0' \oplus K_2' \oplus K_0 \oplus K_2 \oplus K_4, & d = 3. \end{cases} \tag{6.21}$$

$$\mathsf{C} = \begin{cases} K_0'(\kappa) + K_0(\mu) + K_2'(A) + \mathsf{K}_4(\mathsf{C}), & d = 2, \\ K_0'(\kappa) + K_0(\mu) + K_2'(A) + K_2(\boldsymbol{B}) + \mathsf{K}_4(\mathsf{C}), & d = 3, \end{cases} \tag{6.22}$$

where $\{A, \boldsymbol{B}\} \subset \mathrm{Sym}_0(\mathbb{R}^d)$ and $\mathsf{K}_4(\mathsf{C}) \in K_4$ is the orthogonal projection of $\mathsf{C} \in \mathbb{E}^d$ onto K_4. Moreover, for a given $\mathsf{C} \in \mathbb{E}^d$ we can obtain the explicit expressions for κ, μ, $\boldsymbol{A}$ and $\boldsymbol{B}$, according to the following formulas (see [1]).

$$\kappa = \frac{1}{d^2}\mathrm{Tr}\,(\mathsf{C}\boldsymbol{I}_d), \qquad \boldsymbol{A} = \frac{1}{d}\mathrm{dev}(\mathsf{C}\boldsymbol{I}_d), \tag{6.23}$$

$$\boldsymbol{B} = \frac{2\mathrm{dev}(d\boldsymbol{M}_{\mathsf{C}} - 2\mathsf{C}\boldsymbol{I}_d)}{(d-2)(d+4)}, \qquad \mu = \frac{\mathrm{Tr}\,(d\boldsymbol{M}_{\mathsf{C}} - \mathsf{C}\boldsymbol{I}_d)}{d(d-1)(d+2)}. \tag{6.24}$$

In fact, these formulas are not hard to derive. For example, formulas (6.23) follow from (6.12), (6.13), if we observe that

$$\mathsf{C}\boldsymbol{I}_d = d(\kappa\boldsymbol{I}_d + \boldsymbol{A}). \tag{6.25}$$

It is also fairly straightforward to compute, using formulas (6.12)–(6.17), that

$$\boldsymbol{M}_{\mathsf{C}} = \kappa\boldsymbol{I}_d + 2\boldsymbol{A} + \mu\left(d + 1 - \frac{2}{d}\right)\boldsymbol{I}_d + \left(\frac{d}{2} + 1 - \frac{4}{d}\right)\boldsymbol{B}. \tag{6.26}$$

Formulas (6.24) then follow, upon using the already established formulas (6.23).

Now, the search for all SO(d)-invariant Jordan multialgebras in $\mathbb{E}^d$ is reduced to SO(d)-invariant subspaces of $\mathbb{E}^d$. We note that formula (6.21) does not mean that there are only a small finite number of such subspaces. In fact, the presence of several irreps of the same weight, such as K_0 and K_0', as well as K_2 and K_2' makes it possible to combine them to create infinite families of subrepresentations. For example, the subspaces

$$K_0^\alpha = \{K_0(\mu) + K_0'(\alpha\mu)\colon \mu \in \mathbb{R}\}, \tag{6.27}$$

$$K_2^\alpha = \{K_2(\boldsymbol{A}) + K_2'(\alpha\boldsymbol{A})\colon \boldsymbol{A} \in \mathrm{Sym}_0(\mathbb{R}^3)\} \tag{6.28}$$

are also rotationally invariant. The list of all rotationally invariant subspaces in $\mathbb{E}^d$ can thus be described as follows.

Theorem 6.2.

(a) *A subspace* $\Pi \subset \mathrm{Sym}(\mathrm{Sym}(\mathbb{R}^2))$ *is rotationally invariant if and only if* $\Pi = U_0 \oplus U_2 \oplus U_4$, *where* U_0 *is either* $\{0\}$, K_0', K_0^α *for some* $\alpha \in \mathbb{R}$, *or* $K_0 \oplus K_0'$; U_2 *is either* $\{0\}$ *or* K_2'; U_4 *is either* $\{0\}$ *or* K_4.

(b) *A subspace* $\Pi \subset \mathrm{Sym}(\mathrm{Sym}(\mathbb{R}^3))$ *is rotationally invariant if and only if* $\Pi = U_0 \oplus U_2 \oplus U_4$, *where* U_0 *is either* $\{0\}$, K_0', K_0^α *or* $K_0 \oplus K_0'$; U_2 *is either* $\{0\}$, K_2', K_2^β *or* $K_2 \oplus K_2'$; U_4 *is either* $\{0\}$ *or* K_4.

We can also characterize all rotationally invariant subspaces in the context of thermoelectricity (see chapter 12), where $\mathscr{T} = \mathbb{R}^d \oplus \mathbb{R}^d$. It will be instructive to generalize this example to an arbitrary number of $\mathbb{R}^d$-valued fields

$$\mathscr{T} = \underbrace{\mathbb{R}^d \oplus \cdots \oplus \mathbb{R}^d}_{n} = \mathbb{R}^N \otimes \mathbb{R}^d. \tag{6.29}$$

If $\boldsymbol{R} \in \mathrm{SO}(d)$, and $\boldsymbol{E} = (\boldsymbol{e}_1, \ldots, \boldsymbol{e}_N) \in \mathscr{T}$, then $\boldsymbol{R} \cdot \boldsymbol{E} = (\boldsymbol{R}\boldsymbol{e}_1, \ldots, \boldsymbol{R}\boldsymbol{e}_N)$. Mathematically, we say that, as a representation of SO(d), $\mathscr{T} \cong \mathbb{R}^N \otimes W_1$, when $d = 2$ or 3. In that case $\mathrm{End}(\mathscr{T}) \cong \mathrm{End}(\mathbb{R}^N) \otimes \mathrm{End}(W_1)$, while

$$\mathrm{Sym}(\mathscr{T}) \cong \mathrm{Sym}(\mathbb{R}^N) \otimes \mathrm{Sym}(W_1) \oplus \mathrm{Skew}(\mathbb{R}^N) \otimes \mathrm{Skew}(W_1).$$

The decomposition (6.10) can be written as $\mathrm{Sym}(W_1) \cong W_0 \oplus W_2$, while

$$\mathrm{Skew}(W_1) \cong \begin{cases} W_0, & d = 2, \\ W_1, & d = 3. \end{cases} \tag{6.30}$$

The meaning of (6.30) is very simple. The set of skew-symmetric 2×2 matrices is one-dimensional, spanned by the 90° rotation matrix $\boldsymbol{R}_\perp = \begin{bmatrix} 0 & -1 \\ 1 & 0 \end{bmatrix}$, which is unchanged by rotations in SO(2):

$$\boldsymbol{R}_\theta \cdot \boldsymbol{R}_\perp = \boldsymbol{R}_\theta \boldsymbol{R}_\perp \boldsymbol{R}_\theta^T = \boldsymbol{R}_\theta \boldsymbol{R}_\theta^T \boldsymbol{R}_\perp = \boldsymbol{R}_\perp .$$

The set of skew-symmetric 3×3 matrices is three-dimensional, where every skew-symmetric matrix $\boldsymbol{S}$ is uniquely characterized by its axial vector $\boldsymbol{\omega}$, via the rule

$$\boldsymbol{S}\boldsymbol{u} = \boldsymbol{\omega} \times \boldsymbol{u}, \qquad \boldsymbol{S} = \begin{bmatrix} 0 & -\omega_3 & \omega_2 \\ \omega_3 & 0 & -\omega_1 \\ -\omega_2 & \omega_1 & 0 \end{bmatrix} : \stackrel{\text{def}}{=} \boldsymbol{\omega} \times , \tag{6.31}$$

where the notation $\boldsymbol{S} = \boldsymbol{\omega}\times$ was defined in (2.17). It is well-known that for any rotation $\boldsymbol{R} \in \mathrm{SO}(3)$

$$\boldsymbol{R} \cdot (\boldsymbol{\omega} \times) = \boldsymbol{R}(\boldsymbol{\omega} \times)\boldsymbol{R}^T = (\boldsymbol{R}\boldsymbol{\omega}) \times .$$

In other words, the action of SO(3) on the space of 3×3 skew-symmetric matrices is equivalent to the action of SO(3) on their axial vectors. This fact underlies the second line in (6.30). We therefore obtain

$$\mathrm{Sym}(\mathbb{R}^N \otimes \mathbb{R}^d) = \begin{cases} \mathrm{End}(\mathbb{R}^N) \otimes W_0 \oplus \mathrm{Sym}(\mathbb{R}^N) \otimes W_2, & d = 2, \\ \mathrm{Sym}(\mathbb{R}^N) \otimes (W_0 \oplus W_2) \oplus \mathrm{Skew}(\mathbb{R}^N) \otimes W_1, & d = 3. \end{cases} \tag{6.32}$$

This permits us to characterize all rotationally invariant subspaces in $\mathrm{Sym}(\mathbb{R}^N \otimes \mathbb{R}^d)$.

Theorem 6.3.

(a) *A subspace* $\Pi \subset \mathrm{Sym}(\mathbb{R}^N \otimes \mathbb{R}^2)$ *is rotationally invariant if and only if* $\Pi = \mathscr{L}_0 \oplus \mathscr{L}_2$, *where* $\mathscr{L}_0$ *is a subspace in* $\mathrm{End}(\mathbb{R}^N)$ *and* $\mathscr{L}_2$ *is a subspace in* $\mathrm{Sym}(\mathbb{C}^N)$, *regarded as a complex vector space.*

(b) *A subspace* $\Pi \subset \mathrm{Sym}(\mathbb{R}^N \otimes \mathbb{R}^3)$ *is rotationally invariant if and only if*

$$\Pi = \mathscr{L}_0 \otimes W_0 \oplus \mathscr{L}_1 \otimes W_1 \oplus \mathscr{L}_2 \otimes W_2, \tag{6.33}$$

where $\mathscr{L}_0$ *and* $\mathscr{L}_2$ *are subspaces in* $\mathrm{Sym}(\mathbb{R}^N)$, *while* $\mathscr{L}_1$ *is a subspace in* $\mathrm{Skew}(\mathbb{R}^N)$.

Since when $d = 3$ we have

$$\mathrm{End}(\mathbb{R}^N \otimes \mathbb{R}^3) = \mathrm{End}(\mathbb{R}^N) \otimes \mathrm{End}(\mathbb{R}^3) \cong \mathrm{End}(\mathbb{R}^N) \otimes (W_0 \oplus W_1 \oplus W_2),$$

where $W_0 = \mathbb{R}\boldsymbol{I}_3$, $W_1 = \mathrm{Skew}(\mathbb{R}^3)$ and $W_2 = \mathrm{Sym}_0(\mathbb{R}^3)$, the interpretation of elements of Π, given by (6.33), as operators on $\mathscr{T} = \mathbb{R}^N \otimes \mathbb{R}^3$, is clear. This is not the case, when $d = 2$. Each matrix $\boldsymbol{X} \in \mathscr{L}_0 \subset \mathrm{End}(\mathbb{R}^N)$ must be interpreted as $\boldsymbol{X}_{\text{sym}} \otimes \boldsymbol{I}_2 + \boldsymbol{X}_{\text{skew}} \otimes \boldsymbol{R}_\perp$, where

$$X_{\text{sym}} = \frac{X + X^T}{2}, \quad X_{\text{skew}} = \frac{X - X^T}{2}.$$

Each matrix $Y \in \mathrm{Sym}(\mathbb{C}^N)$ must be interpreted as $\mathfrak{R}(Y) \otimes e_1 + \mathfrak{I}(Y) \otimes e_2$, where $\{e_1, e_2\}$ is one of the special bases of W_2 in the definition of the SO(2) irreps W_m. In the case of $W_2 = \mathrm{Sym}_0(\mathbb{R}^2)$ we can choose

$$e_1 = \begin{bmatrix} 1 & 0 \\ 0 & -1 \end{bmatrix}, \quad e_2 = \begin{bmatrix} 0 & 1 \\ 1 & 0 \end{bmatrix}.$$

6.1.3 The subspace $\mathscr{A}$ for polycrystalline composites

We begin by showing that for the polycrystalline composites the rotationally invariant subspace $\mathscr{A}$, given by (4.10) has a special structure as a representation of SO(d). Specifically, we are going to show here that the decomposition of $\mathscr{A}$ into the direct sum of irreps is unique, i.e., $\mathscr{A}$ contains no more than a single copy of each irrep of any particular positive weight, while it cannot contain any irreps of weight zero.

Lemma 6.4. *If $W \subset \mathscr{A}$ and $W' \subset \mathscr{A}$ are distinct irreps then they must have different positive weights. In other words, the decomposition of $\mathscr{A}$ into a direct sum of irreps contains at most one irrep of each weight.*

Proof. According to lemma 4.15 the subspace $\mathscr{A}$ is generated by a single vector $\tilde{\boldsymbol{\Gamma}}$ via (4.17). Let us first show that $\mathscr{A}$ cannot contain any nonzero isotropic tensors. Indeed, suppose $\mathsf{A}_0 \in \mathscr{A}$ is isotropic. Then, there exist rotations $\boldsymbol{R}_j \in \mathrm{SO}(3)$ and real numbers α_j, such that

$$\mathsf{A}_0 = \sum_{j=1}^{N} \alpha_j \boldsymbol{R}_j \cdot \tilde{\boldsymbol{\Gamma}}.$$

By assumption of isotropy of A_0 we also have

$$\mathsf{A}_0 = \sum_{j=1}^{N} \alpha_j (\boldsymbol{R}\boldsymbol{R}_j) \cdot \tilde{\boldsymbol{\Gamma}}, \quad \forall \boldsymbol{R} \in SO(3).$$

Integrating this identity over SO(3) with respect to the Haar measure (We used it in the proof of lemma 4.15.) we obtain

$$\mathsf{A}_0 = \sum_{j=1}^{N} \alpha_j \int_{SO(3)} (\boldsymbol{R}\boldsymbol{R}_j) \cdot \tilde{\boldsymbol{\Gamma}} d\mu(\boldsymbol{R}).$$

Using the right-invariance of the Haar measure we conclude that

$$\mathsf{A}_0 = \sum_{j=1}^{N} \alpha_j \int_{SO(3)} \boldsymbol{R} \cdot \tilde{\boldsymbol{\Gamma}} d\mu(\boldsymbol{R}) = 0,$$

according to (4.20).

To understand the representation theoretic structure of the smallest SO(d)-invariant subspace of Sym($\mathscr{T}$) containing $\tilde{\Gamma}$ we first observe that the decomposition of Sym($\mathscr{T}$) in the orthogonal sum of irreps can be written as

$$\mathrm{Sym}(\mathscr{T}) \cong \bigoplus_{k=0}^{M} \mathbb{R}^{m_k} \otimes W_k, \tag{6.34}$$

meaning that Sym($\mathscr{T}$) contains an orthogonal sum of exactly m_k irreps, each isomorphic to W_k. If $d = 2$, one can choose a special basis in each irrep W_k, so that a vector in W_k with coordinates (x, y), encoded by a complex number $z = x + iy$ is rotated by $\boldsymbol{R}_\theta$ to a vector in W_k, whose complex representation is $e^{ik\theta}z$. If such a basis is chosen in each and every irrep in Sym($\mathscr{T}$), then the decomposition (6.34) can be written as

$$\mathrm{Sym}(\mathscr{T}) \cong \bigoplus_{k=0}^{M} \mathbb{C}^{m_k},$$

so that

$$\boldsymbol{R}_\theta \cdot (\boldsymbol{u}_0, \boldsymbol{u}_1, \ldots, \boldsymbol{u}_M) = (\boldsymbol{u}_0, e^{i\theta}\boldsymbol{u}_1, \ldots, e^{iM\theta}\boldsymbol{u}_M),$$

where $\boldsymbol{u}_k \in \mathbb{C}^{m_k}$. If in this encoding $\tilde{\Gamma} = (0, \boldsymbol{u}_1, \ldots, \boldsymbol{u}_M)$. Then the subspace

$$\mathscr{A} = \bigoplus_{k=1}^{M} \mathbb{C}\boldsymbol{u}_k$$

is obviously the smallest SO(2)-invariant subspace in Sym($\mathscr{T}$) containing $\tilde{\Gamma}$. Moreover, it contains a single irrep $\mathbb{C}\boldsymbol{u}_k$ of weight k, provided $\boldsymbol{u}_k \neq 0$.

Now, consider the case $d = 3$. We saw in the proof of lemma 4.15 that $\tilde{\Gamma} = \Gamma_0(\boldsymbol{n}_0) - \bar{\Gamma}$, where $\boldsymbol{n}_0$ can be chosen arbitrarily. This representation clearly shows that $\tilde{\Gamma}$ is a zero weight vector (fixed point of a copy of SO(2) $\subset$ SO(3) of all rotations around $\boldsymbol{n}_0$). Let w_k be a zero weight vector in an irrep W_k. Then every zero weight vector in Sym($\mathscr{T}$) would have the form

$$\tilde{\Gamma} = \bigoplus_{k=1}^{M} \boldsymbol{u}_k \otimes w_k.$$

But then the smallest SO(3)-invariant subspace containing $\tilde{\Gamma}$ would be

$$\mathscr{A} = \bigoplus_{k=1}^{M} \mathbb{R}\boldsymbol{u}_k \otimes W_k.$$

In other words, $\mathscr{A}$ would contain exactly one copy of an irrep of weight k, unless $\boldsymbol{u}_k = 0$. □

We can now show that the representation-theoretic structure of the subspace $\mathscr{A}$ is independent of the choice of the isotropic reference medium L_0.

Corollary 6.5. *Suppose that* $\dim \mathscr{A}$ *is the same*[3] *for any positive definite, symmetric isotropic tensor* L_0, *used in the definition of* $\mathscr{A}$. *Then all subspaces* $\mathscr{A}$ *are isomorphic as representations of* $SO(d)$, $d = 2$ *or 3.*

Proof. The key observation is that the generating vector $\tilde{\Gamma}$ of $\mathscr{A}$ depends continuously on L_0. We will denote $\mathscr{A}[\mathsf{L}]$ the subspace $\mathscr{A}$ corresponding to the isotropic reference material L. Lemma 6.4 tells us that each $\mathscr{A}[\mathsf{L}]$ may contain at most one irrep of each weight k. The continuous dependence of $\tilde{\Gamma}$ on L implies that if an irrep of weight k is present in $\mathscr{A}[\mathsf{L}_0]$, then it will be present in all $\mathscr{A}[\mathsf{L}]$, if L lies in a sufficiently small neighborhood of L_0. Let G be the set of all isotropic material tensors L, such that $\mathscr{A}[\mathsf{L}]$ is isomorphic to $\mathscr{A}[\mathsf{L}_0]$. Then G must be an open subset of $\mathrm{Sym}^+_{SO(d)}(\mathscr{T})$, and for any $\mathsf{L} \in \partial G$ there will be one or more weights k, such that $\mathscr{A}[\mathsf{L}]$ will no longer contain an irrep of weight k. But then $\dim \mathscr{A}[\mathsf{L}] < \dim \mathscr{A}[\mathsf{L}_0]$ in contradiction with our assumption. Therefore, $G = \mathrm{Sym}^+_{SO(d)}(\mathscr{T})$, i.e., the set of weights present in the subrepresentation $\mathscr{A}[\mathsf{L}]$ will be the same for all isotropic positive definite reference tensors L. □

We continue to use two and three-dimensional elasticity to illustrate the process of computation of all exact relations and links listed in part III of the book. Once all rotationally invariant subspaces in $\mathrm{Sym}(\mathscr{T})$ have been identified, the subspace $\mathscr{A}$, given by (4.10) can be computed with little extra effort. Let $\mathsf{C}_0 \in \mathbb{E}^d_+$ be an arbitrary isotropic positive definite (this is what subscript '+' in $\mathbb{E}^d_+$ indicates) reference medium. Then C_0 has the form[4]

$$\mathsf{C}_0\varepsilon = \kappa_0 \,\mathrm{Tr}\,(\varepsilon)\boldsymbol{I} + 2\mu_0 \mathrm{dev}(\varepsilon), \qquad \kappa_0 > 0,\ \mu_0 > 0.$$

It is not difficult to compute [15] that

$$\boldsymbol{\Gamma}_0(\boldsymbol{n})\varepsilon = \frac{1}{\mu_0}((\varepsilon\boldsymbol{n}) \odot \boldsymbol{n} - \rho_0(\varepsilon\boldsymbol{n} \cdot \boldsymbol{n})\boldsymbol{n} \otimes \boldsymbol{n}), \tag{6.35}$$

where $\boldsymbol{a} \odot \boldsymbol{n} = \frac{1}{2}(\boldsymbol{a} \otimes \boldsymbol{n} + \boldsymbol{n} \otimes \boldsymbol{a})$, and

$$\rho_0 = \frac{\lambda_0 + \mu_0}{\lambda_0 + 2\mu_0} = \frac{d\kappa_0 + (d-2)\mu_0}{d\kappa_0 + 2(d-1)\mu_0}, \qquad d = 2, 3. \tag{6.36}$$

Formula (4.17) shows that in order to compute the subspace $\mathscr{A}$ we need to find the projections of $\boldsymbol{\Gamma}_0(\boldsymbol{n})$ onto K_2', K_2 and K_4. Using explicit formulas (6.23), (6.24) we obtain

$$\boldsymbol{A} = \frac{\mathrm{dev}(\boldsymbol{n} \otimes \boldsymbol{n})}{d(\lambda_0 + 2\mu_0)}, \qquad \boldsymbol{B} = \frac{7 - 4\rho_0}{14\mu_0}\mathrm{dev}(\boldsymbol{n} \otimes \boldsymbol{n}). \tag{6.37}$$

[3] In fact, we conjecture that this is always the case. The proof of that seems to be out of reach for now.

[4] An arbitrary isotropic tensor for any three-dimensional material property is a positive linear combination of orthogonal projections onto the SO(3)-irreps in $\mathscr{T}$.

We also verify that the projection of $\Gamma_0(\boldsymbol{n})$ onto K_4 is nonzero (for any $|\boldsymbol{n}| = 1$), if $\rho_0 > 0$, so that $\mathscr{A}_{2d} = K_2' \oplus K_4$, and

$$\mathscr{A}_{3d} = K_2^{\alpha_0} \oplus K_4, \qquad \alpha_0 = \frac{14(1 - \rho_0)}{3(7 - 4\rho_0)} = \frac{14\mu_0}{3(3\kappa_0 + 8\mu_0)},$$

where K_2^{α} is given by (6.28). We see that in three-dimensional elasticity the subspace $\mathscr{A}_{3d}$ depends on C_0, while in the two-dimensional elasticity the subspace $\mathscr{A}_{2d}$ does not. We therefore follow our strategy from section 6.1.1 to simplify the subspace $\mathscr{A}_{3d}$ by means of a covariance transformation (6.2). In order to preserve both the SO(3)-invariance of all subspaces and the symmetry of tensors, we can only use the transformation (6.2), where $\mathsf{C}_2 = \mathsf{C}_1^T$ is isotropic. In fact, we can achieve our goal with symmetric isotropic tensors $\mathsf{C}_1 = \mathsf{C}_2 = \bar{\mathsf{C}}$:

$$\bar{\mathsf{C}}\varepsilon = \bar{\kappa}\,\mathrm{Tr}\,(\varepsilon)\boldsymbol{I}_3 + 2\bar{\mu}\mathrm{dev}(\varepsilon), \qquad \varepsilon \in \mathrm{Sym}(\mathbb{R}^3).$$

In that case, it is easy to check that

$$\bar{\mathsf{C}}K_2'(\boldsymbol{A})\bar{\mathsf{C}} = 6\bar{\kappa}\bar{\mu}K_2'(\boldsymbol{A}), \qquad \bar{\mathsf{C}}K_2(\boldsymbol{B})\bar{\mathsf{C}} = 4\bar{\mu}^2K_2(\boldsymbol{B}).$$

Hence,

$$\bar{\mathsf{C}}K_2^{\beta}\bar{\mathsf{C}} = K_2^{\gamma}, \qquad \gamma = \frac{3\bar{\kappa}}{2\bar{\mu}}\beta.$$

This shows that in order to eliminate the dependence of $\mathscr{A}_0$ on C_0 we may choose $\bar{\kappa}/\bar{\mu} = \alpha_0$, in which case

$$\mathscr{A}_0 = \bar{\mathsf{C}}^{-1}\mathscr{A}_{3d}\bar{\mathsf{C}}^{-1} = K_4 \oplus K_2^{2/3}. \tag{6.38}$$

The choice of the 'slope' 2/3 in $\mathscr{A}_0$ is convenient because

$$\left(K_2(\boldsymbol{A}) + \frac{2}{3}K_2'(\boldsymbol{A})\right)\varepsilon = \varepsilon\boldsymbol{A} + \boldsymbol{A}\varepsilon, \qquad \varepsilon \in \mathrm{Sym}(\mathbb{R}^3),\ \boldsymbol{A} \in \mathrm{Sym}_0(\mathbb{R}^3). \tag{6.39}$$

6.1.4 Subspace multiplication

We are now ready to discuss the algebraic problem of finding all rotationally invariant Jordan $\mathscr{A}_0$-multialgebras. For 3D elastic composites the straightforward approach would be to test each of the 32 possibilities for Π listed in theorem 6.2, on whether it satisfies (4.9). Instead of verifying (4.9) directly, we observe that it can be rewritten in the language of subspace multiplication. For two subspaces U_1 and U_2 of $\mathrm{End}(\mathscr{T})$ we define

$$U_1U_2 = \mathrm{Span}\{\mathsf{K}_1\mathsf{K}_2\colon \mathsf{K}_1 \in U_1, \mathsf{K}_2 \in U_2\}. \tag{6.40}$$

With this notation, a subspace $\Pi \subset \mathrm{Sym}(\mathscr{T})$ is a Jordan $\mathscr{A}$-multialgebra if and only if

$$(\Pi\mathscr{A}\Pi)_{\mathrm{sym}} \subset \Pi, \tag{6.41}$$

where

$$U_{\text{sym}} = U \bigcap \operatorname{Sym}(\mathscr{T}).$$

We remark that for $\Pi \subset \operatorname{Sym}(\mathscr{T})$ and $\mathscr{A} \subset \operatorname{Sym}(\mathscr{T})$ the subspace $U = \Pi\mathscr{A}\Pi$ is *symmetric*, i.e., $\mathsf{K} \in U \Rightarrow \mathsf{K}^T \in U$. In that case we also have

$$U_{\text{sym}} = \{\mathsf{K} + \mathsf{K}^T : \mathsf{K} \in U\}.$$

It is easy to check that if U_1 and U_2 are rotationally invariant, then so is $U_1 U_2$. Moreover, the usual space addition and space multiplication (6.40) satisfy the distributive laws

$$(U_1 + U_2)U_3 = U_1 U_3 + U_2 U_3, \qquad U_3(U_1 + U_2) = U_3 U_1 + U_3 U_3.$$

Hence, in order to compute the product $U_1 U_2$ of any two rotationally invariant subspaces we only need to know the products of all of the irreps in $\operatorname{End}(\mathscr{T})$. Representation theory of SO(3) permits us to derive explicit formulas for such products, which in turn makes it possible to use computer algebra software to compute all the SO(3)-invariant Jordan $\mathscr{A}$-multialgebras.

We begin the analysis by noting the similarity between a product of subspaces in $\operatorname{End}(\mathscr{T})$, given by (6.40), and a *tensor product* of the same subspaces. Specifically, the map

$$\Phi: U_1 \otimes U_2 \to U_1 U_2, \qquad \Phi(\mathsf{K}_1 \otimes \mathsf{K}_2) = \mathsf{K}_1\mathsf{K}_2 \tag{6.42}$$

is SO(d) equivariant, i.e.,

$$\boldsymbol{R} \cdot \Phi(\mathsf{K}_1 \otimes \mathsf{K}_2) = \Phi((\boldsymbol{R} \cdot \mathsf{K}_1) \otimes (\boldsymbol{R} \cdot \mathsf{K}_2)).$$

That means that Φ will either 'transfer' an irrep in $U_1 \otimes U_2$ to an isomorphic irrep in $U_1 U_2$, or it will 'kill' it. The decomposition of a tensor product of irreps into a direct sum of irreps is one of the basic problems studied in representation theory. In the case of SO(3) this problem has arisen in quantum mechanics in order to understand angular momentum in a hydrogen atom [18]. The decomposition of a tensor product of irreps into a direct sum of irreps is given by the following formulas, called the Clebsch–Gordan formulas.

$$W_m \otimes W_0 \cong W_m, \quad m \geqslant 0 \tag{6.43}$$

$$W_m \otimes W_m \cong \begin{cases} W_0 \oplus W_0 \oplus W_{2m}, & d = 2, \\ W_0 \oplus W_1 \oplus \cdots \oplus W_{2m}, & d = 3, \end{cases} \quad m \geqslant 1 \tag{6.44}$$

$$W_m \otimes W_n \cong \begin{cases} W_{|m-n|} \oplus W_{m+n}, & d = 2, \\ W_{|m-n|} \oplus \cdots \oplus W_{m+n}, & d = 3, \end{cases} \quad m \neq n,\, m \geqslant 1,\, n \geqslant 1. \tag{6.45}$$

We can use these formulas to identify all of the irreps in $\operatorname{End}(\mathscr{T}) \cong \mathscr{T} \otimes \mathscr{T}$. According to (6.10), for elasticity

$$\mathscr{T} = \operatorname{Sym}(\mathbb{R}^d) = \mathbb{R}\boldsymbol{I}_d \oplus \operatorname{Sym}_0(\mathbb{R}^d) \cong W_0 \oplus W_2.$$

Then

$$\mathrm{End}(\mathscr{T}) \cong \mathrm{End}(W_0) \oplus \mathrm{Hom}(W_0, W_2) \oplus \mathrm{Hom}(W_2, W_0) \oplus \mathrm{End}(W_2),$$

where $\mathrm{Hom}(U, V)$ denotes the space of linear maps from U to V. It is well-known that $\mathrm{Hom}(U, V) \cong U \otimes V$, as a representation of SO(d). Therefore, according to (6.43), (6.44)

$$\mathrm{End}(\mathrm{Sym}(\mathbb{R}^d)) \cong \begin{cases} W_0 \oplus W_2 \oplus W_2 \oplus W_0 \oplus W_0 \oplus W_4, & d = 2, \\ W_0 \oplus W_2 \oplus W_2 \oplus W_0 \oplus W_1 \oplus W_2 \oplus W_3 \oplus W_4, & d = 3. \end{cases} \tag{6.46}$$

We also recall from section 6.1.2 that

$$\mathrm{Sym}(\mathrm{Sym}(\mathbb{R}^d)) = \begin{cases} K_0' \oplus K_2' \oplus K_0 \oplus K_4, & d = 2, \\ K_0' \oplus K_2' \oplus K_0 \oplus K_2 \oplus K_4, & d = 3. \end{cases} \tag{6.47}$$

This shows that for $d = 3$, the terms in (6.46) with odd weights comprise the space of all antisymmetric operators on $\mathrm{Sym}_0(\mathbb{R}^3)$. For $d = 2$, we see that all antisymmetric operators on $\mathrm{Sym}_0(\mathbb{R}^2)$ are rotationally invariant (isotropic).

Surprisingly the theory of invariant subspace multiplication is more tractable when $d = 3$. Hence, in the foregoing discussion we will be dealing only with SO(3)-invariant subspaces. In the most general case $\mathscr{T}$ can be written as a direct sum of k_0 irreps of weight 0, k_1 irreps of weight 1, and so on:

$$\mathscr{T} \cong \mathbb{R}^{k_0} \otimes W_0 \oplus \mathbb{R}^{k_1} \otimes W_1 \oplus \cdots \oplus \mathbb{R}^{k_n} \otimes W_n.$$

It is convenient to think of $\mathscr{T}$ as a subspace in $\mathbb{R}^p \otimes \mathscr{W}$, where

$$\mathscr{W} = W_0 \oplus \cdots \oplus W_n.$$

So that

$$\mathrm{End}(\mathscr{T}) \subset \mathrm{End}(\mathbb{R}^p \otimes \mathscr{W}) = \mathrm{End}(\mathbb{R}^p) \otimes \mathrm{End}(\mathscr{W}).$$

Our first task is to identify (coordinatize) all irreps in

$$\mathrm{End}(\mathbb{R}^p) \otimes \mathrm{End}(\mathscr{W}) = \mathrm{End}(\mathbb{R}^p) \otimes \left(\bigoplus_{j,k=0}^{n} \mathrm{Hom}(W_j, W_k) \right).$$

It is a special property of SO(3) that, according to (6.45),

$$\mathrm{Hom}(W_j, W_k) \cong W_j \otimes W_k \cong \bigoplus_{\alpha=|j-k|}^{j+k} W_\alpha,$$

so that *there is at most one irrep of each weight in* $\mathrm{Hom}(W_j, W_k)$. We still need to deal with the fact that there could be multiple irreps of the same weight in various $\mathrm{Hom}(W_j, W_k)$ blocks. For that purpose we select specific zero weight vectors $\mathsf{E}_{jk\alpha}$ (fixed points of a particular copy of SO(2) inside SO(3)) in each irrep $W_\alpha \subset \mathrm{Hom}(W_j, W_k)$. We may think of this choice as the choice of a basis in a vector space for solving linear equations numerically. Choosing specific zero weight vectors $\mathsf{E}_{jk\alpha}$ becomes essential only if there are multiple irreps of weight α in

$\mathrm{Sym}(\mathscr{T})$. What makes SO(3) theory especially simple is the fact that all SO(3) irreps are of real type. That means that, if W_α is an irrep in $\mathrm{End}(\mathbb{R}^p) \otimes \mathrm{End}(\mathscr{W})$, then its zero weight vector can be written as

$$\sum_{j,k=0}^{n} \boldsymbol{X}_{jk} \otimes \mathsf{E}_{jk\alpha},$$

where $\boldsymbol{X}_{jk} \in \mathrm{End}(\mathbb{R}^p)$. We will denote such a zero weight vector $\mathsf{E}(\mathbb{X})$, where $\mathbb{X}$ is the $(n+1) \times (n+1)$ block-matrix, whose (jk) th block is the $p \times p$ matrix $\boldsymbol{X}_{jk}$, where block numbering runs from 0 to n. The unique irrep of weight α containing $\mathsf{E}(\mathbb{X})$ will be accordingly denoted by $W_\alpha(\mathbb{X})$. We call the block matrix $\mathbb{X}$ the homogeneous coordinates of $W_\alpha \subset \mathrm{End}(\mathbb{R}^p) \otimes \mathrm{End}(\mathscr{W})$, since $\lambda\mathbb{X}$, $\lambda \neq 0$, defines the same irrep. The fundamental question is to identify the product $W_\beta(\mathbb{Y}) W_\alpha(\mathbb{X})$ as a direct sum of irreps in $\mathrm{End}(\mathscr{T})$ with specified homogeneous coordinates. The map (6.42) is SO(3)-equivariant, and hence, there are block matrices $\mathbb{Z}^\gamma$, such that

$$W_\beta(\mathbb{Y}) W_\alpha(\mathbb{X}) = \bigoplus_{\gamma=|\alpha-\beta|}^{\alpha+\beta} W_\gamma(\mathbb{Z}^\gamma),$$

with the convention that $\mathbb{Z}^\gamma = 0$, if there is no weight γ irrep in $W_\beta(\mathbb{Y}) W_\alpha(\mathbb{X})$. In chapter 17 we prove the explicit formula for block-matrices $\mathbb{Z}^\gamma$:

$$\boldsymbol{Z}_{jr}^{\gamma} = \frac{1}{\hat{K}_{jr\gamma}} \sum_{k=0}^{n} \boldsymbol{Y}_{kr} \boldsymbol{X}_{jk} \hat{R}_{jkr}^{\alpha\beta\gamma} \hat{K}_{kr\beta} \hat{K}_{jk\alpha}, \tag{6.48}$$

where the Racah coefficients $\hat{R}_{jkr}^{\alpha\beta\gamma}$, given by (17.10), are absolute numbers, while the numbers $\hat{K}_{jr\gamma}$, given by (17.9) involve the choice of specific zero weight vectors $w_j \in W_j \subset \mathscr{W}$, as well as a choice of an SO(3)-invariant inner product on $\mathscr{W}$, even though the resulting matrices $\mathbb{Z}^\gamma$ are independent of these choices. We set $\boldsymbol{Z}_{jr}^{\gamma} = 0$, whenever $\hat{K}_{jr\gamma} = 0$. Formula (6.48) permits us to compute a product of any two SO(3)-invariant subspaces. In particular, we can use it to build a 'Jordan $\mathscr{A}_0$-multiplication table'. The Maple implementation of subspace multiplication, as well as the 'Jordan $\mathscr{A}_0$-multiplication table' for piezoelectricity ($n = 2, p = 1$), which includes elasticity, as a particular case, was built by an undergraduate researcher Daniel Lapsley, working under author's supervision in the summer of 2016.

For elasticity, $\mathscr{A}_0$ is given by (6.38). Tables 6.1 and 6.2 show the 'Jordan products' $(U_1 \mathscr{A}_0 U_2)_{\mathrm{sym}}$ for almost all pairs of irreps U_1 and U_2 in $\mathrm{Sym}(\mathrm{Sym}(\mathbb{R}^d))$, $d = 2, 3$. Since, Jordan multiplication is commutative, we do not need to list the product $(U_2 \mathscr{A}_0 U_1)_{\mathrm{sym}}$ if the product $(U_1 \mathscr{A}_0 U_2)_{\mathrm{sym}}$ has already been listed. We also use the notation γV, for a real number γ and a subspace V to denote the subspace equal to V, if $\gamma \neq 0$ and to $\{0\}$, if $\gamma = 0$. We note that table 6.2 omits K_2^α (or rather contains the results for $\alpha = 0$ only). For the purposes of identifying all SO(3)-invariant Jordan $\mathscr{A}_0$-multialgebras it will be sufficient to have the following results:

$$K_4 \subset (K_2^\alpha \mathscr{A}_0 K_2^\beta)_{\mathrm{sym}}, \qquad K_4 \subset (K_2^\alpha \mathscr{A}_0 K_0^\beta)_{\mathrm{sym}}. \tag{6.49}$$

Table 6.1. 2D Jordan $\mathscr{A}_{2d}$-multiplication table.

	K_0'	K_2'	K_0^α	K_4
K_0'	0	K_0'	K_2'	K_2'
K_2'		$K_0' \oplus K_2'$	$K_2' \oplus K_0^{2\alpha} \oplus K_4$	$K_2' \oplus K_0 \oplus K_4$
K_0^β			$K_4 \oplus (\alpha + \beta)K_2'$	$K_0 + \beta K_2'$
K_4				K_4

Table 6.2. 3D Jordan $\mathscr{A}_0$-multiplication table.

	K_0'	K_2'	K_0^α	K_2	K_4
K_0'	0	K_0'	K_2'	K_2'	K_2'
K_2'		$K_0' \oplus K_2'$	$K_2' \oplus K_0^{5\alpha} \oplus K_2 \oplus K_4$	$K_2' \oplus K_0 \oplus K_2 \oplus K_4$	$K_2' \oplus K_0 \oplus K_2 \oplus K_4$
K_0^β			$K_2^{(\alpha+\beta)/2} \oplus K_4$	$K_0 \oplus K_2 \oplus K_4 \oplus \beta K_2'$	$K_0 \oplus K_2 \oplus K_4 \oplus \beta K_2'$
K_2				$K_0 \oplus K_2 \oplus K_4$	$K_0 \oplus K_2 \oplus K_4$
K_4					$K_0 \oplus K_2 \oplus K_4$

For the sake of completeness, using Daniel Lapsley's Maple program we give the missing formulas

$$(K_2^\alpha \mathscr{A}_0 K_0')_{\text{sym}} = \alpha K_0' \oplus K_2', \qquad (K_2^\alpha \mathscr{A}_0 K_2')_{\text{sym}} = \alpha K_0' \oplus K_2' \oplus K_0 \oplus K_2 \oplus K_4.$$

$$(K_2^\alpha \mathscr{A}_0 K_0^\beta)_{\text{sym}} = K_4 \oplus \begin{cases} K_0' \oplus K_2^{-49/24}, & \alpha = -\dfrac{7}{6}, \beta = \dfrac{49}{8}, \\ \beta K_0' \oplus K_2' \oplus K_2, & \alpha = -\dfrac{7}{6}, \beta \neq \dfrac{49}{8}, \\ K_0^{30\alpha\beta/(6\alpha+7)} \oplus K_2^{7\alpha/4}, & \alpha \neq -\dfrac{7}{6}, \beta = 3\alpha^2 - \dfrac{7\alpha}{4}, \\ K_0^{30\alpha\beta/(6\alpha+7)} \oplus K_2' \oplus K_2, \alpha \neq -\dfrac{7}{6}, & \beta \neq 3\alpha^2 - \dfrac{7\alpha}{4}. \end{cases}$$

$$(K_2^\alpha \mathscr{A}_0 K_2^\beta)_{\text{sym}} = (\alpha^2 + \beta^2)K_2' \oplus K_2 \oplus K_4 \oplus \begin{cases} K_0^{15\alpha\beta}, & \alpha + \beta = \dfrac{7}{12}, \\ K_0 \oplus \alpha\beta K_0', \alpha + \beta \neq \dfrac{7}{12}. \end{cases}$$

$$(K_2^\alpha \mathscr{A}_0 K_4)_{\text{sym}} = K_0 \oplus \alpha K_2' \oplus K_2 \oplus K_4.$$

6.1.5 Computing all multialgebras

Once $\mathscr{A}$-products of irreducibles have been computed, it is then a fairly simple matter, in most cases, to characterize all SO(d)-invariant Jordan $\mathscr{A}$-multialgebras. Let us again turn to the cases of 2D and 3D elasticity for illustration. The main observation is that in addition to 'coordinate irreps' denoted by K_j and K_j' we also have 'askew' irreps K_0^α and K_2^α. The hardest part of the analysis is to identify all 'askew' SO(d)-invariant Jordan $\mathscr{A}$-multialgebras, i.e., multialgebras containing K_0^α or K_2^α, but not $K_0 \oplus K_0'$ or $K_2 \oplus K_2'$. The following theorem resolves this issue.

Theorem 6.6. *There are no proper rotationally invariant Jordan $\mathscr{A}$-multialgebras containing either K_0^α or K_2^α for $\alpha \neq 0$.*

Proof. First, consider 2D elasticity. According to the multiplication table 6.1, if $\alpha \neq 0$ and $K_0^\alpha \subset \Pi$, then $K_4 \oplus K_2' \subset \Pi$. But then $K_0' \subset \Pi$ and $K_0 \subset \Pi$. This proves that $\Pi = \mathbb{E}^2$.

Next, consider 3D elasticity. In that case table 6.2 shows that if $K_0^\alpha \subset \Pi$, $\alpha \neq 0$, then $K_2^\alpha \subset \Pi$. The result would follow, when we prove that $K_2^\alpha \subset \Pi$ implies that $\Pi = \mathbb{E}^3$.

According to (6.49), if $K_2^\alpha \subset \Pi$, $\alpha \neq 0$, then $K_4 \subset \Pi$. This in turn implies that $K_0 \oplus K_2 \subset \Pi$. In particular, $K_2^\alpha + K_2 \subset \Pi$. But, obviously,

$$K_2^\alpha + K_2 = \begin{cases} K_2, & \alpha = 0, \\ K_2 \oplus K_2', & \alpha \neq 0. \end{cases}$$

Finally, $K_2' \subset \Pi$ implies that $K_0' \subset \Pi$ and so, $\Pi = \mathbb{E}^3$. □

It remains now to examine only a small finite number of cases. The results are given in the following complete lists of SO(d)-invariant Jordan $\mathscr{A}$-multialgebras.

1. $\Pi = \mathrm{Sym}(\mathrm{Sym}_0(\mathbb{R}^d))$. We observe that $\Pi = \mathrm{Ann}(\mathbb{R}\boldsymbol{I}_d) \cap \mathbb{E}^d$. Hence, this solution corresponds to a uniform field relation via (4.24). As such, it can be written as a set of all elastic tensors C with the property $\mathsf{C}\boldsymbol{I}_d = d\kappa_0 \boldsymbol{I}_d$ for some constant bulk modulus $\kappa_0 > 0$. This exact relation was discovered in [12].
2. $\Pi = K_0' \oplus K_2'$. We will see that this Jordan multialgebra corresponds to the exact relation (9.13) in section 9.2. This is the first exact relation obtained in [9] by the application of the general theory of exact relations. In fact, no other proof exists in the three-dimensional case.
3. $\Pi = K_0'$. We observe that $\Pi = \mathrm{Ann}(\mathrm{Sym}_0(\mathbb{R}^d))$. Hence, this solution corresponds to a uniform field relation via (4.24). This is the Hill exact relation (4.2).
4. If $d = 2$, then $\Pi = K_4$. This solution corresponds to the Lurie-Cherkaev exact relation (9.7).

Tables 6.1 and 6.2 allow fast calculations of Jordan 'squares' (5.28). We observe that $\Pi^2 = \Pi$, except for $\Pi = K_0'$, in which case $\Pi^2 = \{0\}$. According to the theory of section 5.3, $\Pi = K_0'$ must have an accompanying volume fraction relation. In the next section we will show that this volume fraction relation is the famous Hill formula (4.1).

In order to verify that the manifolds $\mathbb{M}$ corresponding to the computed Jordan multialgebras are exact relations, we need to show that they are reflexive in the sense of definition 4.10. The reflexivity of uniform field relations is automatic (see theorem 4.17). In order to establish reflexivity of the remaining multialgebras we construct explicit associative multialgebras, whose set of symmetric operators coincides with the given Jordan multialgebra. It is easy to verify that when $d=2$, $\Pi = K_4$ is also an associative $\mathscr{A}$-multialgebra. For $\Pi = K_0' \oplus K_2'$, the subspace

$$\Pi' = \{\mathsf{C} \in \mathrm{End}(\mathrm{Sym}(\mathbb{R}^d)) \colon \mathsf{C}\mathrm{Sym}_0(\mathbb{R}^d) \subset \mathbb{R}\boldsymbol{I}_d\}.$$

is an associative $\mathscr{A}$-multialgebra, whose symmetric elements comprise Π. This becomes apparent, if we recall that all elements of $\mathscr{A}$ map $\boldsymbol{I}_d$ into symmetric trace-free matrices, so that $\mathsf{C}_1\mathsf{A}\mathsf{C}_2 \in \Pi'$, whenever $\{\mathsf{C}_1, \mathsf{C}_2\} \subset \Pi'$ and $\mathsf{A} \in \mathscr{A}$.

6.2 Computing exact relations

The final task is to compute explicit representations of the manifold $\mathbb{M}$, given by (4.4), for each Jordan $\mathscr{A}$-multialgebra. Using formula (4.4) for this purpose is inconvenient, since it involves a unit vector $\boldsymbol{n}_0$, which, according to the general theory, the manifold $\mathbb{M}$ does not depend on. In this section we will show that we can compute the explicit representation of $\mathbb{M}$ via (4.4), where instead of the transformation $W_{\boldsymbol{n}}$, given by (3.21) and (3.37), we use a simplified transformation W_{M}, given by (3.21), provided M has the property

$$\mathsf{K}(\bar{\boldsymbol{\Gamma}} - \mathsf{M})\mathsf{K} \in \Pi, \quad \forall \mathsf{K} \in \Pi, \tag{6.50}$$

where the fixed tensor $\bar{\boldsymbol{\Gamma}}$ can be chosen arbitrarily among the points on the affine hull $\mathscr{G}$ of $\{\boldsymbol{\Gamma}_0(\boldsymbol{n}) \colon |\boldsymbol{n}| = 1\}$ (smallest affine hyperplane containing all matrices $\boldsymbol{\Gamma}(\boldsymbol{n})$, $|\boldsymbol{n}| = 1$). In particular, $\bar{\boldsymbol{\Gamma}}$, given by (4.18), or $\boldsymbol{\Gamma}_0(\boldsymbol{n})$, for any $|\boldsymbol{n}| = 1$ are all valid choices.

Definition 6.7. *A tensor* M *satisfying (6.50) is called an inversion key for the Jordan multialgebra* Π.

Choosing the inversion key is the same as recognizing that the Jordan $\mathscr{A}$-multialgebra Π is also a Jordan $\mathscr{A}'$-multialgebra, where

$$\mathscr{A}' = \mathrm{Span}\{\mathscr{A}, \bar{\boldsymbol{\Gamma}} - \mathsf{M}\}. \tag{6.51}$$

It is easy to see that if M satisfies (6.50) with one choice of $\bar{\boldsymbol{\Gamma}} \in \mathscr{G}$ it will also satisfy (6.50) with any other choice of $\bar{\boldsymbol{\Gamma}} \in \mathscr{G}$. Indeed, if $\bar{\boldsymbol{\Gamma}}' \in \mathscr{G}$, then

$$\mathsf{K}(\bar{\boldsymbol{\Gamma}}' - \mathsf{M})\mathsf{K} = \mathsf{K}(\bar{\boldsymbol{\Gamma}} - \mathsf{M})\mathsf{K} + \mathsf{K}(\bar{\boldsymbol{\Gamma}}' - \bar{\boldsymbol{\Gamma}})\mathsf{K} \in \Pi,$$

since $\bar{\boldsymbol{\Gamma}}' - \bar{\boldsymbol{\Gamma}} \in \mathscr{A}$.

To choose the simplest inversion key we first check if $\mathsf{M} = 0$ satisfies (6.50). If not, then other possibilities become apparent if we use the isotropic $\bar{\boldsymbol{\Gamma}}$, given by (4.18). In this case $\bar{\boldsymbol{\Gamma}}$ is a sum of scalar multiples of orthogonal projections onto $SO(d)$-invariant subspaces of $\mathscr{T}$. If $\mathsf{M} = 0$ does not satisfy (6.50), we can then try M to be the sum of one or more of the terms in the expression for $\bar{\boldsymbol{\Gamma}}$. In applications we were always able to find a much simpler choice for M than $\bar{\boldsymbol{\Gamma}}$.

Let us prove now that we can compute an exact relation submanifold $\mathbb{M}$ using the simplified transformation W_{M}. It is therefore necessary that the operator $\mathsf{I} + (\mathsf{L} - \mathsf{L}_0)\mathsf{M}$ be invertible. However, our criterion (6.50) does not guarantee that M will have the universal invertibility property (see definition 3.14). For this reason we denote

$$\mathbb{P}_{\mathsf{M}} = \{\mathsf{L} \in \mathrm{End}^{+}(\mathscr{T}): \mathsf{I} + (\mathsf{L} - \mathsf{L}_0)\mathsf{M} \text{ is invertible}\}.$$

We also define

$$\mathbb{M}' = \mathbb{P}_{\mathsf{M}} \bigcap \mathbb{M}. \tag{6.52}$$

Lemma 6.8. *Suppose that* $\mathsf{L}_0 \in \mathbb{M}$, *where* $\mathbb{M}$ *is given by (4.4) in terms of a Jordan* $\mathscr{A}$*-multialgebra* Π. *Then the set* $\mathbb{M}'$ *is an open and dense subset of* $\mathbb{M}$.

Proof. The idea is to obtain an explicit parametrization of $\mathbb{M}$ by $m = \dim \mathbb{M}$ real parameters. Recall that $\mathbb{K} = W_{\boldsymbol{n}_0}(\mathbb{M})$ is an open subset of a Jordan $\mathscr{A}$-multialgebra Π, containing $0 \in \mathrm{End}(\mathscr{T})$. Let $\{\mathsf{K}_1,\ldots, \mathsf{K}_m\}$ be a basis of Π. For any $\boldsymbol{u} \in \mathbb{R}^m$ we denote $\mathsf{K}(\boldsymbol{u}) = u_1\mathsf{K}_1 + \ldots + u_m\mathsf{K}_m$. Then, $\mathscr{O} = \{\boldsymbol{u} \in \mathbb{R}^m: \mathsf{K}(\boldsymbol{u}) \in \mathbb{K}\}$ is an open subset of $\mathbb{R}^m$, containing $0 \in \mathbb{R}^m$. The inverse of $W_{\boldsymbol{n}_0}$, can be computed by (3.42), where $\mathsf{M} = \boldsymbol{\Gamma}_0(\boldsymbol{n})$ has the universal invertibility property, according to corollary 3.16. This formula defines a diffeomorphism $\Phi: \mathscr{O} \to \mathbb{M}$

$$\Phi(\boldsymbol{u}) = \mathsf{L}_0 + (\mathsf{I} - \mathsf{K}(\boldsymbol{u})\boldsymbol{\Gamma}_0(\boldsymbol{n}_0))^{-1}\mathsf{K}(\boldsymbol{u}), \quad \boldsymbol{u} \in \mathscr{O}.$$

Thus, in order to prove the lemma we need to prove that the operator

$$\mathsf{P}(\boldsymbol{u}) = \mathsf{I} + (\Phi(\boldsymbol{u}) - \mathsf{L}_0)\mathsf{M}$$

is invertible on an open and dense subset of $\mathscr{O}$. Substituting the expression for $\Phi(\boldsymbol{u})$ into the formula for $\mathsf{P}(\boldsymbol{u})$ we obtain

$$\mathsf{P}(\boldsymbol{u}) = (\mathsf{I} - \mathsf{K}(\boldsymbol{u})\boldsymbol{\Gamma}_0(\boldsymbol{n}_0))^{-1}(\mathsf{I} - \mathsf{K}(\boldsymbol{u})(\boldsymbol{\Gamma}_0(\boldsymbol{n}_0) - \mathsf{M})).$$

It remains to observe that the polynomial

$$p(\boldsymbol{u}) = \det(\mathsf{I} - \mathsf{K}(\boldsymbol{u})(\boldsymbol{\Gamma}_0(\boldsymbol{n}_0) - \mathsf{M}))$$

is nonzero on a sufficiently small neighborhood of $0 \in \mathbb{R}^m$, since $\mathsf{K}(0) = 0$. Hence, $p(\boldsymbol{u})$ is a nonzero polynomial on $\mathbb{R}^m$. We conclude that $p(\boldsymbol{u}) \neq 0$ on an open and dense subset $\mathscr{O}' \subset \mathscr{O}$. Thus, $\mathsf{P}(\boldsymbol{u})$ must be invertible for all $\boldsymbol{u} \in \mathscr{O}'$. The lemma is proved now. □

Lemma 6.8 implies that the subset $\mathbb{M}' \subset \mathbb{M}$, given by (6.52), describes the submanifold $\mathbb{M}$ completely, since $\overline{\mathbb{M}'} = \mathbb{M}$, where the closure is taken in the relative topology of $\mathrm{End}^+(\mathscr{T})$. We can now prove our main inversion theorem.

Theorem 6.9. *Suppose* M *is an inversion key for the Jordan multialgebra* Π *in the sense of definition 6.7. Then*

$$\mathbb{M}' = \{\mathsf{L} \in \mathbb{P}_{\mathsf{M}}: W_{\mathsf{M}}(\mathsf{L}) \in \Pi\}, \tag{6.53}$$

where the submanifold $\mathbb{M}'$ *is given by (6.52) and* W_{M}*, by (3.21).*

Proof. Let $\tilde{\mathbb{M}}$ denote the right-hand side of (6.53). To establish the theorem we will show that $\mathbb{M}' \subset \tilde{\mathbb{M}}$ and $\tilde{\mathbb{M}} \subset \mathbb{M}'$.

To prove the first inclusion we assume that $\mathsf{L} \in \mathbb{M}'$ and prove that $\mathsf{L} \in \tilde{\mathbb{M}}$. We have

$$\mathsf{I} + (\mathsf{L} - \mathsf{L}_0)\mathsf{M} = \mathsf{I} + (\mathsf{L} - \mathsf{L}_0)\boldsymbol{\Gamma}_0(\boldsymbol{n}_0) - (\mathsf{L} - \mathsf{L}_0)\mathsf{A}', \quad \mathsf{A}' = \boldsymbol{\Gamma}_0(\boldsymbol{n}_0) - \mathsf{M}.$$

Multiplying by $(\mathsf{I} + (\mathsf{L} - \mathsf{L}_0)\boldsymbol{\Gamma}_0(\boldsymbol{n}_0))^{-1}$ on the left, and using definition (3.37) of $W_{\boldsymbol{n}_0}(\mathsf{L})$, we obtain

$$(\mathsf{I} + (\mathsf{L} - \mathsf{L}_0)\boldsymbol{\Gamma}_0(\boldsymbol{n}_0))^{-1}(\mathsf{I} + (\mathsf{L} - \mathsf{L}_0)\mathsf{M}) = \mathsf{I} - W_{\boldsymbol{n}_0}(\mathsf{L})\mathsf{A}'. \tag{6.54}$$

The left-hand side of (6.54) is invertible, since $\mathsf{L} \in \mathbb{P}_{\mathsf{M}}$. If we take inverse of both sides in (6.54) and then multiply by $W_{\boldsymbol{n}_0}(\mathsf{L})$ on the right we obtain

$$W_{\mathsf{M}}(\mathsf{L}) = (\mathsf{I} - W_{\boldsymbol{n}_0}(\mathsf{L})\mathsf{A}')^{-1}W_{\boldsymbol{n}_0}(\mathsf{L}) \in \Pi,$$

by lemma 4.5, which applies, since $W_{\boldsymbol{n}_0}(\mathsf{L}) \in \Pi$ and $\mathsf{A}' \in \mathscr{A}'$, where $\mathscr{A}'$ is defined in (6.51). The fact that M is an inversion key implies that Π is a Jordan $\mathscr{A}'$-multialgebra. This proves the inclusion $\mathbb{M}' \subset \tilde{\mathbb{M}}$. The other inclusion is proved in a symmetric way by computing that

$$W_{\boldsymbol{n}_0}(\mathsf{L}) = (\mathsf{I} + W_{\mathsf{M}}(\mathsf{L})\mathsf{A}')^{-1}W_{\mathsf{M}}(\mathsf{L}) \in \Pi.$$

□

The simplification afforded by a choice of the inversion key can also be combined with the covariance transformation that was used to simplify the subspace $\mathscr{A}$.

Theorem 6.10. *Suppose* $\mathscr{A}_0$ *is given by (6.3) and* Π_0 *is a Jordan* $\mathscr{A}_0$*-multialgebra. Suppose that the operator* M_0 *satisfies*

$$\mathsf{K}(\bar{\boldsymbol{\Gamma}}^0 - \mathsf{M}_0)\mathsf{K} \in \Pi_0 \quad \forall \mathsf{K} \in \Pi_0, \tag{6.55}$$

where $\bar{\boldsymbol{\Gamma}}^0 = \mathsf{C}_2^{-1}\bar{\boldsymbol{\Gamma}}\mathsf{C}_1^{-1}$*. Then the exact relation corresponding to* Π_0 *can be computed by the formula*

$$\mathbb{M}' = \{\mathsf{L} \in \mathrm{End}^+(\mathscr{T}): \mathsf{L} = \mathsf{L}_0 + \mathsf{C}_1^{-1}(\mathsf{I} - \mathsf{K}\mathsf{M}_0)^{-1}\mathsf{K}\mathsf{C}_2^{-1},\ \mathsf{K} \in \Pi_0\}. \tag{6.56}$$

Example 6.11. Let us compute the exact relation corresponding to the SO(d)-invariant Jordan $\mathscr{A}_0$-multialgebra $\Pi_0 = K_0' \oplus K_2'$, from section 6.1.5. Recall that for elasticity $\Gamma_0(\boldsymbol{n})$ is given by (6.35). Then we easily compute, using (6.23) and (6.24), that $\bar{\Gamma}$ is an isotropic elastic tensor with bulk and shear moduli, given respectively by

$$\kappa_\Gamma = \begin{cases} \dfrac{1}{4(\kappa_0 + \mu_0)}, & d = 2, \\ \dfrac{1}{3(3\kappa_0 + 4\mu_0)}, & d = 3, \end{cases} \qquad \mu_\Gamma = \begin{cases} \dfrac{\kappa_0 + 2\mu_0}{8\mu_0(\kappa_0 + \mu_0)}, & d = 2, \\ \dfrac{3(\kappa_0 + 2\mu_0)}{10\mu_0(3\kappa_0 + 4\mu_0)}, & d = 3. \end{cases} \tag{6.57}$$

Recall that we have simplified the subspace $\mathscr{A}$ by (6.38). That means that in theorem 6.10 we need to set $\mathsf{C}_1 = \mathsf{C}_2 = \bar{\mathsf{C}}$, where the bulk and shear moduli $\bar{\kappa}$ and $\bar{\mu}$, respectively, of $\bar{\mathsf{C}}$ can be chosen arbitrarily for 2D elasticity, and are constrained by

$$\frac{\bar{\kappa}}{\bar{\mu}} = \alpha_0 = \frac{14\mu_0}{3(3\kappa_0 + 8\mu_0)}, \tag{6.58}$$

for 3D elasticity. This gives us some freedom to simplify $\bar{\Gamma}^0$ in (6.55). In two space dimensions we can choose

$$\bar{\kappa} = \frac{1}{2}\sqrt{\kappa_\Gamma}, \qquad \bar{\mu} = \frac{1}{2}\sqrt{\mu_\Gamma},$$

so that both the bulk and the shear moduli of $\bar{\Gamma}^0 = \bar{\mathsf{C}}^{-1}\bar{\Gamma}\bar{\mathsf{C}}^{-1}$ are 1. For three-dimensional elasticity we may only choose

$$\bar{\kappa} = \frac{1}{3}\sqrt{\kappa_\Gamma},$$

making the bulk modulus of $\bar{\Gamma}^0$ equal to 1. The shear modulus $\bar{\mu}$ is then uniquely determined by (6.58).

Let us show that for $\Pi = K_0' \oplus K_2'$ we can choose $\mathsf{M}_0 = \boldsymbol{I}_d \otimes \boldsymbol{I}_d$, $d = 2, 3$, i.e.,

$$\mathsf{M}_0\varepsilon = (\mathrm{Tr}\,\varepsilon)\boldsymbol{I}_d, \qquad \varepsilon \in \mathrm{Sym}(\mathbb{R}^d).$$

Indeed, this choice implies that $(\bar{\Gamma}^0 - \mathsf{M}_0)\boldsymbol{I}_d = 0$, since in both two and three-dimensional elasticity the bulk modulus of $\bar{\Gamma}^0$ is 1. Observing that every $\mathsf{K} \in \Pi$ maps $\mathrm{Sym}_0(\mathbb{R}^d)$ into multiples of the identity, it follows that $(\bar{\Gamma}^0 - \mathsf{M}_0)\mathsf{K}$, and consequently $\mathsf{K}(\bar{\Gamma}^0 - \mathsf{M}_0)\mathsf{K}$ annihilate $\mathrm{Sym}_0(\mathbb{R}^d)$. Thus,

$$\mathsf{K}(\bar{\Gamma}^0 - \mathsf{M}_0)\mathsf{K} \subset K_0' \subset \Pi.$$

We can now use the inversion key M_0 to compute the corresponding exact relation. We begin by parametrizing Π using maps (6.12)–(6.17). So, if $\mathsf{K} \in \Pi$, then $\mathsf{K} = K_0'(\theta) + K_2'(\boldsymbol{A})$, $\theta \in \mathbb{R}$, $\boldsymbol{A} \in \mathrm{Sym}_0(\mathbb{R}^d)$. Observe that

$$\mathsf{K}\boldsymbol{I}_d = d\theta\boldsymbol{I}_d + d\boldsymbol{A} = \boldsymbol{B} \in \mathrm{Sym}(\mathbb{R}^d).$$

We note that $\boldsymbol{B}$ also uniquely determines parameters θ and $\boldsymbol{A}$. In fact,

$$\mathsf{K} = \frac{1}{d}\left(\boldsymbol{I}_d \otimes \boldsymbol{B} + \boldsymbol{B} \otimes \boldsymbol{I}_d - \frac{1}{d}(\mathrm{Tr}\, \boldsymbol{B})\boldsymbol{I}_d \otimes \boldsymbol{I}_d\right). \tag{6.59}$$

We compute

$$(\mathsf{I} - \mathsf{K}\boldsymbol{I}_d \otimes \boldsymbol{I}_d)^{-1} = (\mathsf{I} - \boldsymbol{B} \otimes \boldsymbol{I}_d)^{-1} = \mathsf{I} + \frac{\boldsymbol{B} \otimes \boldsymbol{I}_d}{1 - \mathrm{Tr}\, \boldsymbol{B}}.$$

Hence,

$$(\mathsf{I} - \mathsf{K}\boldsymbol{I}_d \otimes \boldsymbol{I}_d)^{-1}\mathsf{K} = \mathsf{K} + \frac{\boldsymbol{B} \otimes \boldsymbol{B}}{1 - \mathrm{Tr}\, \boldsymbol{B}}.$$

Using (6.59) we obtain

$$(\mathsf{I} - \mathsf{K}\mathsf{M}_0)^{-1}\mathsf{K} = \frac{1}{d}\left(\boldsymbol{I}_d \otimes \boldsymbol{B} + \boldsymbol{B} \otimes \boldsymbol{I}_d - \frac{1}{d}(\mathrm{Tr}\, \boldsymbol{B})\boldsymbol{I}_d \otimes \boldsymbol{I}_d\right) + \frac{\boldsymbol{B} \otimes \boldsymbol{B}}{1 - \mathrm{Tr}\, \boldsymbol{B}}.$$

Substituting this expression into (6.56) gives a (rather ugly) parametrization of $\mathbb{M}$. It is a remarkable property of exact relations that they invariably possess beautiful representations. Indeed, the ugly expression above can also be written as

$$(\mathsf{I} - \mathsf{K}\mathsf{M}_0)^{-1}\mathsf{K} = -\frac{\boldsymbol{I}_d \otimes \boldsymbol{I}_d}{d^2} + \boldsymbol{C} \otimes \boldsymbol{C}, \qquad \boldsymbol{C} = \frac{\boldsymbol{B} + (1 - \mathrm{Tr}\, \boldsymbol{B})\boldsymbol{I}_d/d}{\sqrt{1 - \mathrm{Tr}\, \boldsymbol{B}}} \tag{6.60}$$

Thus,

$$\mathsf{C} = \mathsf{C}_0 - \frac{1}{d^4\bar{\kappa}^2}\boldsymbol{I}_d \otimes \boldsymbol{I}_d + \bar{\mathsf{C}}^{-1}\boldsymbol{C} \otimes \bar{\mathsf{C}}^{-1}\boldsymbol{C}. \tag{6.61}$$

We see now that $\mathbb{M}$ is parametrized by $\boldsymbol{D} = \bar{\mathsf{C}}^{-1}\boldsymbol{C} \in \mathrm{Sym}(\mathbb{R}^d)$ and has the form

$$\mathsf{C} = \tilde{\mathsf{C}}_0 + \boldsymbol{D} \otimes \boldsymbol{D}, \tag{6.62}$$

where

$$\tilde{\mathsf{C}}_0 = \mathsf{C}_0 - \frac{1}{d^2\kappa_\Gamma}\boldsymbol{I}_d \otimes \boldsymbol{I}_d$$

is an isotropic elastic tensor with shear modulus μ_0 and bulk modulus

$$\tilde{\kappa} = \kappa_0 - \frac{1}{d^2\kappa_\Gamma} = \begin{cases} -\mu_0, & d = 2, \\ -4\mu_0/3, & d = 3. \end{cases}$$

Thus, $\tilde{\mathsf{C}}_0 = 2\mu_0\mathsf{T}$, where T is the isotropic 'null-Lagrangian', given by the formula

$$\mathsf{T}\varepsilon = \varepsilon - (\mathrm{Tr}\, \varepsilon)\boldsymbol{I}_d, \qquad \varepsilon \in \mathrm{Sym}(\mathbb{R}^d). \tag{6.63}$$

This shows that all elastic tensors of materials in the exact relation corresponding to $\Pi = K_0' \oplus K_2'$ have a 'rank-one plus a null-Lagrangian' structure, given by

$$\mathsf{C} = 2\mu_0\mathsf{T} + \boldsymbol{C} \otimes \boldsymbol{C}\colon \boldsymbol{C} \in \mathrm{Sym}(\mathbb{R}^3). \tag{6.64}$$

6.3 Computing volume fraction relations

Recall that additional formulas involving volume averages accompany an exact relation, if the Jordan square (5.28) is strictly smaller than the original Jordan multialgebra. In order to compute such formulas explicitly we derive the rule for finding the inversion key M, so that the transformation W_{M} can be used in (5.33) instead of $W_{\boldsymbol{n}}$.

Theorem 6.12. *Let* Π *be a Jordan* $\mathscr{A}$*-multialgebra, and let* $\mathscr{P}_{(\Pi^2)^c}$ *denote the orthogonal projection onto the orthogonal complement of* Π^2 *in* Π*. Then*

$$\mathscr{P}_{(\Pi^2)^c} W_{\mathsf{M}}(\mathsf{L}_*) = \mathscr{P}_{(\Pi^2)^c} \langle W_{\mathsf{M}}(\mathsf{L}(z)) \rangle, \tag{6.65}$$

where W_{M} *is given by (3.21) and* M *satisfies*

$$\mathsf{K}(\bar{\Gamma} - \mathsf{M})\mathsf{K} \in \Pi^2 \quad \forall \mathsf{K} \in \Pi. \tag{6.66}$$

Whether or not M *satisfies (6.66) does not depend on the choice of* $\bar{\Gamma} \in \mathscr{G}$*—the affine hull of* $\{\Gamma_0(\boldsymbol{n}) \colon |\boldsymbol{n}| = 1\}$.

Proof. In our theory the formulas involving volume averages are particular instances of links, as was discussed in section 5.3. The links, according to section 5.1, can be interpreted as exact relations. Thus, formulas (6.55) and (6.56) must apply to volume fraction relations. Therefore, in order to prove the theorem we need to show that these formulas reduce to (6.66) and (6.65), respectively, for Jordan $\hat{\mathscr{A}}$-multialgebra $\hat{\Pi}$, defined in (5.29), where $\hat{\mathscr{A}}$ is given by (5.23). Now, let $\hat{\mathsf{M}} = [\mathsf{M}, 0]$ be such that (6.55) is satisfied for $\hat{\Pi}$. That means that

$$[\mathsf{K}(\bar{\Gamma} - \mathsf{M})\mathsf{K}, 0] \in \hat{\Pi}, \quad \forall \mathsf{K} \in \Pi.$$

Hence, according to (5.29) we conclude that (6.66) must hold. Then, by theorem 6.9, $\hat{\mathbb{M}}$, given by (5.32) can also be described as the closure of

$$\hat{\mathbb{M}}' = \{[\mathsf{L}, \tilde{\mathsf{L}}] \in \mathbb{M}' \times \mathrm{End}^+(\mathbb{R}^N) \colon \tilde{\mathsf{L}}_0 - \tilde{\mathsf{L}} = \mathscr{P}_{(\Pi^2)^c} W_{\mathsf{M}}(\mathsf{L})\}.$$

Formula (6.65) follows from $[\mathsf{L}, \tilde{\mathsf{L}}]_* = [\mathsf{L}_*, \langle \tilde{\mathsf{L}} \rangle]$ and the fact that the closure of $\hat{\mathbb{M}}'$ is an exact relation. □

According to (6.66) the volume fraction inversion key can also serve as the inversion key for the corresponding multialgebra Π. The converse is not always true. It is important to emphasize that formulas (6.65) and (6.66) use the original (unsimplified) versions of $\Gamma_0(\boldsymbol{n})$ and $\mathscr{A}$. If we apply the covariance transformation (6.2) and use $\mathscr{A}_0$, given by (6.3), instead of $\mathscr{A}$, then formula (6.65) becomes

$$\mathscr{P}_{(\Pi_0^2)^c} \mathsf{K}(\mathsf{L}_*) = \mathscr{P}_{(\Pi_0^2)^c} \langle \mathsf{K}(\mathsf{L}(z)) \rangle. \tag{6.67}$$

where

$$\mathsf{K}(\mathsf{L}) = \mathsf{C}_1 W_{\mathsf{C}_2 \mathsf{M}_0 \mathsf{C}_1}(\mathsf{L}) \mathsf{C}_2$$

is the inverse of the transformation in (6.56), and Π_0 is a Jordan $\mathscr{A}_0$-multialgebra. However, the inversion key M_0 is required to satisfy

$$\mathsf{K}(\bar{\boldsymbol{\Gamma}}^0 - \mathsf{M}_0)\mathsf{K} \in \Pi_0^2 \quad \forall \mathsf{K} \in \Pi_0. \tag{6.68}$$

We note that in cases when the simplest inversion key M_0 for Π_0 also satisfies (6.68), the calculations in (6.56) usually also yield the explicit formulas for $\mathsf{K} = \mathsf{K}(\mathsf{L})$. In this case the seemingly more complicated formula (6.67) is more convenient than the more compact (6.65).

Let us now apply this theory for computing the volume fraction relation associated to the Jordan $\mathscr{A}_0$-multialgebra $\Pi = K_0'$, obtained in section 6.1.5. It has the property that $\Pi^2 = \{0\}$. The analysis in the example in section 6.2 shows that we should choose $\mathsf{M}_0 = \boldsymbol{I}_d \otimes \boldsymbol{I}_d$, since $(\bar{\boldsymbol{\Gamma}}^0 - \mathsf{M}_0)\boldsymbol{I}_d = 0$. Indeed, all operators in K_0' map $\mathbb{E}^d$ to multiples of the identity, and hence, $(\bar{\boldsymbol{\Gamma}}^0 - \mathsf{M}_0)\mathsf{K} = 0$. Consequently, $\mathsf{K}(\bar{\boldsymbol{\Gamma}}^0 - \mathsf{M}_0)\mathsf{K} = 0$, as required by (6.68). Thus, all calculations from section 6.2 apply with $\boldsymbol{B} = d\theta\boldsymbol{I}_d$. This gives us

$$(\mathsf{I} - \mathsf{K}\mathsf{M}_0)^{-1}\mathsf{K} = \mathsf{K} + \frac{\boldsymbol{B} \otimes \boldsymbol{B}}{1 - \operatorname{Tr} \boldsymbol{B}} = \frac{\theta\boldsymbol{I}_d \otimes \boldsymbol{I}_d}{1 - d^2\theta}$$

Hence, C is isotropic with the shear modulus μ_0 and the bulk modulus

$$\kappa = \frac{1}{d}\mathsf{C}\boldsymbol{I}_d = \kappa_0 + \frac{\theta}{d\bar{\kappa}^2(1 - d^2\theta)} = \kappa_0 + \frac{\theta}{\kappa_\Gamma(1 - d^2\theta)}.$$

Solving for θ we obtain

$$\theta = \theta(\kappa) = \frac{\kappa_\Gamma(\kappa - \kappa_0)}{1 + d^2\kappa_\Gamma(\kappa - \kappa_0)}.$$

According to (6.67) the volume fraction relation reads $\theta(\kappa_*) = \langle\theta(\kappa)\rangle$. Hence, after some simple algebra, we conclude that

$$\frac{1}{\kappa_\Gamma^{-1} - d^2\kappa_0 + d^2\kappa_*} = \langle\frac{1}{\kappa_\Gamma^{-1} - d^2\kappa_0 + d^2\kappa}\rangle.$$

This gives

$$\begin{cases} \dfrac{1}{\mu_0 + \kappa_*} = \langle\dfrac{1}{\mu_0 + \kappa}\rangle, & d = 2, \\ \dfrac{1}{4\mu_0 + 3\kappa_*} = \langle\dfrac{1}{4\mu_0 + 3\kappa}\rangle, & d = 3. \end{cases} \tag{6.69}$$

These are the well-known Hill's formulas for the effective bulk modulus of two and three-dimensional composites made of isotropic materials with a constant shear modulus.

6.4 Finding Jordan $\hat{\mathscr{A}}$-multialgebras

In section 5.1 we showed that links can be regarded as exact relations in general. In this section we will focus on the most important subcase of polycrystalline links when $\mathscr{T}_1 = \mathscr{T}_2 = \mathscr{T}$. Here we will be able to say a lot more about the structure of the subspace $\hat{\mathscr{A}}$ and rotationally invariant Jordan $\hat{\mathscr{A}}$-multialgebras.

Lemma 6.13. *Suppose* $\mathscr{T}_1 = \mathscr{T}_2 = \mathscr{T}$ *and* $\hat{\mathsf{L}}_0 = [\mathsf{L}_0^{(1)}, \mathsf{L}_0^{(1)}]$ *is isotropic. Assume that* $\mathscr{A}_1$ *and* $\mathscr{A}_2$ *are isomorphic as representations of SO(d), d = 2 or 3. Then the subspace* $\hat{\mathscr{A}}$ *satisfies (5.9).*

Proof. According to (5.8), $\hat{\mathscr{A}}$ is isomorphic to $\mathfrak{A}_1 \oplus (\mathscr{A}_1/\mathfrak{A}_1) \oplus \mathfrak{A}_2$ as a representation of SO(d). It is also isomorphic to $\mathfrak{A}_1 \oplus (\mathscr{A}_2/\mathfrak{A}_2) \oplus \mathfrak{A}_2$. Now, the assumption that $\mathscr{A}_1$ and $\mathscr{A}_2$ are isomorphic as representations of SO(d) combined with

$$\mathfrak{A}_1 \oplus (\mathscr{A}_1/\mathfrak{A}_1) \oplus \mathfrak{A}_2 \cong \hat{\mathscr{A}} \cong \mathfrak{A}_1 \oplus (\mathscr{A}_2/\mathfrak{A}_2) \oplus \mathfrak{A}_2 \tag{6.70}$$

implies that $\mathfrak{A}_1 \cong \mathfrak{A}_2$ as representations of SO(d). But this contradicts the fact that $\hat{\mathscr{A}}$ contains no more than one copy of each irrep, unless $\mathfrak{A}_1 = \mathfrak{A}_2 = \{0\}$, since, according to (6.70) there is a subrepresentation of $\hat{\mathscr{A}}$, containing $\mathfrak{A}_1 \oplus \mathfrak{A}_2$. □

Corollary 6.5 says that the assumption in lemma 6.13 that $\mathscr{A}_1$ and $\mathscr{A}_2$ are isomorphic as representations of SO(d) will be satisfied whenever subspaces $\mathscr{A}$ have the same dimension for any isotropic reference media L. We conjecture that this should be the case in general, as it holds in all of our examples. Lemma 6.13 already simplifies the general structure of links described in section 5.2, as it implies that

$$\hat{\mathscr{A}} = \{[\mathsf{A}, \Xi(\mathsf{A})] \in \mathscr{A}_1 \oplus \mathscr{A}_2 \colon \mathsf{A} \in \mathscr{A}_1\}, \tag{6.71}$$

where $\Xi\colon \mathscr{A}_1 \mapsto \mathscr{A}_2$ is a linear SO(d)-equivariant isomorphism, i.e.

$$\boldsymbol{R} \cdot \Xi(\mathsf{A}) = \Xi(\boldsymbol{R} \cdot \mathsf{A}), \qquad \boldsymbol{R} \in SO(d).$$

Then, the key factor determining the number and structure of links is the existence of a linear SO(d)-equivariant isomorphism $\Phi_0\colon \mathrm{Sym}(\mathscr{T}) \to \mathrm{Sym}(\mathscr{T})$ satisfying

$$\Phi_0(\mathsf{KAK}) = \Phi_0(\mathsf{K})\Xi(\mathsf{A})\Phi_0(\mathsf{K}), \qquad \forall\, \mathsf{K} \in \mathrm{End}(\mathscr{T}), \forall\, \mathsf{A} \in \mathscr{A}_1. \tag{6.72}$$

Then, the transformation

$$\mathfrak{F}_0\colon \mathrm{Sym}(\mathscr{T}) \oplus \mathrm{Sym}(\mathscr{T}) \to \mathrm{Sym}(\mathscr{T}) \oplus \mathrm{Sym}(\mathscr{T}), \qquad \mathfrak{F}_0([\mathsf{K}_1, \mathsf{K}_2]) = [\mathsf{K}_1, \Phi_0^{-1}(\mathsf{K}_2)]$$

has the property

$$\mathfrak{F}_0(\hat{\mathsf{K}}[\mathsf{A}, \Xi(\mathsf{A})]\hat{\mathsf{K}}) = \mathfrak{F}_0(\hat{\mathsf{K}})[\mathsf{A}, \mathsf{A}]\mathfrak{F}_0(\hat{\mathsf{K}}), \qquad \hat{\mathsf{K}} \in \mathrm{Sym}(\mathscr{T}) \oplus \mathrm{Sym}(\mathscr{T}), \mathsf{A} \in \mathscr{A}_1.$$

In other words, $\mathfrak{F}_0$ establishes a one-to-one correspondence between Jordan $\hat{\mathscr{A}}$-multialgebras $\hat{\Pi}$ and Jordan $\hat{\mathscr{A}}_0$-multialgebras $\hat{\Pi}_0 = \mathfrak{F}_0(\hat{\Pi})$, where

$$\hat{\mathscr{A}}_0 = \{[\mathsf{A}, \mathsf{A}]\colon \mathsf{A} \in \mathscr{A}_1\}. \tag{6.73}$$

The set of all Φ_0, satisfying (6.72) is described explicitly by theorem 6.1, applied to $\tilde{\Gamma}_1$ that generates $\mathscr{A}_1$ via (4.17) and $\tilde{\Gamma}_2 = \Xi(\tilde{\Gamma}_1)$ that generates $\mathscr{A}_2$.

When the transformation Φ_0, satisfying (6.72) exists, the structure of all links acquires a much nicer description. Indeed, property (5.14) becomes

$$\Phi(\bar{\mathsf{K}}\mathsf{A}\bar{\mathsf{K}}) = \Phi(\bar{\mathsf{K}})\mathsf{A}\Phi(\bar{\mathsf{K}}), \quad \forall\, \bar{\mathsf{K}} \in \Pi_1/\mathscr{I}_1, \forall\, \mathsf{A} \in \mathscr{A}_1, \tag{6.74}$$

making Φ: $\Pi_1/\mathscr{I}_1 \to \Pi_2/\mathscr{I}_2$ a Jordan factor-multialgebra isomorphism. Thus, the computation of all links requires us to classify ideals and corresponding factor-multialgebras, up to an SO(d)-equivariant Jordan factor-multialgebra isomorphism. If such isomorphism classes contain more than one factor-multialgebra, then we only need to find one (simplest) isomorphism between the factor-multialgebras $\Pi_1/\mathscr{I}_1$ and $\Pi_2/\mathscr{I}_2$ in the same isomorphism class. Once these simple isomorphisms are established the entire collection of links will be described by single representatives of each factor-multialgebras isomorphism class, together with the group of all of their Jordan automorphisms.

In the case, when the global isomorphism (6.72) does not exist, it is theoretically possible that such an isomorphism exists on a subalgebra Π:

$$\Phi(\mathsf{KAK}) = \Phi(\mathsf{K})\Xi_0(\mathsf{A})\Phi(\mathsf{K}), \qquad \forall\, \mathsf{K} \in \Pi, \forall\, \mathsf{A} \in \mathscr{A}, \tag{6.75}$$

where Φ: $\Pi \to \Pi$ is an SO(d)-invariant linear isomorphism.

We illustrate these observations in the contexts of two and three-dimensional conductivity in section 6.4.1. The application of this methodology to two and three-dimensional elasticity is illustrated in the subsequent sections 6.4.2–6.4.5. The results are also summarized in chapter 8 for conductivity, and chapter 9 for elasticity. However, it is in the context of fiber-reinforced elastic and two-dimensional thermoelectric composites that the tools developed here become essential for calculation and effective classification of all exact relations and links listed in sections 9.3 and 12.4, respectively, in part III of the book (see [10, 11] and [8]).

6.4.1 Computing $\hat{\mathscr{A}}$ multialgebras in conductivity

Example 6.14. *Two and three-dimensional conductivity.*

In the context of polycrystalline conducting composites we take $\hat{\mathsf{L}}_0 = [\sigma_1^0 \boldsymbol{I}_d, \sigma_2^0 \boldsymbol{I}_d]$.

$$\boldsymbol{\Gamma}_0^{(i)}(\boldsymbol{n}) = \frac{\boldsymbol{n} \otimes \boldsymbol{n}}{\sigma_i^0}.$$

Therefore,

$$\hat{\mathscr{A}} = \{[\boldsymbol{A}, \alpha\boldsymbol{A}]\colon \boldsymbol{A} \in \mathrm{Sym}_0(\mathbb{R}^d)\}, \qquad \alpha = \frac{\sigma_1^0}{\sigma_2^0},$$

where $\mathrm{Sym}_0(\mathbb{R}^d)$ was defined in (4.22). This shows that $\Xi(\boldsymbol{A}) = \alpha\boldsymbol{A}$, and therefore, the map $\Phi_0(\boldsymbol{K}) = \boldsymbol{K}/\alpha$ satisfies (6.72). Hence, the map $\mathfrak{F}_0([\boldsymbol{K}_1, \boldsymbol{K}_2]) = [\boldsymbol{K}_1, \alpha\boldsymbol{K}_2]$

transforms SO(d)-invariant Jordan $\hat{\mathscr{A}}$-multialgebras into SO(d)-invariant Jordan $\hat{\mathscr{A}}_0$-multialgebras, where

$$\hat{\mathscr{A}}_0 = \{[\boldsymbol{A}, \boldsymbol{A}]: \boldsymbol{A} \in \mathrm{Sym}_0(\mathbb{R}^d)\}.$$

Now, to classify all links in the context of two and three-dimensional conductivity, we need a complete list of SO(d)-invariant Jordan $\mathscr{A}$-multialgebras, where $\mathscr{A} = \mathrm{Sym}_0(\mathbb{R}^d)$. When $d = 3$, the list is $\{0\}$ and $\mathrm{Sym}(\mathbb{R}^3)$. When $d = 2$, the list is $\{0\}$, $\mathrm{Sym}_0(\mathbb{R}^2)$ and $\mathrm{Sym}(\mathbb{R}^2)$. It is easy to check that $\mathrm{Sym}_0(\mathbb{R}^2)$ is not an ideal in $\mathrm{Sym}(\mathbb{R}^2)$. Hence, the complete lists of Jordan factor multialgebras are

$$\{0\}/\{0\}, \quad \mathrm{Sym}(\mathbb{R}^3)/\{0\},$$

for three-dimensional conductivity, and

$$\{0\}/\{0\}, \quad \mathrm{Sym}_0(\mathbb{R}^2)/\{0\}, \quad \mathrm{Sym}(\mathbb{R}^2)/\{0\}$$

for the two-dimensional one. Since all listed factor-algebras are of different dimensions, no two of them could possibly be isomorphic.

To classify all links it remains to compute all rotationally equivariant Jordan $\mathscr{A}$-multialgebra *automorphisms*. The requirement of rotational equivariance severely limits the possibilities. In addition, the trivial factor algebra $\{0\}/\{0\}$ is always present regardless of context and carries no useful information. The only rotationally equivariant linear automorphism of $\mathrm{Sym}_0(\mathbb{R}^2)$ must have the form $\Phi(\boldsymbol{K}) = \lambda \boldsymbol{K}$. It is easy to check that only $\lambda = 1$ is compatible with the homomorphism property (6.74), resulting in a trivial Jordan $\mathscr{A}$-multialgebra automorphism.

The only rotationally invariant linear automorphism of $\mathrm{Sym}(\mathbb{R}^d)$ must have the form $\Phi(\boldsymbol{K}) = \lambda \boldsymbol{I}_d \,\mathrm{Tr}\, \boldsymbol{K} + \mu \boldsymbol{K}$. Then, property (6.74) becomes

$$\lambda \boldsymbol{I}_d \,\mathrm{Tr}\,(\boldsymbol{KAK}) + \mu \boldsymbol{KAK} = (\lambda \boldsymbol{I}_d \,\mathrm{Tr}\, \boldsymbol{K} + \mu \boldsymbol{K})\boldsymbol{A}(\lambda \boldsymbol{I}_d \,\mathrm{Tr}\, \boldsymbol{K} + \mu \boldsymbol{K}).$$

Taking $\boldsymbol{K} = \boldsymbol{I}_d$ we get the equation $\mu = (d\lambda + \mu)^2$. If we take $\boldsymbol{K} \in \mathrm{Sym}_0(\mathbb{R}^d)$ we obtain

$$(\mu^2 - \mu)\boldsymbol{KAK} = \lambda \boldsymbol{I}_d \,\mathrm{Tr}\,(\boldsymbol{KAK}).$$

Taking traces of both sides, we obtain

$$\mu^2 - \mu = 3\lambda,$$

when $d = 3$, since it is easy to find $\{\boldsymbol{K}, \boldsymbol{A}\} \subset \mathrm{Sym}_0(\mathbb{R}^3)$, such that $\mathrm{Tr}\,(\boldsymbol{KAK}) \neq 0$. Hence, when $d = 3$, the system of equations

$$\mu = (3\lambda + \mu)^2, \quad \mu^2 - \mu = 3\lambda$$

implies $\mu = \mu^4$, so that $\mu = 1$, $\lambda = 0$ is the only solution (aside from $\lambda = \mu = 0$ corresponding to the zero map $\Phi(\boldsymbol{K}) = 0$). Hence, when $d = 3$, the $\mathscr{A}$-multialgebra $\mathrm{Sym}(\mathbb{R}^d)$ has only the trivial automorphism.

The situation is different when $d = 2$. In this case, $\mathrm{Tr}\,(\boldsymbol{KAK}) = 0$ *for any* $\{\boldsymbol{K}, \boldsymbol{A}\} \subset \mathrm{Sym}_0(\mathbb{R}^2)$ and therefore,

$$(\mu^2 - \mu)\boldsymbol{KAK} = \lambda \boldsymbol{I}_2 \operatorname{Tr}(\boldsymbol{KAK}) = 0$$

implies that we must have $\mu = 1$, yielding $2\lambda + 1 = \pm 1$. The '+' sign gives $\lambda = 0$ and the trivial automorphism $\Phi(\boldsymbol{K}) = \boldsymbol{K}$. The '−' sign gives $\lambda = -1$, and one can verify, either by hand, or with a computer algebra system, that the map $\Phi(\boldsymbol{K}) = \boldsymbol{K} - \boldsymbol{I}_2 \operatorname{Tr} \boldsymbol{K}$ is, in fact, a Jordan $\mathscr{A}$-multialgebra automorphism of $\mathrm{Sym}(\mathbb{R}^2)$. The resulting Jordan $\hat{\mathscr{A}}_0$-multialgebra

$$\hat{\Pi} = \{[\boldsymbol{K}, \boldsymbol{K} - \boldsymbol{I}_2 \operatorname{Tr} \boldsymbol{K}]: \boldsymbol{K} \in \mathrm{Sym}(\mathbb{R}^2)\}$$

corresponds to the Mendelson's link (5.1).

The transformation $\mathfrak{F}_0([\boldsymbol{K}_1, \boldsymbol{K}_2]) = [\boldsymbol{K}_1, \alpha \boldsymbol{K}_2]$ gives a recipe of how to go from links passing through $\hat{\mathsf{L}}_0 = [\boldsymbol{I}_d, \boldsymbol{I}_d]$ to the ones passing through $\hat{\mathsf{L}}_0 = [\sigma_1^0 \boldsymbol{I}_d, \sigma_2^0 \boldsymbol{I}_d]$. In our example of conducting composites, it is just the homogeneity property of the effective tensors: $\tilde{\boldsymbol{\sigma}}(z) = \alpha \boldsymbol{\sigma}(z)$ implies $\tilde{\boldsymbol{\sigma}}_* = \alpha \boldsymbol{\sigma}_*$.

We have now proved that there are no nontrivial exact relations and links for three-dimensional conducting polycrystals, and that, aside from the Keller-Dykhne exact relation (4.3) and Mendelson's link (5.1), there are no other microstructure-independent formulas for two-dimensional conducting polycrystals.

6.4.2 Computing $\hat{\mathscr{A}}$ in elasticity

Let us compute the subspace $\hat{\mathscr{A}}$ for 2D and 3D elasticity. Using formulas (6.37) and noting that the first term in (6.35) has no component isomorphic to W_4 we obtain

$$\boldsymbol{\Gamma}_0(\boldsymbol{n}) - \boldsymbol{\Gamma}_0(\boldsymbol{n}_0) = \begin{cases} aK_2'(\boldsymbol{G}) + \dfrac{\rho_0}{\mu_0}\mathsf{K}_4, & d = 2, \\ aK_2'(\boldsymbol{G}) + bK_2(\boldsymbol{G}) + \dfrac{\rho_0}{\mu_0}\mathsf{K}_4, & d = 3, \end{cases}$$

where $\boldsymbol{G} = \boldsymbol{n} \otimes \boldsymbol{n} - \boldsymbol{n}_0 \otimes \boldsymbol{n}_0$, ρ_0 is defined in (6.36) and

$$a = \frac{1 - \rho_0}{\mu_0 d} = d\kappa_\Gamma, \qquad b = \frac{7 - 4\rho_0}{14\mu_0} \tag{6.76}$$

are the coefficients in (6.37). The precise expression of the tensor $\mathsf{K}_4 \in K_4$ is unimportant for computing $\hat{\mathscr{A}}$. It is only important that K_4 is independent of the elastic moduli κ_0, μ_0. We can simplify $\boldsymbol{\Gamma}_0(\boldsymbol{n}) - \boldsymbol{\Gamma}_0(\boldsymbol{n}_0)$ by a covariance transformation $\boldsymbol{\Gamma}_0 \mapsto \bar{\mathsf{C}}^{-1}\boldsymbol{\Gamma}_0\bar{\mathsf{C}}^{-1}$, where $\bar{\mathsf{C}}$ is an isotropic elastic tensor with bulk and shear moduli $\bar{\kappa}$, $\bar{\mu}$, respectively. Using the formulas

$$\bar{\mathsf{C}}^{-1}K_2'(\boldsymbol{A})\bar{\mathsf{C}}^{-1} = \frac{K_2'(\boldsymbol{A})}{2d\bar{\kappa}\bar{\mu}}, \quad \bar{\mathsf{C}}^{-1}K_2(\boldsymbol{B})\bar{\mathsf{C}}^{-1} = \frac{K_2(\boldsymbol{B})}{4\bar{\mu}^2}, \quad \bar{\mathsf{C}}^{-1}\mathsf{K}_4\bar{\mathsf{C}}^{-1} = \frac{\mathsf{K}_4}{4\bar{\mu}^2}, \tag{6.77}$$

we choose, for isotropic reference media $\hat{\mathsf{C}}_0 = [\mathsf{C}_1^0, \mathsf{C}_2^0]$ with bulk and shear moduli κ_j and μ_j, respectively, $j = 1, 2$,

$$\bar{\mu}_j = \begin{cases} m_0\sqrt{\dfrac{\rho_j}{\mu_j}}, d=2, \\ k_0\sqrt{b_j}, \ d=3, \end{cases} \qquad \bar{\kappa}_j = \begin{cases} k_0 a_j\sqrt{\dfrac{\mu_j}{\rho_j}}, d=2, \\ \dfrac{k_0 a_j}{\sqrt{b_j}}, \qquad d=3, \end{cases} \qquad j=1,2. \tag{6.78}$$

Here m_0 and k_0 are arbitrary positive constants that do not affect the resulting subspace

$$\hat{\mathscr{A}}_0 = \begin{cases} \{[\mathsf{A}, \mathsf{A}]: \mathsf{A} \in \mathscr{A}_{2d}\}, \qquad d=2, \\ \{[\mathsf{A}, \Xi_0(\mathsf{A})]: \mathsf{A} \in \mathscr{A}_0\}, d=3, \end{cases} \qquad \mathscr{A}_{2d} = K_2' \oplus K_4, \tag{6.79}$$

where $\mathscr{A}_0$ is given by (6.38). The map Ξ_0: $\mathscr{A}_0 \to \mathscr{A}_0$ is defined in terms of the unique representation $\mathsf{A} = \mathsf{A}_2 + \mathsf{A}_4$, where $\mathsf{A}_2 \in K_2^{2/3}$, and $\mathsf{A}_4 \in K_4$. Then

$$\Xi_0(\mathsf{A}) = \mathsf{A}_2 + \xi\mathsf{A}_4, \qquad \xi = \frac{(3\kappa_1 + 8\mu_1)(3\kappa_2 + \mu_2)}{(3\kappa_2 + 8\mu_2)(3\kappa_1 + \mu_1)}.$$

It is easy to see that $\xi = 1$ if and only if $\mathsf{C}_2^0 = s\mathsf{C}_1^0$ for some scalar $s > 0$. In particular, we see that if $d=3$ and $\xi \neq 1$ there are no transformations satisfying (6.72). This implies that in 3D elasticity we should expect to have very few links, unlike 2D elasticity.

6.4.3 Ideals and factor-multialgebras in elasticity

The next step in our program of obtaining a complete list of all links is to identify all Jordan multialgebra-ideal pairs. In the case of elasticity, there are so few Jordan multialgebras that it is a simple matter (using multiplication tables 6.1 and 6.2) to produce the following list of all Jordan multialgebra-ideal pairs. In this regard we make two simple observations. The first, is that $\mathscr{I} = \{0\}$ is an ideal in every Jordan multialgebra. The second is that any subspace $\mathscr{I} \subset \Pi$, such that $\Pi^2 \subset \mathscr{I}$ is an ideal. It is then easy to verify that in the setting of elasticity (both two and three-dimensional), the only multialgebra-ideal pair, not covered by the above observations, is $\Pi = K_0' \oplus K_2'$, $\mathscr{I} = K_0'$. Hence, the complete list of multialgebra-ideal pairs for elastic composites is

List 6.15.

1. $\Pi = \mathbb{E}^d = \mathrm{Sym}(\mathrm{Sym}(\mathbb{R}^d))$, $\mathscr{I} = \{0\}$.
2. $\Pi = \mathrm{Sym}(\mathrm{Sym}_0(\mathbb{R}^d))$, $\mathscr{I} = \{0\}$.
3. $\Pi = K_0' \oplus K_2'$, $\mathscr{I} = K_0'$ and $\mathscr{I} = \{0\}$.
4. $\Pi = K_0'$, $\mathscr{I} = \{0\}$.
5. If $d=2$, then $\Pi = K_4$, $\mathscr{I} = \{0\}$.

A simple inspection shows that different factor-multialgebras are not isomorphic even as representations of SO(d), let alone as Jordan factor-multialgebras.

6.4.4 Jordan isomorphisms in 3D elasticity

Now that all the factor-multialgebras are computed, it remains to find all SO(3)-invariant Jordan factor-multialgebra isomorphisms Φ described in theorem 5.1. It is here that the different structures of subspaces $\hat{\mathscr{A}}$ for the 2D and 3D elasticity will result in very different set of links for the two cases.

Theorem 6.16. *Suppose that $\xi \neq 1$. Then*

(i) *There are no SO(3)-equivariant linear isomorphisms Φ of $\mathbb{E}^3 = \mathrm{Sym}(\mathrm{Sym}(\mathbb{R}^3))$, satisfying (6.72).*

(ii) *There are no SO(3)-equivariant linear isomorphisms Φ of $K_0 \oplus K_2 \oplus K_4$, satisfying (6.75).*

(iii) *There are no SO(3)-equivariant linear isomorphisms Φ of $K_0' \oplus K_2'$, satisfying (6.75).*

(iv) *The only SO(3)-equivariant linear isomorphism of $\mathscr{F} = (K_0' \oplus K_2')/K_0' \cong K_2'$, satisfying (6.74) is $\Phi(\bar{\mathsf{K}}) = \bar{\mathsf{K}}$.*

(v) *All maps $\Phi(\mathsf{K}) = \lambda\mathsf{K}$, $\mathsf{K} \in K_0'$ satisfy (6.75).*

Proof. Before we prove part (i), let us first prove parts (ii)–(v). In each case (ii)–(v) an SO(3)-equivariant linear isomorphism Φ must be a multiple of the identity on each irrep. (This is known as Schur's lemma in representation theory.) Then, for the map Φ in part (ii), there exist nonzero scalars λ_0, λ_2, λ_4, such that $\Phi(\mathsf{K}) = \lambda_i\mathsf{K}$ for every $\mathsf{K} \in K_i$, $i = 0, 2, 4$. Recall that $\mathscr{A}_0 = K_2^{2/3} \oplus K_4$. Let us test (6.75) by setting $\mathsf{K} = K_0(\mu)$, $\mathsf{A} = (2/3)K_2'(A) + K_2(A) + \mathsf{A}_4$. Then, according to (6.77) (see also table 6.2)

$$\mathsf{KAK} = 4\mu^2(K_2(A) + \mathsf{A}_4).$$

Thus,

$$\Phi(\mathsf{KAK}) - 4\mu^2(\lambda_2 K_2(A) + \lambda_4\mathsf{A}_4).$$

We also have

$$\Phi(\mathsf{K})\Xi_0(\mathsf{A})\Phi(\mathsf{K}) = 4\mu^2\lambda_0^2(K_2(A) + \xi\mathsf{A}_4).$$

It follows that

$$\lambda_2 = \lambda_0^2, \qquad \lambda_4 = \xi\lambda_0^2.$$

Now let us take $\mathsf{K} \in K_j$, $j = 2, 4$ and $\mathsf{A} \in K_2^{3/2}$, such that

$$\mathsf{KAK} = \mathsf{B}_0 + \mathsf{B}_2 + \mathsf{B}_4, \qquad \mathsf{B}_i \in K_i \setminus \{0\},\ i = 0, 2, 4.$$

This is possible, because $(K_j K_2^{3/2} K_j)_{\mathrm{sym}} = K_0 \oplus K_2 \oplus K_4$, $j = 2, 4$, and the sets $\{(\mathsf{K}, \mathsf{A}) \in K_j \oplus K_2^{3/2}\colon \mathsf{KAK} = \mathsf{B}_0 + \mathsf{B}_2 + \mathsf{B}_4,\ \mathsf{B}_k \neq 0\}$, $k = 0, 2, 4$ arc opcn and dense in $K_j \oplus K_2^{3/2}$, $j = 2, 4$. Then, we must have

$$\Phi(\mathsf{KAK}) = \lambda_0 \mathsf{B}_0 + \lambda_2 \mathsf{B}_2 + \lambda_4 \mathsf{B}_4.$$

We also have

$$\Phi(\mathsf{K})\Xi_0(\mathsf{A})\Phi(\mathsf{K}) = \lambda_j^2 \mathsf{KAK} = \lambda_j^2(\mathsf{B}_0 + \mathsf{B}_2 + \mathsf{B}_4), \quad j = 2, 4,$$

since $\Xi_0(\mathsf{A}) = \mathsf{A}$ for $\mathsf{A} \in K_2^{3/2}$. This implies that $\lambda_0 = \lambda_2 = \lambda_4 = 1$. It is now clear that (6.75) cannot hold, unless $\xi = 1$.

Now let us prove part (iii). We first note that for $\mathsf{K}_0 \in K_0'$, $\mathsf{K}_2 \in K_2'$, $\mathsf{A}_2 \in K_2^{3/2}$, $\mathsf{A}_4 \in K_4$ we have

$$\mathsf{K}_0\mathsf{A}_4\mathsf{K}_2 = \mathsf{K}_2\mathsf{A}_4\mathsf{K}_0 = 0, \qquad \mathsf{K}_0\mathsf{A}_2\mathsf{K}_2 + \mathsf{K}_2\mathsf{A}_2\mathsf{K}_0 = \mathsf{B}_0 \in K_0'$$

Thus, (6.75) implies

$$\lambda_0 \mathsf{B}_0 = \lambda_0 \lambda_2 \mathsf{B}_0.$$

It follows that $\lambda_2 = 1$. Now let $\mathsf{K} \in K_2'$. Then

$$\mathsf{KA}_4\mathsf{K} = \mathsf{B}_0 \in K_0'.$$

Hence, (6.75) implies

$$\lambda_0 \mathsf{B}_0 = \lambda_2^2 \xi \mathsf{B}_0 = \xi \mathsf{B}_0.$$

This implies that $\lambda_0 = \xi$. To obtain a contradiction we compute for $\mathsf{K} = K_2'(\boldsymbol{B})$ and $\mathsf{A} \in K_2^{2/3}$, given by (6.39)

$$K_2'(\boldsymbol{B})\mathsf{A}K_2'(\boldsymbol{B})\varepsilon = 2\,\mathrm{Tr}\,(\boldsymbol{AB})K_2'(\boldsymbol{B})\varepsilon + K_0'(2\,\mathrm{Tr}\,(\boldsymbol{BAB}))\varepsilon. \tag{6.80}$$

Choosing non-singular $\boldsymbol{A} = \boldsymbol{B} \in \mathrm{Sym}_0(\mathbb{R}^3)$, we obtain

$$\mathsf{KAK} = \mathsf{B}_0 + \mathsf{B}_2, \qquad \mathsf{B}_i \in K_i' \setminus \{0\},\ i = 0, 2.$$

Then, for (6.75) to hold we must have

$$\lambda_0 \mathsf{B}_0 + \lambda_2 \mathsf{B}_2 = \lambda_2^2(\mathsf{B}_0 + \mathsf{B}_2), \tag{6.81}$$

which cannot hold for $\lambda_0 = \xi \neq 1$, $\lambda_2 = 1$.

Part (iv) is very simple, since we only need to test the maps $\Phi(\bar{\mathsf{K}}) = \lambda\bar{\mathsf{K}}$. Factoring (6.81) by K_0' we conclude that we must have $\lambda^2 = \lambda$, while $\Phi(\bar{\mathsf{K}}) = \bar{\mathsf{K}}$ is easily seen to satisfy (6.75), since

$$\bar{\mathsf{K}}\mathsf{A}_4\bar{\mathsf{K}} = \bar{0}, \quad \forall \bar{\mathsf{K}} \in K_2'.$$

Part (v) is obvious, since $\mathsf{KAK} = 0$ for all $\mathsf{K} \in K_0'$ and all $\mathsf{A} \in \mathscr{A}_0$.

We are now ready to prove part (i). Let us assume that such Φ exists. Then for $\Pi = K_0 \oplus K_2 \oplus K_4$ its Φ image $\Phi(\Pi)$ must be isomorphic to Π as a representation of SO(3). It also has to be a Jordan $\mathscr{A}_0$ multialgebra. But Π is the only SO(3)-invariant $\mathscr{A}_0$ multialgebra isomorphic to Π as a representation of SO(3). This implies that $\Phi(\Pi) = \Pi$, which does not exist due to part (ii). □

We conclude that in the case $\xi \neq 1$ there is only one nontrivial link, corresponding to

$$\hat{\Pi} = \{[\mathsf{K}, \mathsf{K} + \mathsf{J}]: \mathsf{J} \in K_0', \mathsf{K} \in K_0' \oplus K_2'\}. \quad (6.82)$$

Moreover, when $\xi = 1$, the arguments in the proof of theorem 6.16 show that $\Phi(\mathsf{K}) = \mathsf{K}$ are the only automorphisms in cases (i)–(iii). We also remark that the link in case (v) does not bring any new information, since we already have fully explicit formulas (4.1) for the effective elastic tensor.

Finally, we need to verify that link (6.82) satisfies sufficient conditions from theorem 5.2. We note that the map Φ in (6.82) is trivial, i.e., $\Phi(\bar{\mathsf{K}}) = \bar{\mathsf{K}}$, so that conditions (5.19), (5.20) follow from (5.17), (5.18) and the already verified 3 and 4-chain properties for $\Pi = K_0' \oplus K_4'$. Conditions (5.17), (5.18) for (6.82) follow from a simple verification that

$$K_0' \mathscr{A}_0 K_2' = K_0', \qquad K_2' \mathscr{A}_0 K_0' = K_0'.$$

6.4.5 Jordan isomorphisms in 2D elasticity

Recall that Ξ_0 is the identity transformation in 2D elasticity. Thus, the goal is to compute all Jordan automorphisms of the factor-multialgebras listed in section 6.4.3. Let us first consider the case $\mathscr{I} = \{0\}$. Using the fact than none of the Jordan $\mathscr{A}_{2d}$-multialgebras are isomorphic as representations of SO(2), we conclude that any Jordan automorphism Φ of $\mathbb{E}^2$ must map $K_0' \oplus K_2'$ into itself and $K_0 \oplus K_4$ into itself. It follows that

$$\Phi(K_0) = K_0, \quad \Phi(K_4) = K_4, \quad \Phi(K_0') = K_0', \quad \Phi(K_2') = K_2'.$$

Let $\mathsf{K}_0' = K_0'(1)$, $\mathsf{K}_0 = K_0(1)$. Then

$$\Phi(\mathsf{K}_0') = \lambda_0' \mathsf{K}_0', \qquad \Phi(\mathsf{K}_0) = \lambda_0 \mathsf{K}_0.$$

According to the multiplication table 6.1

$$\mathsf{K}_0 \mathsf{A} \mathsf{K}_0 = \mathsf{K}_4(\mathsf{A}) \in K_4, \qquad \mathsf{A} \in \mathscr{A}_{2d}.$$

Applying Φ and using (6.75) we obtain

$$\Phi(\mathsf{K}_4(\mathsf{A})) = \lambda_0^2 \mathsf{K}_4(\mathsf{A}).$$

As A ranges over $\mathscr{A}_{2d}$, $\mathsf{K}_4(\mathsf{A})$ must range over K_4 and thus,

$$\Phi(\mathsf{K}) = \lambda_0^2 \mathsf{K}, \quad \mathsf{K} \in K_4.$$

Similarly,

$$\mathsf{K}_0' \mathsf{A} \mathsf{K}_0 + \mathsf{K}_0 \mathsf{A} \mathsf{K}_0' = \mathsf{K}_2'(\mathsf{A}) \in K_2', \qquad \mathsf{A} \in \mathscr{A}_{2d}.$$

Applying Φ we conclude that $\Phi(\mathsf{K}_2'(\mathsf{A})) = \lambda_0 \lambda_0' \mathsf{K}_2'(\mathsf{A})$, so that

$$\Phi(\mathsf{K}) = \lambda_0 \lambda_0' \mathsf{K}, \quad \mathsf{K} \in K_2'.$$

Using these results we apply Φ to

$$\mathsf{K}_0'\mathsf{A}\mathsf{K} + \mathsf{K}\mathsf{A}\mathsf{K}_0' = \theta(\mathsf{K}, \mathsf{A})\mathsf{K}_0', \qquad \mathsf{A} \in \mathscr{A}_{2d}, \ \mathsf{K} \in K_2'$$

and to

$$\mathsf{K}_0'\mathsf{A}\mathsf{K} + \mathsf{K}\mathsf{A}\mathsf{K}_0' = \mathsf{K}_2'(\mathsf{K}, \mathsf{A}), \qquad \mathsf{A} \in \mathscr{A}_{2d}, \ \mathsf{K} \in K_4$$

we conclude that we must have $\lambda_0 = \lambda_0' = 1$. Thus, $\Phi(\mathsf{K}) = \mathsf{K}$ for all $\mathsf{K} \in \mathbb{E}^2$.

Let us now compute all Jordan automorphisms of $K_0 \oplus K_4$. As before,

$$\Phi(\mathsf{K}_0) = \lambda_0\mathsf{K}_0, \qquad \Phi(\mathsf{K}) = \lambda_0^2\mathsf{K}, \quad \mathsf{K} \in K_4.$$

Applying Φ to the remaining two relations

$$\mathsf{K}\mathsf{A}\mathsf{K} = \mathsf{P}(\mathsf{K}, \mathsf{A}) \in K_4, \quad \mathsf{K}_0\mathsf{A}\mathsf{K} + \mathsf{K}\mathsf{A}\mathsf{K}_0 = \theta(\mathsf{K}, \mathsf{A})\mathsf{K}_0, \quad \mathsf{A} \in \mathscr{A}_{2d}, \ \mathsf{K} \in K_4$$

we obtain $\lambda_0^2 = 1$, which in addition to the trivial automorphism gives a new one:

$$\Phi(\tau\mathsf{K}_0 + \mathsf{K}_4) = \mathsf{K}_4 - \tau\mathsf{K}_0, \qquad \mathsf{K}_4 \in K_4, \tag{6.83}$$

giving rise to the corresponding Jordan $\hat{\mathscr{A}}_0$-multialgebra

$$\hat{\Pi} = \{[\mathsf{K}, \Phi(\mathsf{K})]\colon \mathsf{K} \in K_0 \oplus K_4\}. \tag{6.84}$$

We have already remarked on the fact that sometimes links contain more information than exact relations. For example, (6.84) implies that K_4 must be a Jordan $\mathscr{A}_{2d}$-multialgebra. Indeed,

$$K_4 = \{\mathsf{K} \in K_0 \oplus K_4\colon \Phi(\mathsf{K}) = \mathsf{K}\},$$

i.e., K_4 is the space of fixed points of an automorphism Φ. Then, for any fixed point K of a Jordan multialgebra automorphism Φ we have

$$\Phi(\mathsf{K}\mathsf{A}\mathsf{K}) = \Phi(\mathsf{K})\mathsf{A}\Phi(\mathsf{K}) = \mathsf{K}\mathsf{A}\mathsf{K},$$

showing that $\mathsf{K}\mathsf{A}\mathsf{K}$ is also a fixed point of Φ. Therefore, the set of fixed points of a Jordan $\mathscr{A}$-multialgebra automorphism is always a Jordan $\mathscr{A}$-multialgebra. We also note that the 3 and 4-chain properties (see definition 4.7) of the fixed point subspace follow from (5.19) and (5.20) that had to be verified in order to check whether the underlying link $\hat{\Pi}$, given by (6.84), holds for all pairs of composites, regardless of the microstructure.

Next, we compute all automorphisms of $K_0' \oplus K_2'$. As before, $\Phi(\mathsf{K}_0') = \lambda_0'\mathsf{K}_0'$. However, to make progress we need to go a little beyond the multiplication table. We compute for $\mathsf{K} = K_2'(\boldsymbol{B})$ and $\mathsf{A} \in K_2^{2/3}$, given by (6.39),

$$\mathsf{K}_0'\mathsf{A}K_2'(\boldsymbol{B}) + K_2'(\boldsymbol{B})\mathsf{A}\mathsf{K}_0' = 4\,\mathrm{Tr}\,(\boldsymbol{A}\boldsymbol{B})\mathsf{K}_0'.$$

Suppose that Φ, acting on K_2' transforms $\boldsymbol{B}$ into $\boldsymbol{B}'$, i.e. $\Phi(K_2'(\boldsymbol{B})) = K_2'(\boldsymbol{B}')$. Then, we must have

$$\mathrm{Tr}\,(\boldsymbol{A}\boldsymbol{B}) = \mathrm{Tr}\,(\boldsymbol{A}\boldsymbol{B}'), \qquad \forall\,\boldsymbol{A} \in \mathrm{Sym}_0(\mathbb{R}^2).$$

But that implies $\boldsymbol{B}' = \boldsymbol{B}$. To complete the calculation we apply Φ to (6.80) and obtain $\lambda_0' = 1$, proving that there are no nontrivial automorphisms of $K_0' \oplus K_2'$. Incidentally, the same calculation also proves that there are no nontrivial automorphisms of the factor-multialgebra $K_2' \cong (K_0' \oplus K_2')/K_0'$.

Finally, we compute all SO(2)-invariant automorphisms of K_4. In order to do that we need to delve a little deeper into the structure of $\mathbb{E}^2$. We begin by recalling that every $\varepsilon \in \mathrm{Sym}(\mathbb{R}^2)$ can be decomposed as

$$\varepsilon = \frac{1}{2}(\mathrm{Tr}\ \varepsilon)\boldsymbol{I}_2 + \mathrm{dev}(\varepsilon).$$

We now make one more step and parametrize $\mathrm{Sym}_0(\mathbb{R}^2)$ by complex numbers as follows

$$\mathbb{C} \ni z = x + iy \mapsto \psi(z) = \begin{bmatrix} x & y \\ y & -x \end{bmatrix}.$$

Thus, we will write

$$\varepsilon = \frac{1}{2}(\mathrm{Tr}\ \varepsilon)\boldsymbol{I}_2 + \psi(e), \quad e \in \mathbb{C}.$$

Then we can parametrize the SO(2) irrep K_4 by complex numbers as follows:

$$\mathbb{C} \ni v \mapsto K_4(v), \qquad K_4(v)\varepsilon = \psi(v\bar{e}).$$

The subspace $\mathscr{A}_0$ can now be parametrized by two complex numbers

$$\mathscr{A}_0 = \{\mathsf{A}(a, b) = \mathsf{A}_2(a) + K_4(b)\colon \{a, b\} \subset \mathbb{C}\}, \qquad \mathsf{A}_2(a)\varepsilon = \psi(a)\varepsilon + \varepsilon\psi(a).$$

It is now easy to check that

$$K_4(v)\mathsf{A}_2(a)K_4(v) = 0, \qquad K_4(v)K_4(b)K_4(v) = K_4(\bar{b}v^2). \tag{6.85}$$

Any SO(2)-equivariant map Φ: $K_4 \to K_4$ must have the form $\Phi(K_4(v)) = K_4(cv)$ for some $c \in \mathbb{C}$. Applying Φ to (6.85) we conclude that $c^2 = c$, i.e. $c = 1$ is the only choice for which Φ satisfies (6.75).

Let us now verify that link (6.84) satisfies sufficient conditions from theorem 5.2. We note that the ideals $\mathscr{I}_1$, $\mathscr{I}_2$ in (6.84) are trivial, so that that conditions (5.17), (5.18) are trivially satisfied. Hence, we only need to verify conditions (5.19), (5.20) for (6.83). To prove (5.19), (5.20) for (6.83) we recall that $K_0 \oplus K_4$ is the set of symmetric operators in the associative $\mathscr{A}_0$-multialgebra $\mathrm{End}(\mathrm{Sym}_0(\mathbb{R}^2))$. For any $\mathsf{K} \in \mathrm{End}(\mathrm{Sym}_0(\mathbb{R}^2))$ there is a unique pair of complex numbers v, w, such that $\mathsf{K} = K_4(v) + P_0(w)$, where $P_0(w)\psi(e) = \psi(we)$. We claim that the transformation

$$\hat{\Phi}(K_4(v) + P_0(w)) = K_4(v) - P_0(w)$$

is the associative $\mathscr{A}_0$-multialgebra automorphism. This claim can be verified directly by applying $\hat{\Phi}$ to

$$\begin{aligned}&(K_4(v) + P_0(w))(\mathsf{A}_2(a) + K_4(b))(K_4(v') + P_0(w'))\psi(e)\\&= \psi((vv'\bar{b} + w\overline{w'}b)\bar{e} + (vw'\bar{b} + w\overline{v'}b)e)\\&= (K_4(vv'\bar{b} + w\overline{w'}b) + P_0(vw'\bar{b} + w\overline{v'}b))\psi(e).\end{aligned}$$

Hence, we see that Φ is a restriction of an associative $\mathscr{A}_0$-multialgebra automorphism to the set of symmetric operators there. Hence, it must satisfy (5.19), (5.20).

6.5 Computing links

We continue to view links as exact relations in $\mathrm{End}(\mathscr{T}_1) \oplus \mathrm{End}(\mathscr{T}_2)$. Hence, the passage from the algebraic representation of links to their formulation in physical variables occurs along the lines of section 6.2, where we explain how to find the simplest possible inversion key $\hat{\mathsf{M}} = [\mathsf{M}_1, \mathsf{M}_2]$ to be used in (6.56). In other words, from the general theoretic point of view there is nothing to be added to methods of section 6.2. Instead we illustrate the procedure using the example of elastic polycrystals. We observe that $\hat{\boldsymbol{\Gamma}} = [\bar{\boldsymbol{\Gamma}}_1, \bar{\boldsymbol{\Gamma}}_2]$, where $\bar{\boldsymbol{\Gamma}}_j$ are isotropic elastic tensors with bulk and shear moduli κ_Γ^j, μ_Γ^j, $i = 1, 2$, given by (6.57) with κ_0 and μ_0, replaced by κ_j, μ_j, respectively. Recall that we have also applied the Milgrom-Shtrikman covariance transformation (6.78) in order to simplify the subspace $\hat{\mathscr{A}}$. Hence, the same covariance transformation applied to $\hat{\boldsymbol{\Gamma}}$ transforms it into $\hat{\boldsymbol{\Gamma}}^0 = [\bar{\boldsymbol{\Gamma}}_1^0, \bar{\boldsymbol{\Gamma}}_2^0]$, where $\bar{\boldsymbol{\Gamma}}_j^0$, $j = 1, 2$, are isotropic tensors with bulk and shear moduli κ_j^0, μ_j^0, $j = 1, 2$, given by

$$\kappa_j^0 = \frac{\kappa_\Gamma^j}{d^2\bar{\kappa}_j^2}, \qquad \mu_j^0 = \frac{\mu_\Gamma^j}{4\bar{\mu}_j^2}, \qquad j = 1, 2, \tag{6.86}$$

where $\bar{\kappa}_j$ and $\bar{\mu}_j$ are defined in (6.78).

6.5.1 Links in 3D elasticity

In 3D elasticity there is a single nontrivial link, corresponding to (6.82). If $\hat{\mathsf{M}}_0 = [\mathsf{M}_1, \mathsf{M}_2]$ is its inversion key, then the defining property (6.55) is equivalent to the following set of conditions on the two isotropic elastic tensors M_1, M_2.

- $\mathsf{K}(\bar{\boldsymbol{\Gamma}}_j^0 - \mathsf{M}_j)\mathsf{K} \in K_0' \oplus K_2'$, for all $\mathsf{K} \in K_0' \oplus K_2'$, $j = 1, 2$.
- $\mathsf{K}(\bar{\boldsymbol{\Gamma}}_j^0 - \mathsf{M}_j)\mathsf{J} + \mathsf{J}(\bar{\boldsymbol{\Gamma}}_j^0 - \mathsf{M}_j)\mathsf{K} \in K_0'$, for all $\mathsf{K} \in K_0' \oplus K_2'$, $\mathsf{J} \in K_0'$, $j = 1, 2$.
- $\mathsf{K}(\bar{\boldsymbol{\Gamma}}_1^0 - \bar{\boldsymbol{\Gamma}}_2^0 - \mathsf{M}_1 + \mathsf{M}_1)\mathsf{K} \in K_0'$, for all $\mathsf{K} \in K_0' \oplus K_2'$.

It is easy to see that $\mathsf{M}_j = \kappa_j^0 \boldsymbol{I}_3 \otimes \boldsymbol{I}_3 = \kappa_j^0 \mathsf{M}_0$ satisfies

$$\mathsf{K}(\bar{\boldsymbol{\Gamma}}_j^0 - \mathsf{M}_j)\mathsf{K} \in K_0', \quad \forall \mathsf{K} \in K_0' \oplus K_2'.$$

Hence, all the requirements in the list above for the inversion key are satisfied. In order to compute the link corresponding to (6.82) we will use the inversion formula (6.56) for each component:

$$\mathsf{C}_j = \mathsf{C}_j^0 + \bar{\mathsf{C}}_j^{-1}(\mathsf{I} - \kappa_j^0\mathsf{K}_j\mathsf{M}_0)^{-1}\mathsf{K}_j\bar{\mathsf{C}}_j^{-1}, \quad j = 1, 2, \tag{6.87}$$

where $\bar{\mathsf{C}}_j$ are isotropic elastic tensors with bulk and shear moduli $\bar{\kappa}_j$, $\bar{\mu}_j$, $j = 1, 2$, respectively, given by (6.78). From (6.62) we already know that

$$\mathsf{C}_j = 2\mu_j\mathsf{T} + \boldsymbol{D}_j \otimes \boldsymbol{D}_j, \quad j = 1, 2. \tag{6.88}$$

The goal is to compute the link between $\boldsymbol{D}_1$ and $\boldsymbol{D}_2$, corresponding to (6.82), which can also be recorded as

$$\operatorname{dev}(\boldsymbol{B}_1) = \operatorname{dev}(\boldsymbol{B}_2), \qquad \boldsymbol{B}_j = \mathsf{K}_j\boldsymbol{I}_d, j = 1, 2. \tag{6.89}$$

We first observe that

$$(\mathsf{I} - \kappa_j^0\mathsf{K}_j\mathsf{M}_0)^{-1}\mathsf{K}_j = \frac{1}{\kappa_j^0}(\mathsf{I} - (\kappa_j^0\mathsf{K}_j)\mathsf{M}_0)^{-1}(\kappa_j^0\mathsf{K}_j).$$

Then we can use formula (6.60) applied to $\kappa_j^0\mathsf{K}_j$ to obtain

$$(\mathsf{I} - \kappa_j^0\mathsf{K}_j\mathsf{M}_0)^{-1}\mathsf{K}_j = -\frac{\boldsymbol{I}_d \otimes \boldsymbol{I}_d}{d^2\kappa_j^0} + \boldsymbol{C}_j \otimes \boldsymbol{C}_j,$$

where

$$\boldsymbol{C}_j = \frac{\kappa_j^0\boldsymbol{B} + (1 - \kappa_j^0 \operatorname{Tr} \boldsymbol{B})\boldsymbol{I}_d/d}{\sqrt{\kappa_j^0(1 - \kappa_j^0 \operatorname{Tr} \boldsymbol{B})}}.$$

Hence,

$$\boldsymbol{D}_j = \bar{\mathsf{C}}_j^{-1}\boldsymbol{C}_j = \frac{\kappa_j^0\bar{\mathsf{C}}_j^{-1}\boldsymbol{B}_j + (1 - \kappa_j^0 \operatorname{Tr} \boldsymbol{B}_j)\boldsymbol{I}_d/(d^2\bar{\kappa}_j)}{\sqrt{\kappa_j^0(1 - \kappa_j^0 \operatorname{Tr} \boldsymbol{B})}}.$$

Taking a trace we get

$$\operatorname{Tr}(\boldsymbol{D}_j) = \frac{1}{d\bar{\kappa}_j\sqrt{\kappa_j^0(1 - \kappa_j^0 \operatorname{Tr} \boldsymbol{B})}}. \tag{6.90}$$

Taking the deviatoric part and using (6.90) and (6.86) we get

$$\operatorname{dev}(\boldsymbol{B}_j) = \frac{2\bar{\mu}_j\operatorname{dev}(\boldsymbol{D}_j)}{d\bar{\kappa}_j\kappa_j^0 \operatorname{Tr}(\boldsymbol{D}_j)} = \frac{2d\bar{\kappa}_j\bar{\mu}_j\operatorname{dev}(\boldsymbol{D}_j)}{\kappa_\Gamma^j \operatorname{Tr}(\boldsymbol{D}_j)}.$$

By (6.78) and (6.76) we find

$$\frac{\bar{\kappa}_j\bar{\mu}_j}{\kappa_\Gamma^j} = \begin{cases} \dfrac{m_0k_0a_j}{\kappa_\Gamma^j}, & d = 2, \\ \dfrac{k_0^2a_j}{\kappa_\Gamma^j}, & d = 3. \end{cases} = \begin{cases} 2m_0k_0, & d = 2, \\ 3k_0^2, & d = 3, \end{cases}$$

which is independent of the moduli of the reference media κ_j and μ_j. We conclude that (6.89) is equivalent to

$$\frac{\operatorname{dev}(\boldsymbol{D}_1)}{\operatorname{Tr}(\boldsymbol{D}_1)} = \frac{\operatorname{dev}(\boldsymbol{D}_2)}{\operatorname{Tr}(\boldsymbol{D}_2)}.$$

which can also be written as

$$\frac{\boldsymbol{D}_1}{\operatorname{Tr}(\boldsymbol{D}_1)} = \frac{\boldsymbol{D}_2}{\operatorname{Tr}(\boldsymbol{D}_2)}. \tag{6.91}$$

The link (6.91) can also be restated in simpler terms, if we observe that (6.91) is satisfied if and only if $\boldsymbol{D}_2$ is a scalar multiple of $\boldsymbol{D}_1$.

What this means physically is that if we have two composites $\mathsf{C}_1(z)$ and $\mathsf{C}_2(z)$ that have the form (6.88), where $\boldsymbol{D}_2(z) = \Lambda(z)\boldsymbol{D}_1(z)$, then $\boldsymbol{D}_{2*}$ is a scalar multiple of $\boldsymbol{D}_{1*}$

$$\boldsymbol{D}_{2*} = \Lambda_* \boldsymbol{D}_{1*},$$

where the scalar Λ_* is microstructure-dependent. We observe that Hill's exact relation (4.2) is obtained from this link, if we choose $\boldsymbol{D}_1 = \boldsymbol{I}_3$.

6.5.2 Links in 2D elasticity

Recall that in addition to (6.91) there are two more links in two-dimensional elasticity, corresponding to Jordan $\hat{\mathscr{A}}_0$-multialgebras (6.84) and

$$\hat{\Pi} = \{[\mathsf{K}, \mathsf{K}] : \mathsf{K} \in \mathbb{E}^2\}. \tag{6.92}$$

We note that the seemingly trivial multialgebra (6.92) may correspond to a nontrivial link since the reference media C_1^0 and C_2^0 are not necessarily scalar multiples of one another. In particular, we cannot choose $\hat{\mathsf{M}}_0 = [0, 0]$ for the inversion key for (6.92), since $\bar{\boldsymbol{\Gamma}}_1^0 \neq \bar{\boldsymbol{\Gamma}}_2^0$, while the inversion key $\hat{\mathsf{M}}_0 = [\mathsf{M}_1, \mathsf{M}_2]$ must satisfy

$$\mathsf{K}(\bar{\boldsymbol{\Gamma}}_1^0 - \mathsf{M}_1)\mathsf{K} = \mathsf{K}(\bar{\boldsymbol{\Gamma}}_2^0 - \mathsf{M}_2)\mathsf{K}, \qquad \forall \mathsf{K} \in \mathbb{E}^2.$$

This is equivalent to

$$\bar{\boldsymbol{\Gamma}}_1^0 - \mathsf{M}_1 = \bar{\boldsymbol{\Gamma}}_2^0 - \mathsf{M}_2. \tag{6.93}$$

Let us now compute the link corresponding to (6.92). In order to simplify (somewhat) the calculations we observe that the first component of $\hat{\mathsf{M}}_0$ can be chosen arbitrarily. Hence, we choose $\mathsf{M}_1 = 0$. In that case $\mathsf{M}_2 = \bar{\boldsymbol{\Gamma}}_2^0 - \bar{\boldsymbol{\Gamma}}_1^0$ and

$$\mathsf{K} = \bar{\mathsf{C}}_1(\mathsf{C}_1 - \mathsf{C}_1^0)\bar{\mathsf{C}}_1.$$

Substituting this expression into (6.56) for C_2 we obtain our initial representation of the link corresponding to (6.92):

$$\mathsf{C}_2 = \mathsf{C}_2^0 + \left[\bar{\mathsf{C}}_2 - \bar{\mathsf{C}}_1(\mathsf{C}_1 - \mathsf{C}_1^0)\bar{\mathsf{C}}_1(\bar{\boldsymbol{\Gamma}}_2^0 - \bar{\boldsymbol{\Gamma}}_1^0)\bar{\mathsf{C}}_2\right]^{-1}\bar{\mathsf{C}}_1(\mathsf{C}_1 - \mathsf{C}_1^0)\bar{\mathsf{C}}_1\bar{\mathsf{C}}_2^{-1}$$

We observe that

$$\bar{\mathsf{C}}_1(\bar{\boldsymbol{\Gamma}}_2^0 - \bar{\boldsymbol{\Gamma}}_1^0)\bar{\mathsf{C}}_2 = \xi\mathsf{T},$$

where T is given by (6.63) and

$$\xi = \frac{\mu_2\kappa_1 - \mu_1\kappa_2}{2\sqrt{\kappa_1\kappa_2\mu_1\mu_2(\kappa_1 + \mu_1)(\kappa_2 + \mu_2)}}.$$

After some matrix algebra we obtain

$$\mathsf{C}_2 = \xi^{-1}\mathsf{T}\left[\mathsf{C}_1^0 + \xi^{-1}\bar{\mathsf{C}}_1^{-1}\bar{\mathsf{C}}_2\mathsf{T} - \mathsf{C}_1\right]^{-1}\left\{(\mathsf{C}_1 - \mathsf{C}_1^0)(\bar{\mathsf{C}}_1\bar{\mathsf{C}}_2^{-1} - \xi\mathsf{T}\mathsf{C}_2^0) + \bar{\mathsf{C}}_1^{-1}\bar{\mathsf{C}}_2\mathsf{C}_2^0\right\}$$

We next compute

$$\mathsf{C}_1^0 + \xi^{-1}\bar{\mathsf{C}}_1^{-1}\bar{\mathsf{C}}_2\mathsf{T} = \eta\mathsf{T}, \qquad \eta = \frac{2\mu_1\kappa_1(\mu_2 + \kappa_2)}{\mu_2\kappa_1 - \mu_1\kappa_2}.$$

$$-\mathsf{C}_1^0(\bar{\mathsf{C}}_1\bar{\mathsf{C}}_2^{-1} - \xi\mathsf{T}\mathsf{C}_2^0) + \bar{\mathsf{C}}_1^{-1}\bar{\mathsf{C}}_2\mathsf{C}_2^0 = 0$$

Hence,

$$\bar{\mathsf{C}}_1\bar{\mathsf{C}}_2^{-1} - \xi\mathsf{T}\mathsf{C}_2^0 = (\bar{\mathsf{C}}_1\mathsf{C}_1^0)^{-1}\bar{\mathsf{C}}_2\mathsf{C}_2^0 = \beta\mathsf{I}, \qquad \beta^2 = \frac{\kappa_2\mu_2(\kappa_1 + \mu_1)}{\kappa_1\mu_1(\kappa_2 + \mu_2)}.$$

Thus, we obtain

$$\mathsf{C}_2 = \frac{\beta}{\xi}\mathsf{T}(\eta\mathsf{T} - \mathsf{C}_1)^{-1}\mathsf{C}_1. \tag{6.94}$$

The most compact form of this two-parameter family of links is obtained if we rewrite (6.94) in terms of compliance tensors $\mathsf{S}_j = \mathsf{C}_j^{-1}$, $j = 1, 2$:

$$\mathsf{S}_2 = b_0(\mathsf{S}_1 - a_0\mathsf{T}). \tag{6.95}$$

where

$$b_0 = \frac{\xi\eta}{\beta} = \frac{\kappa_1\mu_1(\kappa_2 + \mu_2)}{\mu_2\kappa_2(\mu_1 + \kappa_1)}, \qquad a_0 = \frac{1}{\eta} = \frac{\mu_2\kappa_1 - \mu_1\kappa_2}{2\mu_1\kappa_1(\mu_2 + \kappa_2)}.$$

This family of links was discovered in [4].

Finally, we want to compute the link, corresponding to $\hat{\mathscr{A}}_0$-multialgebra (6.84). Before we plunge into computations let us mention another important simplifying device. We have seen how linear independence of isotropic reference media tensors C_1^0 and C_2^0 in $\hat{\mathsf{C}}^0 = [\mathsf{C}_1^0, \mathsf{C}_2^0]$ complicated our calculations. In some sense we have overcome this difficulty when we computed the global link (6.95), since it has the property that for any $\hat{\mathsf{C}}^0 = [\mathsf{C}_1^0, \mathsf{C}_2^0]$ we can choose values of a_0 and b_0, so that the link (6.95) passes through $\hat{\mathsf{C}}^0$. Now it is sufficient to compute only the link corresponding to (6.84) that passes through $\hat{\mathsf{C}}^0 = [\mathsf{I}, \mathsf{I}]$. Indeed, let $\mathsf{C}_2 = \Lambda(\mathsf{C}_1)$,

$\mathsf{C}_1 \in \mathbb{M}$ be such a link, where $\mathbb{M}$ is the set of all elastic tensors $\mathsf{C} > 0$ with the property $\mathsf{C}\boldsymbol{I}_d = d\kappa_0 \boldsymbol{I}_d$ for some constant bulk modulus $\kappa_0 > 0$. Let $\mathsf{C}_0' = \Lambda(\mathsf{C}_0)$. Let Ψ be the global link (6.95), such that $\Psi(\mathsf{I}) = \mathsf{C}_0$ and let Ψ' be the global link, satisfying $\Psi'(\mathsf{C}_0') = \mathsf{I}$. Consider now the map

$$\tilde{\Lambda}(\mathsf{C}) = \Psi'(\Lambda(\Psi(\mathsf{C}))).$$

Then, $\tilde{\Lambda}(\mathsf{I}) = \mathsf{I}$ and $\tilde{\Lambda}(\mathsf{C})$ is also a link. Indeed, let $\mathsf{C}_1(z)$ be a composite and $\mathsf{C}_2(z) = \tilde{\Lambda}(\mathsf{C}_1(z))$. Introduce related composites $\mathsf{C}_3(z) = \Psi(\mathsf{C}_1(z))$ and $\mathsf{C}_4(z) = \Lambda(\mathsf{C}_3(z))$. Then $\mathsf{C}_2(z) = \Psi'(\mathsf{C}_4(z))$. If C_{j*}, is the effective tensor of $\mathsf{C}_j(z)$, $j = 1, 2, 3, 4$, then, since Ψ and Ψ' are links, $\mathsf{C}_{3*} = \Psi(\mathsf{C}_{1*})$ and $\mathsf{C}_{2*} = \Psi'(\mathsf{C}_{4*})$. By assumption Λ is a link, and therefore $\mathsf{C}_{4*} = \Lambda(\mathsf{C}_{3*})$. But then $\mathsf{C}_{2*} = \tilde{\Lambda}(\mathsf{C}_{1*})$, proving that $\tilde{\Lambda}(\mathsf{C})$ is a link. Conversely, every link Λ can be obtained as a $[\Psi, \Psi']$-image of a link $\tilde{\Lambda}$, passing through $\hat{\mathsf{C}}^0 = [\mathsf{I}, \mathsf{I}]$. In this case

$$\bar{\boldsymbol{\Gamma}}_1^0 = \bar{\boldsymbol{\Gamma}}_2^0 = K_0'\left(\frac{1}{4}\right) + K_0\left(\frac{3}{8}\right).$$

We claim that

$$\hat{\mathsf{M}} = \left[K_0\left(\frac{3}{8}\right), K_0\left(\frac{3}{8}\right)\right]$$

is an inversion key. Indeed, it follows from an obvious relation

$$\mathsf{K}\mathsf{K}_0'\mathsf{K} = 0, \qquad \forall\, \mathsf{K} \in \mathrm{Sym}(\mathrm{Sym}_0(\mathbb{R}^2)).$$

The inversion formulas (6.56) read

$$\mathsf{C}_1 = \mathsf{I} - \left(\mathsf{I} - \frac{3}{8}\mathsf{K}K_0(1)\right)^{-1}\mathsf{K}, \quad \mathsf{C}_2 = \mathsf{I} - \left(\mathsf{I} - \frac{3}{8}\Phi(\mathsf{K})K_0(1)\right)^{-1}\Phi(\mathsf{K}),$$

where Φ is given by (6.83). It is clear that $\boldsymbol{I}_2$ is an eigenstrain for both C_1 and C_2 (with eigenvalue 1). Therefore, the action needs to be restricted to the two-dimensional space of symmetric, trace-free 2×2 matrices. On that space $K_0(1) = 2\mathsf{I}$ and hence,

$$\mathsf{C}_1 = \mathsf{I} - \left(\mathsf{I} + \frac{3}{4}\mathsf{K}\right)^{-1}\mathsf{K}, \quad \mathsf{C}_2 = \mathsf{I} - \left(\mathsf{I} + \frac{3}{4}\Phi(\mathsf{K})\right)^{-1}\Phi(\mathsf{K}). \tag{6.96}$$

This shows that C_1 is a rational function of K on $\mathrm{Sym}_0(\mathbb{R}^2)$. A crucial observation is that the transformation Φ preserves eigenstrains of K. Indeed, on $\mathrm{Sym}_0(\mathbb{R}^2)$

$$\Phi(\mathsf{K}) = \mathsf{K} - (\mathrm{Tr}\,\mathsf{K})\mathsf{I},$$

where the trace is taken on $\mathrm{Sym}_0(\mathbb{R}^2)$. Thus, C_1 and C_2 have the same eigenstrains, while the relation between their eigenvalues can be computed from (6.96). Suppose that K has eigenvalues α_1, α_2. Then the eigenvalues of C_1 are

$$2\mu_j = 1 - \frac{\alpha_j}{1 + 3\alpha_j/4}, \qquad j = 1, 2. \tag{6.97}$$

The eigenvalues of $\Phi(\mathsf{K})$ will be $-\alpha_2$, $-\alpha_1$, i.e., $-\alpha_2$ is the eigenvalue of $\Phi(K)$ corresponding to the α_1 eigenstrain of K. Thus, the eigenvalues of C_2 are

$$2\mu'_j = 1 + \frac{\alpha_{j'}}{1 - 3\alpha_{j'}/4}, \qquad j = 1, 2,$$

where

$$j' = 3 - j = \begin{cases} 2, & j = 1, \\ 1, & j = 2. \end{cases}$$

Equivalently,

$$2\mu'_{j'} = 1 + \frac{\alpha_j}{1 - 3\alpha_j/4}, \qquad j = 1, 2, \tag{6.98}$$

Eliminating α_j from (6.97), (6.98) we obtain

$$\left(\frac{1}{\mu_j} + 2\right)\left(\frac{1}{\mu'_{j'}} + 2\right) = 16, \qquad j = 1, 2.$$

In order to obtain the general form of this link we apply the global link (6.95), and after somewhat tedious calculations obtain

$$\left(\frac{1}{\mu_j} + \frac{1}{\kappa_0}\right)\left(\frac{1}{\mu'_{j'}} + \frac{1}{\kappa'_0}\right) = a_0^2, \qquad j = 1, 2, \tag{6.99}$$

where a_0 is an arbitrary constant and κ_0 and κ'_0 are the constant bulk moduli of $\mathsf{C}_1(z)$ and $\mathsf{C}_2(z)$, respectively.

References

[1] Auffray N, Kolev B and Petitot M 2014 On anisotropic polynomial relations for the elasticity tensor *J. Elast.* **115** 77–103

[2] Biedenharn L C and Louck J D 1981 *Angular Momentum in Quantum Physics Encyclopedia of Mathematics and Its Applications* **vol 8** (Reading, MA: Addison-Wesley) Theory and application, With a foreword by P A Carruthers

[3] Bröcker T and Dieck T 1985 *Representations of Compact Lie Groups Graduate Texts in Mathematics* **vol 98** (New York: Springer)

[4] Cherkaev A V and Lurie K A 1992 Invariant properties of the stress in plane elasticity and equivalence classes of composites *Proc. R. Soc. Lond.* A **438** 519–29

[5] Edmonds A R 1996 *Angular Momentum in Quantum Mechanics* (Princeton, NJ: Princeton University Press)

[6] Grabovsky Y 2004 Algebra, geometry and computations of exact relations for effective moduli of composites *Advances in Multifield Theories of Continua with Substructure* Modeling and Simulation in Science, Engineering and Technology G Capriz and P M Mariano (Boston, MA: Birkhäuser) pp 167–97

[7] Grabovsky Y 2009 Exact relations for effective conductivity of fiber-reinforced conducting composites with the Hall effect via a general theory *SIAM J. Math Anal.* **41** 973–1024
[8] Grabovsky Y 2020 Exact relations and links for two-dimensional thermoelectric composites arXiv:2010.05051
[9] Grabovsky Y, Milton G W and Sage D S 2000 Exact relations for effective tensors of polycrystals: necessary conditions and sufficient conditions *Commun. Pure Appl. Math.* **53** 300–53
[10] Hegg M 2012 Exact relations and links for fiber-reinforced elastic composites *PhD Thesis* Temple University, Philadelphia, PA https://scholarshare.temple.edu/items/cc8db7cf-664e-467e-997e-eeebd70cf092
[11] Hegg M 2013 Links between effective tensors for fiber-reinforced elastic composites *C. R. Méc.* **341** 520–32
[12] Lurie K A, Cherkaev A V and Fedorov A V 1984 On the existence of solutions to some problems of optimal design for bars and plates *J. Optim. Theor. Appl.* **42** 247–81
[13] Milgrom M and Shtrikman S 1989 Linear response of polycrystals to coupled fields: exact relations among the coefficients *Phys. Rev. B (Solid State)* **40** 5991–4
[14] Milgrom M and Shtrikman S 1989 Linear response of two-phase composites with cross moduli: exact universal relations *Phys. Rev.* **40** 1568–75
[15] Milton G W 2002 *The Theory of Composites Cambridge Monographs on Applied and Computational Mathematics* (Cambridge: Cambridge University Press)
[16] To H 2004 Homogenization of dynamic materials *PhD Thesis* Temple University, Philadelphia, PA
[17] Weyl H 1931 *Theory of Groups and Quantum Mechanics* (New York: E.P. Dutton) Translated from the second (revised) German editionH P Robertson
[18] Wigner E P 1931 *Gruppentheorie und ihre Anwendung auf die Quantenmechanik der Atomspektren* (Braunschweig: Friedr. Vieweg & Sohn)

Part III

Case studies

Chapter 7

Introduction

This part of the book is a compendium of results on exact relations and links for effective properties of conducting, elastic, piezoelectric, thermoelastic and thermoelectric composites, given without proof, but with references to published derivations, when they exist. If no references are given, then the author is not aware of a prior publication, where a given formula is derived. Many results in the literature, such as [2, 7, 11, 19], for example, give formulas for effective moduli of two phase composites. Such formulas will not be listed explicitly among exact relations, because they do not hold for general multiphase composites. Instead, these results can be derived using links that map the two constituents into one of the general exact relations in our lists, as is done for conductivity in Section 8.1.2 and thermoelectricity in section 12.4.4. See [5, 6, 8] for details and additional examples.

The results listed here are general. They are valid for *all* composites: multiphase, polycrystalline, periodic, random, functionally graded, etc. Moreover, composites themselves do not have to be homogeneous, as is the case for functionally graded composites. The effective tensors of such composites are also inhomogeneous, i.e., spatially-dependent, and all our formulas apply pointwise. The applicability of listed results depends on the physical validity of three assumptions:

- the response of each material in a composite is linear (linear constitutive law);
- there are no residual stresses, net charges or thermal imbalances;
- the constituents are in perfect mechanical/electric/thermal contact.

Even though the latter two assumptions might not be valid for real composites, one can use formulas in this part of the book to estimate the extent to which the failure of these assumptions has affected the effective behavior of composites, by providing benchmarks for comparison between theory and experiment.

Exact relations are sets of explicit formulas satisfied by the effective parameters of a composite involving properties and volume fractions of constituents that are valid

doi:10.1088/978-0-7503-6249-8ch7

for *all microstructures*. By contrast, *links* relate effective moduli of two composites with the same microstructure, but different properties of the constituents.

One of the earliest example of an exact relation was encountered by Keller in [12], and was proved in full generality by Dykhne [4]. It states that any two-dimensional conducting composite, all of whose constituents have the same determinant d_0 of their conductivity tensors, must have the effective conductivity tensor $\boldsymbol{\sigma}_*$, satisfying $\det \boldsymbol{\sigma}_* = d_0$, *regardless of the microstructure*. In fact, the microstructure, or texture, as it is called for polycrystals, affects all components of the effective conductivity tensor, and only their specific combination, the determinant, remains completely insensitive to the microstructure. In our list of exact relations, this result would be listed as a single equation

$$\det \boldsymbol{\sigma} = d_0, \tag{7.1}$$

to be interpreted as described above.

In order to explain the concept of a link, let us assume that we somehow know the effective tensor L_* of a composite whose microscopic properties are described by the local tensor $\mathsf{L}(\boldsymbol{x})$, specifying material properties at each point $\boldsymbol{x}$ in a composite. Now, let us imagine that at every point in our composite we have replaced the material $\mathsf{L}(\boldsymbol{x})$ with the material $\mathsf{L}'(\boldsymbol{x}) = \lambda_0 \mathsf{L}(\boldsymbol{x})$, where λ_0 is a positive scalar, independent of $\boldsymbol{x}$. Then the effective tensor L'_* of the new composite will be $\lambda_0 \mathsf{L}_*$. This obvious, for experts, statement is an example of a link, which in our notation will be recorded as $\mathsf{L}' = \lambda_0 \mathsf{L}$. An example of a nontrivial link in the context of two-dimensional conductivity is due to Mendelson [15, 16]. Once again we assume that we somehow know the effective conductivity tensor $\boldsymbol{\sigma}_*$ of a two-dimensional conducting composite whose microscopic properties are described by the local tensor field $\boldsymbol{\sigma}(\boldsymbol{x})$. Now, let us imagine that at every point $\boldsymbol{x}$ in this composite we have replaced the material $\boldsymbol{\sigma}(\boldsymbol{x})$ with the material

$$\boldsymbol{\sigma}'(\boldsymbol{x}) = \frac{\boldsymbol{\sigma}(\boldsymbol{x})}{\det \boldsymbol{\sigma}(\boldsymbol{x})}. \tag{7.2}$$

Then, according to [15, 16], the effective tensor $\boldsymbol{\sigma}'_*$ of the new composite must be

$$\boldsymbol{\sigma}'_* = \frac{\boldsymbol{\sigma}_*}{\det \boldsymbol{\sigma}_*}. \tag{7.3}$$

We emphasize that in this result, no constraints whatsoever were placed on the microstructure of the original composite $\boldsymbol{\sigma}(\boldsymbol{x})$. In our notation, this result will be recorded as a single equation

$$\boldsymbol{\sigma}' = \frac{\boldsymbol{\sigma}}{\det \boldsymbol{\sigma}}. \tag{7.4}$$

In general links contain a wealth of information. For example, they can simplify and even obviate specific exact relations. For instance, we can use the 'trivial' link $\boldsymbol{\sigma}' = \lambda_0 \boldsymbol{\sigma}$ applied to the two-dimensional conductivity to simplify the Keller–Dykhne exact relation (7.1), by eliminating the parameter d_0:

$$\det \boldsymbol{\sigma} = 1. \tag{7.5}$$

That means that we can always recover (7.1) by combining (7.5) and the link $\boldsymbol{\sigma}' = \lambda_0 \boldsymbol{\sigma}$. In turn, the simplified exact relation (7.5) is a consequence of (7.4). To see this, we observe that the set of all tensors left invariant by any link must be an exact relation. Clearly, the set of conductivity tensors $\boldsymbol{\sigma}$ left invariant by (7.4) is (7.5).

In our examples of links above, the first composite in the description of the link was arbitrary. Links like that are called *global links.* In general, links may be defined only for composites belonging to a specific exact relation. For example, in two-dimensional elasticity there is a link discovered by Lurie, Cherkaev and Fedorov [14] that says that composite polycrystals whose constituents have a square symmetry and one and the same bulk modulus κ_0 must always have square symmetry (even for highly anisotropic microstructures) and that same bulk modulus κ_0. All two-dimensional elasticity tensors belonging to this exact relation have a common eigenvalue $2\kappa_0$, corresponding to the common eigenvector $\boldsymbol{I}_2$, and two other eigenvalues $2\mu_1$, $2\mu_2$. There is a link (as it turns out, not previously known) defined only for elastic tensors belonging to the Lurie–Cherkaev–Fedorov exact relation that maps C to C' with the same eigenvectors and eigenvalues $2\kappa_0'$, $2\mu_1'$, $2\mu_2'$, determined by the equations

$$\left(\frac{1}{\kappa_0} + \frac{1}{\mu_1}\right)\left(\frac{1}{\kappa_0'} + \frac{1}{\mu_2'}\right) = \left(\frac{1}{\kappa_0} + \frac{1}{\mu_2}\right)\left(\frac{1}{\kappa_0'} + \frac{1}{\mu_1'}\right) = c_0, \tag{7.6}$$

where $c_0 > (\kappa_0 \kappa_0')^{-1}$ is an arbitrary constant. Those familiar with the Lurie–Cherkaev exact relation [13] will immediately recognize that tensors belonging to it are exactly the ones left invariant by the link (7.6). A final remark is that links need not be *functions* mapping L into L', they can relate (link) only parts of L and L', leaving remaining parts unrelated. For example, in the setting of fibrous conducting composites, the effective conductivity in the plane transversal to the fibers depends only on the local transversal conductivity. That means that two composites, whose local transversal conductivity tensors are the same, while other components of the full conductivity tensor are unrelated, will have effective conductivity tensors with the same transversal component. In the case studies in this part of the book we will encounter links of all of the described types, that range from simple and natural to unexpected and strange.

We find it convenient (and elegant) to think of exact relations geometrically, as submanifolds $\mathbb{M}$ (of positive codimension) in the space of material tensors. In our example of two-dimensional conductivity, the set of all material tensors is an open subset $\mathrm{Sym}^+(\mathbb{R}^2)$ of all positive definite matrices in the three dimensional space of all 2×2 symmetric matrices. The two-dimensional submanifold $\mathbb{M} = \{\boldsymbol{\sigma} \in \mathrm{Sym}^+(\mathbb{R}^2)\colon \det \boldsymbol{\sigma} = d_0\}$, corresponding to the Keller–Dykhne exact relation (7.1), can be visualized as an entire sheet of a two-sheeted hyperboloid. Very often it will be convenient to define the exact relation submanifolds $\mathbb{M}$ by parametric equations, instead of implicit equations, like (7.1). An example of such a description is a well-know Hill's exact relation [10, 11], that says that if a composite is made of any number of isotropic materials with the same shear modulus μ_0

then, regardless of the microstructure, its effective tensor will always be isotropic and have the same shear modulus μ_0. Geometrically, this exact relation is a one-dimensional submanifold $\mathbb{M}$ in the 21-dimensional space $\mathrm{Sym}^+(\mathrm{Sym}(\mathbb{R}^3))$ of elastic tensors, which are symmetric positive definite operators on the six-dimensional space $\mathrm{Sym}(\mathbb{R}^3)$ of symmetric 3×3 matrices. The submanifold $\mathbb{M} \subset \mathrm{Sym}^+(\mathrm{Sym}(\mathbb{R}^3))$ corresponding to the Hill exact relation is given by parametric equations

$$\mathbb{M}_{\mu_0} = \{\mathsf{C} \in \mathrm{Sym}^+(\mathrm{Sym}(\mathbb{R}^3)) \colon \mathsf{C} = 2\mu_0 \mathsf{I} + \lambda \boldsymbol{I}_3 \otimes \boldsymbol{I}_3, \;\; \lambda \in \mathbb{R}\}. \tag{7.7}$$

The action of $\mathsf{C} \in \mathbb{M}_{\mu_0}$ on a strain tensor ε is given by

$$(2\mu_0 \mathsf{I} + \lambda \boldsymbol{I}_3 \otimes \boldsymbol{I}_3)\varepsilon = 2\mu_0 \varepsilon + \lambda(\mathrm{Tr}\, \varepsilon)\boldsymbol{I}_3.$$

The Lamé modulus λ plays the role of a variable parameter, determining the point on $\mathbb{M}_{\mu_0}$, while the constant shear modulus μ_0 labels a specific exact relation in an infinite family $\mathbb{M}_{\mu_0}$ of exact relations. Hill's theorem that each submanifold $\mathbb{M}_{\mu_0}$, $\mu_0 > 0$ represents an exact relation can be stated as a formula $\mathsf{C}_* = 2\mu_0 \mathsf{I} + \lambda_* \boldsymbol{I}_3 \otimes \boldsymbol{I}_3$ for the effective elastic tensor of a composite $\mathsf{C}(\boldsymbol{x}) = 2\mu_0 \mathsf{I} + \lambda(\boldsymbol{x}) \boldsymbol{I}_3 \otimes \boldsymbol{I}_3$, where λ_* depends on the microstructure. As it turns out, the dependence of λ_* on the microstructure is fairly simple, given by

$$\frac{1}{2\mu_0 + \lambda_*} = \left\langle \frac{1}{2\mu_0 + \lambda} \right\rangle, \tag{7.8}$$

where $\langle \cdot \rangle$ denotes an average, understood in the context of the composite class under consideration. For periodic composites, it is the average over a period cell, for random composites it is an ensemble average, or an average over a representative volume element. For more mathematically inclined readers $\langle \cdot \rangle$ stands for the L^∞ weak-* limit of a sequence of oscillating fields, in which case (7.8) represents an equality between objects obtained via the G or H-limit (denoted by stars) and objects obtained via a much simpler L^∞ weak-* limit. In our list of exact relations, Hill's exact relation (7.7), (7.8) will be listed as

$$\mathsf{C} = 2\mu_0 \mathsf{I} + \lambda \boldsymbol{I}_3 \otimes \boldsymbol{I}_3, \qquad \lambda \in \mathbb{R}, \qquad \left\langle \frac{1}{2\mu_0 + \lambda} \right\rangle. \tag{7.9}$$

An essential part of our system of notation is the consistent use of zero subscript to denote constants, which do not vary throughout the composite, and parameters without the zero subscript that are allowed to vary from point to point in a composite *arbitrarily*, while the explicit dependence on $\boldsymbol{x}$ is not shown in our notation.

The theory described in part II of the book makes it possible to compute *complete* lists of exact relations and links for conducting, elastic, piezoelectric, thermoelastic, and thermoelectric composites. In cases where anisotropic constituents are used in a composite, it is often practically impossible to control the orientation in which each anisotropic crystallite occurs within a composite. We therefore focus almost exclusively on *polycrystalline* exact relations and links, whereby all anisotropic constituents are allowed to be used in arbitrary orientations. A partial exception to this rule occurs in fibrous composites, where anisotropic cylindrical fibers (of arbitrary cross-section) can be injected into an isotropic matrix. In this case, it is natural to assume that we can keep the orientation of the fiber axis fixed, while rotations of fibers around the fiber axis cannot be controlled. In our theory such composites are also regarded as polycrystalline, since in the idealized setting of perfectly parallel fibers the microstructure is essentially two-dimensional, determined by any cross-section perpendicular to the fibers. Hence, mathematically, such composites can be treated as two-dimensional polycrystalline. Another exception to the polycrystalline rule occurs in the study of conducting composites with Hall effect, where the anisotropy is induced by a fixed orientation of the ambient magnetic field, which also breaks the symmetry of conductivity tensors. In this specific case all exact relations and links, without restriction, are computed.

The actual derivation of complete lists of exact relations and links, while feasible, is often very labor-intensive. In many cases it required combined efforts of many students, both graduate and undergraduate, working over several academic years with the author to accomplish the task, as in some cases the number of different families of exact relations and links number into hundreds. In order to make information contained in these results presentable, it was expedient to eliminate logically redundant facts. For example, we could have chosen to omit the Keller–Dykhne exact relation (7.1) from the list of formulas for two-dimensional conducting composites because, as we have shown above, it is a logical corollary of Mendelson's link (7.4) (and the universal homogeneity link $\mathsf{L}' = \lambda_0 \mathsf{L}$). After elimination of logical redundancies the remaining several dozens essential results, could then be effectively presented. Thus, one has to keep in mind that completeness of the lists in this part of the book is understood not in the sense that every single exact relation and link is listed, but in the sense that every exact relation and link can be obtained, if desired, as a logical consequence of listed results. As a rule, no effort has been made to make all of the lists maximally logically reduced, as long as they can be conveniently and concisely presented.

It is natural to ask if there is a simple way to understand or explain exact relation formulas in terms of standard physical principles. The answer is 'no' in general, in the sense that the mathematical theory developed in this book delivering all exact relations and links in a unified way cannot be interpreted in terms of equally unified general physical principles. However, there are classes of exact relations and links that can be obtained by particular methods (e.g., [17]). A most well-known example of such a class of exact relations is called, following Milton [18], *uniform field relations* (UFR). These relations are consequences of the observation, that if all constituents in a composite respond identically to a particular applied field, then any composite, made with these materials, will also respond in exactly the same way to that applied field. Physically, this is obvious, since, as far as that applied field is concerned, all constituent materials are indistinguishable from one another. This idea originated in [9, 11], in fact, Hill's exact relation (7.7) is an example of a uniform field relation. It was subsequently formalized in [3, 14], among others. This simple idea becomes more and more effective as the number of coupled physical properties grow. It results in a very large class of exact relations in the context of piezothermoelectroelastic fibrous composites exhibiting pyroelectricity [1]. All uniform field relations are also easily identifiable in the framework of our general theory. In each of our case studies the uniform field relations are identified, by the 'UFR' label.

References

[1] Benveniste Y 1993 Exact results in the micromechanics of fibrous piezoelectric composites exhibiting pyroelectricity *Proc. R. Soc. Lond.* A **441** 59–81

[2] Dunn M L 1993 Exact relations between the thermoelectroelastic moduli of heterogeneous materials *Proc. R. Soc. Lond.* A **441** 549–57

[3] Dvorak G J 1990 On uniform fields in heterogeneous media *Proc. R. Soc. Lond.* A **431** 89–110

[4] Dykhne A M 1971 *Sov. Phys. JETP* **32** 63–5 [Zh. Eksp. Teor. Fiz., 59 (1970) 110–5.]

[5] Grabovsky Y 2009 An application of the general theory of exact relations to fiber-reinforced conducting composites with Hall effect *Mech. Mater.* **41** 456–62

[6] Grabovsky Y 2020 Exact relations and links for two-dimensional thermoelectric composites (arXiv preprint arXiv:2010.05051)

[7] Hashin Z and Rosen B W 1964 The elastic moduli of fiber-reinforced materials *J. Appl. Mech.* **31** 223–32

[8] Hegg M 2013 Links between effective tensors for fiber-reinforced elastic composites *C. R. Méc.* **341** 520–32

[9] Hill R 1952 The elastic behavior of a crystalline aggregate *Proc. Phys. Soc. Lond.* A **65** 349–54

[10] Hill R 1963 Elastic properties of reinforced solids: some theoretical principles *J. Mech. Phys. Solids* **11** 357–72

[11] Hill R 1964 Theory of mechanical properties of fibre-strengthened materials: I. Elastic behaviour *J. Mech. Phys. Solids* **12** 199–212

[12] Keller J B 1964 A theorem on the conductivity of a composite medium *J. Math. Phys.* **5** 548–9

[13] Lurie K A and Cherkaev A V 1984 G-closure of some particular sets of admissible material characteristics for the problem of bending of thin plates *J. Opt. Theor. Appl.* **42** 305–16

[14] Lurie K A, Cherkaev A V and Fedorov A V 1984 On the existence of solutions to some problems of optimal design for bars and plates *J. Optim. Theor. Appl.* **42** 247–81

[15] Mendelson K S 1975 Effective conductivities of two-phase material with cylindrical phase boundaries *J. Appl. Phys.* **46** 917–8

[16] Mendelson K S 1975 A theorem on the conductivity of two-dimensional heterogeneous medium *J. Appl. Phys.* **46** 4740–1

[17] Milton G W 1997 Composites: a myriad of microstructure independent relations *Theoretical and applied mechanics (Proc. of the XIX Int. Congress of Theoretical and Applied Mechanics, Kyoto, 1996)* ed T Tatsumi, E Watanabe and T Kamb (Amsterdam: Elsevier)) 443–59

[18] Milton G W 2002 *The Theory of Composites Cambridge Monographs on Applied and Computational Mathematics* (Cambridge: Cambridge University Press)

[19] Schulgasser K 1992 Relationships between the effective properties of transversely isotropic piezoelectric composites *J. Mech. Phys. Solids* **40** 473–9

Chapter 8

Conductivity with Hall effect

Consider composites made with conductors exhibiting the Hall effect. That means that the electric field $\boldsymbol{e}$ and the current field $\boldsymbol{j}$ are related by a linear transformation $\boldsymbol{L}$, whose symmetric part $\boldsymbol{\sigma}$ represents the conductivity tensor of the material, and whose antisymmetric part depends (linearly) on the applied magnetic field. The system combining Maxwell equations with the linear constitutive law

$$\nabla \times \boldsymbol{e} = 0, \qquad \nabla \cdot \boldsymbol{j} = 0, \qquad \boldsymbol{j} = \boldsymbol{L}\boldsymbol{e}, \tag{8.1}$$

augmented with appropriate boundary conditions determines the electric and the current field inside the material uniquely.

8.1 2D conductivity with Hall effect

Physically, 2D conductivity results would apply to fibrous conducting composites placed in a uniform magnetic field directed along the fibers, while both the electric field and the current field are perpendicular to the fibers. The arrangement and cross-section of fibers can be arbitrary, including the case where different fibers have different cross-sections. All exact relations and links in this section can be found elsewhere in the literature (see, e.g., [4, 5, 11, 14]). The calculations using the theory of exact relations can be found in [6, 8]. The Hall conductivity tensors in this context will be represented as $\boldsymbol{L} = \boldsymbol{\sigma} + r\boldsymbol{R}_{\perp}$, where $\boldsymbol{\sigma}$ denotes the symmetric part of the conductivity tensor, and

$$\boldsymbol{R}_{\perp} = \begin{bmatrix} 0 & -1 \\ 1 & 0 \end{bmatrix}.$$

The Hall coefficient r depends linearly on the applied magnetic field. However, since we will not be studying the dependence of Hall conductivity on the magnetic field, the dependence of r on the magnetic field will not be made explicit.

doi:10.1088/978-0-7503-6249-8ch8

8.1.1 Exact relations and links

Exact relations

1. (UFR) If $\boldsymbol{La}_0 = \boldsymbol{b}_0$
2. $r = r_0$
3. $(r - r_0)^2 + \det \boldsymbol{\sigma} = c_0$

Links

1. $\boldsymbol{L}' = \boldsymbol{L}^T$
2. $\boldsymbol{L}' = \boldsymbol{L} + r_0 \boldsymbol{R}_\perp$.
3. $\boldsymbol{L}' = \boldsymbol{\sigma}' + r'\boldsymbol{R}_\perp$, where

$$\boldsymbol{\sigma}' = \frac{c_0 \boldsymbol{\sigma}}{(r_0 - r)^2 + \det \boldsymbol{\sigma}}, \qquad r' = \frac{c_0(r_0 - r)}{(r_0 - r)^2 + \det \boldsymbol{\sigma}} + r_0'. \tag{8.2}$$

Technically speaking, there are other results, but all of them are consequences of the ones listed above. Our theory also guarantees that there are no other genuinely different microstructure-independent equalities.

8.1.2 Applications

As an example of application of these formulas we consider a polycrystalline composite made of a single crystal with conductivity tensor $\boldsymbol{L}_0 = \boldsymbol{\sigma}_0 + r_0 \boldsymbol{R}_\perp$. Then the effective tensor of such a polycrystal must have the form $\boldsymbol{L}_* = \boldsymbol{\sigma}_* + r_0 \boldsymbol{R}_\perp$, where $\det \boldsymbol{\sigma}_* = \det \boldsymbol{\sigma}_0$. In particular, if the polycrystalline texture of the composite is statistically isotropic, then $\boldsymbol{L}_* = \sqrt{\det \boldsymbol{\sigma}_0}\, \boldsymbol{I}_2 + r_0 \boldsymbol{R}_\perp$.

A more powerful example is provided by composites made of two isotropic constituents

$$\boldsymbol{L}_i = \sigma_i \boldsymbol{I}_2 + r_i \boldsymbol{R}_\perp, \quad i = 1, 2. \tag{8.3}$$

Then all components of the effective tensor $\boldsymbol{L}_*$ can be expressed in terms of a single microstructure-dependent function $\Sigma_*(h)$, which is the effective conductivity of the original composite in which materials $\boldsymbol{L}_1$ and $\boldsymbol{L}_2$ are replaced with materials $\boldsymbol{I}_2$ and $h\boldsymbol{I}_2$, respectively. Assuming $r_1 \neq r_2$, we first use the global link (8.2) to 'eliminate' the Hall effect. In other words, we need to choose the constants r_0, r_0' and c_0 in (8.2), such that $r_1' = r_2' = 0$ and $\boldsymbol{\sigma}_1' = \boldsymbol{I}_2$. Thus, we need to solve the quadratic equation

$$\frac{r_0 - r_1}{(r_0 - r_1)^2 + \sigma_1^2} = \frac{r_0 - r_2}{(r_0 - r_2)^2 + \sigma_2^2} \tag{8.4}$$

for r_0, and then set

$$c_0 = \frac{(r_0 - r_1)^2 + \sigma_1^2}{\sigma_1}, \qquad r_0' = -\frac{c_0(r_0 - r_1)}{(r_0 - r_1)^2 + \sigma_1^2}.$$

It is easy to see that equation (8.4) is equivalent to a quadratic equation with two distinct real roots (assuming $\boldsymbol{L}_1 \neq \boldsymbol{L}_2$). Let r_0 denote one of the roots. (Choosing either root for r_0 does not affect the final answer.) Then, $\boldsymbol{L}_1$ gets mapped to $\boldsymbol{I}_2$ and $\boldsymbol{L}_2$ gets mapped to $h\boldsymbol{I}_2$, where

$$h = \frac{\sigma_2}{\sigma_1} \cdot \frac{(r_0 - r_1)^2 + \sigma_1^2}{(r_0 - r_2)^2 + \sigma_2^2} > 0.$$

We note that equation (8.4) stays the same if we interchange indices 1 and 2. However, h gets transformed to $1/h$. Hence, we may assume, without loss of generality, that $h > 1$. In fact, we may derive the equation for h in terms of σ_i, r_i, $i = 1, 2$:

$$\sigma_1\sigma_2 h^2 - (\sigma_1^2 + \sigma_2^2 + (r_1 - r_2)^2)h + \sigma_1\sigma_2 = 0, \quad h > 1. \tag{8.5}$$

Then, we compute

$$r_0' = \frac{\sigma_2 h - \sigma_1}{r_1 - r_2}, \qquad r_0 = r_1 - \sigma_1 r_0', \qquad c_0 = \sigma_2 h + \sigma_1 + r_0' \frac{\sigma_2^2 - \sigma_1^2}{r_1 - r_2}. \tag{8.6}$$

Finally, solving the equations

$$\frac{c_0 \boldsymbol{\sigma}^*}{(r_0 - r^*)^2 + \det \boldsymbol{\sigma}^*} = \boldsymbol{\Sigma}_*(h\boldsymbol{I}_2), \qquad \frac{c_0(r_0 - r^*)}{(r_0 - r^*)^2 + \det \boldsymbol{\sigma}^*} + r_0' = 0 \tag{8.7}$$

for $\boldsymbol{\sigma}_*$ and r_*, we obtain the effective Hall conductivity of the composite in terms of the properties of the constituents and the microstructure-dependent symmetric 2×2 matrix $\boldsymbol{\Sigma}_* = \boldsymbol{\Sigma}_*(h\boldsymbol{I}_2)$:

$$\boldsymbol{\sigma}_* = \frac{c_0 \boldsymbol{\Sigma}_*}{(r_0')^2 + \det \boldsymbol{\Sigma}_*}, \qquad r_* = r_0 - \frac{c_0 r_0'}{(r_0')^2 + \det \boldsymbol{\Sigma}_*}, \tag{8.8}$$

where c_0, r_0, and r_0' are given by (8.6), and h is the larger root of (8.5). Since the G and G_θ-closure of two isotropic conductors $\boldsymbol{I}_2$ and $h\boldsymbol{I}_2$ is known, formulas (8.8) give an exact description of the G and G_θ-closures of two isotropic conductors (8.3). In the three-dimensional space $(\lambda_1, \lambda_2, r)$, where λ_1, λ_2 are the eigenvalues of $\boldsymbol{\sigma}$, these G-closures lie on a two-dimensional surface (a hyperboloid of one sheet, whose equation is $(8.7)_2$).

8.2 3D conductivity with Hall effect

Here we are assuming that a composite is placed in a uniform magnetic field, whose direction is arbitrary, but fixed. In three space dimensions the constitutive relation $\boldsymbol{j} = \boldsymbol{L}\boldsymbol{e}$ can be written as follows

$$\boldsymbol{L}\boldsymbol{e} = \boldsymbol{\sigma}\boldsymbol{e} + \boldsymbol{r} \times \boldsymbol{e},$$

where $\boldsymbol{\sigma}$ is the symmetric part of $\boldsymbol{L}$ and the map $\boldsymbol{e} \mapsto \boldsymbol{r} \times \boldsymbol{e}$ constitutes its antisymmetric part. There are very few exact relations and links in the context of three-dimensional conductivity with Hall effect. In fact the only link is

$$L' = L^T. \tag{8.9}$$

All of the exact relations can be obtained by means of intersections and applications of the transposition link (8.9) from a family of uniform field relations

$$L e_0 = j_0,$$

and a single non-UFR exact relation $L^T = L$ (the set of fixed points of (8.9)), saying that the Hall effect cannot be obtained from non-Hall conductors by means of making anisotropic composites. These results were obtained in PhD dissertation of Hansun To [15].

8.3 Fibrous conducting composites with Hall effect

By 'fibrous composites' we mean that there exists a direction $\hat{f}$ (fiber direction) so that all cross-sections of a composite by planes orthogonal to $\hat{f}$ have the same microstructure. That means that we consider a class of composites that include all fibrous ones, but is broader in the sense that we do not place any geometric connectivity constraints on each phase. No other restrictions are placed on the microstructure. We would like to emphasize here that the direction of the magnetic field is not assumed to be aligned in any way either with the crystallographic axes of the constituent materials nor with the fiber direction. By contrast with two and three-dimensional conductivity there is a very large number of exact relations and links, some of which are highly non-trivial. This happens because we are working in a significantly larger space of three-dimensional material tensors, while the microstructure is two-dimensional. The results of this section were the eventual outcome of three consecutive Summer REU programs, involving a total of 14 undergraduate students. A complete report is published in [8].

The presence of the special fiber direction suggests that we should split the electric and the current fields into their transversal and fiber components:

$$e = e_\top + e_\| \hat{f}, \qquad j = j_\top + j_\| \hat{f},$$

where $\hat{f}$ is a unit vector directed along the fibers. The conductivity tensors will then we represented as 2×2 block matrices

$$L = \begin{bmatrix} \Lambda_\top & p_{\top\|} \\ q_{\|\top} & \sigma_\| \end{bmatrix},$$

effecting the constitutive relation

$$j_\top = \Lambda_\top e_\top + e_\| p_{\top\|}, \quad j_\| = q_{\|\top} \cdot e_\top + \sigma_\| e_\|.$$

We formulate all exact relations and links in terms of the block-components $\Lambda_\top$, $p_{\top\|}$, $q_{\|\top}$, $\sigma_\|$.

Exact relations

1. $\boldsymbol{p}_{\top\|} = \boldsymbol{p}_0$, $\langle \sigma_\| \rangle$ (see, e.g. [9])
2. $\boldsymbol{\Lambda}_\top = \boldsymbol{\Lambda}_0$, $\langle \boldsymbol{p}_{\top\|} \rangle$, $\langle \boldsymbol{q}_{\|\top} \rangle$

Links

1. $\boldsymbol{L}' = \begin{bmatrix} a_0 \boldsymbol{I}_2 & 0 \\ \boldsymbol{q}_0 & \alpha_0 \end{bmatrix} \boldsymbol{L} \begin{bmatrix} b_0 \boldsymbol{I}_2 & \boldsymbol{p}_0 \\ 0 & \beta_0 \end{bmatrix} + \begin{bmatrix} r_0 \boldsymbol{R}_\perp & \boldsymbol{p}_0' \\ \boldsymbol{q}_0' & \sigma_0 \end{bmatrix}$
2. $\boldsymbol{L}' = \begin{bmatrix} \dfrac{\boldsymbol{\Lambda}_\top^T}{\det \boldsymbol{\Lambda}_\top} & \dfrac{\boldsymbol{\Lambda}_\top^T \boldsymbol{p}_{\top\|}^\perp}{\det \boldsymbol{\Lambda}_\top} \\ \dfrac{\boldsymbol{\Lambda}_\top \boldsymbol{q}_{\|\top}^\perp}{\det \boldsymbol{\Lambda}_\top} & \boldsymbol{\Lambda}^{-1} \boldsymbol{p}_{\top\|} \cdot \boldsymbol{q}_{\|\top} - \sigma_\| + \sigma_\|^0 \end{bmatrix}$
3. $\boldsymbol{L}' = \boldsymbol{L}^T$
4. $\boldsymbol{\Lambda}_{\top *}$ depends only on $\boldsymbol{\Lambda}_\top$. In particular, restricting $\boldsymbol{\Lambda}_\top$ to an exact relation for two-dimensional Hall effect conductivity, described in section 8.1.1, while keeping $\boldsymbol{p}_{\top\|}$, $\boldsymbol{q}_{\|\top}$ and $\sigma_\|$ arbitrary produces an exact relation. An observation to this effect was made in [10].
5. $\boldsymbol{p}_{\top\| *}$ depends only on $\boldsymbol{\Lambda}_\top$ and $\boldsymbol{p}_{\top\|}$
6. $$\boldsymbol{L} = \begin{bmatrix} \boldsymbol{\sigma} & \boldsymbol{p}_{\top\|} \\ \boldsymbol{q}_{\|\top} & \sigma_\| \end{bmatrix}, \qquad \boldsymbol{L}' = \begin{bmatrix} \boldsymbol{\sigma} & \boldsymbol{p}_{\top\|} + \boldsymbol{q}_{\|\top} \\ \boldsymbol{q}_{\|\top} & \sigma_\| \end{bmatrix} \tag{8.10}$$

   where $\boldsymbol{\sigma}^T = \boldsymbol{\sigma}$. This link is an example of a link defined only on a specific exact relation, and not on the entire space of conductivity tensors.

Combining results in this section one can obtain many interesting corollaries. We refer the reader to a long series of papers by Bergman, Strelniker [1–3, 12, 13] for a taste of what is possible. Another application of these results is given in [7].

References

[1] Bergman D J, Li X and Strelniker Y M 2005 Macroscopic conductivity tensor of a three-dimensional composite with a one- or two-dimensional microstructure *Phys. Rev.* **71** 035120

[2] Bergman D J and Strelniker Y M 1998 Duality transformation in a three dimensional conducting medium with two dimensional heterogeneity and an in-plane magnetic field *Phys. Rev. Lett.* **80** 3356–9

[3] Bergman D J and Strelniker Y M 1999 Magnetotransport in conducting composite films with a disordered columnar microstructure and an in-plane magnetic field *Phys. Rev.* B **60** 13016–27

[4] Dykhne A M 1971 Anomalous plasma resistance in a strong magnetic field *Sov. Phys. JETP* **32** 348–51 [Zh. Eksp. Teor. Fiz., 59 (1970) 641–7.]

[5] Dykhne A M and Ruzin I M 1994 On the theory of the fractional quantum Hall effect: the two-phase model *J. Phys. Rev.* B **50** 2369–79

[6] Grabovsky Y 2004 Algebra, geometry and computations of exact relations for effective moduli of composites *Advances in Multifield Theories of Continua with Substructure*

Modeling and Simulation in Science, Engineering and Technology G Capriz and P M Mariano (Boston, MA: Birkhäuser) 167–97
[7] Grabovsky Y 2009 An application of the general theory of exact relations to fiber-reinforced conducting composites with Hall effect *Mech. Mater.* **41** 456–62
[8] Grabovsky Y 2009 Exact relations for effective conductivity of fiber-reinforced conducting composites with the Hall effect via a general theory *SIAM J. Math Anal.* **41** 973–1024
[9] Hashin Z 1968 Assessment of the self consistent scheme approximation: conductivity of particulate composites *J. Compos. Mater.* **2** 284–300
[10] Hashin Z 1983 Analysis of composite materials–a survey *J. Appl. Mech.* **50** 481–505
[11] Milton G W 1988 Classical Hall effect in two-dimensional composites: a characterization of the set of realizable effective conductivity tensors *Phys. Rev.* B **38** 11296–303
[12] Strelniker Y M and Bergman D J 2000 Exact relations between magnetoresistivity tensor components of conducting composites with a columnar microstructure *Phys. Rev.* B **61** 6288–97
[13] Strelniker Y M and Bergman D J 2003 Exact relations between macroscopic moduli of composite media in three dimensions: application to magnetoconductivity and magneto-optics of three-dimensional composites with related columnar microstructures *Phys. Rev.* B **67** 184416
[14] Stroud D and Bergman D J 1984 New exact results for the Hall-coefficient and magneto-resistance of inhomogeneous two-dimensional metals *Phys. Rev.* B **30** 447–9
[15] To H 2004 Homogenization of dynamic materials *PhD Thesis* Temple University, Philadelphia, PA

IOP Publishing

Composite Materials (Second Edition)
Mathematical theory and exact relations
Yury Grabovsky

Chapter 9

Elasticity

Elastic properties of materials, in the context of linear elasticity, are described by the elasticity tensor $\mathsf{C} \in \mathrm{Sym}(\mathrm{Sym}(\mathbb{R}^d))$, $d = 2, 3$, relating the strain field $\varepsilon(\boldsymbol{x})$ to the stress field $\boldsymbol{\sigma}(\boldsymbol{x})$

$$\boldsymbol{\sigma}(\boldsymbol{x}) = \mathsf{C}\varepsilon(\boldsymbol{x}). \tag{9.1}$$

Both fields take values in the space of symmetric $d \times d$ matrices. The strain field is a symmetric part of the gradient of the displacement vector $\boldsymbol{u}$:

$$\varepsilon = e(\boldsymbol{u}) = \frac{1}{2}(\nabla \boldsymbol{u} + (\nabla \boldsymbol{u})^T), \tag{9.2}$$

while the stress field satisfies the equation of equilibrium.

$$\nabla \cdot \boldsymbol{\sigma} = 0. \tag{9.3}$$

We equip the stress-strain space $\mathrm{Sym}(\mathbb{R}^d)$, with thc Frobenius inner product

$$(\boldsymbol{A}, \boldsymbol{B})_F = \mathrm{Tr}\,(\boldsymbol{AB}), \qquad \{\boldsymbol{A}, \boldsymbol{B}\} \subset \mathrm{Sym}(\mathbb{R}^d),\ d = 2, 3,$$

so that the tensor product $\boldsymbol{C} \otimes \boldsymbol{C}$, $\boldsymbol{C} \subset \mathrm{Sym}(\mathbb{R}^d)$ can be regarded as an operator on $\mathrm{Sym}(\mathbb{R}^d)$, acting by the rule

$$(\boldsymbol{C} \otimes \boldsymbol{C})\varepsilon = (\boldsymbol{C}, \varepsilon)_F \boldsymbol{C}.$$

For example, an isotropic elastic tensor with shear modulus μ and Lamé modulus λ can be written as

$$\mathsf{C} = 2\mu\mathsf{I} + \lambda \boldsymbol{I}_d \otimes \boldsymbol{I}_d,$$

where I is the identity operator on $\mathrm{Sym}(\mathbb{R}^d)$.

We note an unfortunate circumstance of the absence of a canonical basis in $\mathrm{Sym}(\mathbb{R}^d)$. Any choice of basis permits one to represent any elastic tensor as a 6×6 matrix. However, a different basis choice will lead to a different representation of the same elastic tensor, making comparisons of results from different papers difficult. Our results are written in coordinate-free form in the sense that they only use invariant (or covariant) objects, like trace, identity matrix, identity operator and a tensor product $\otimes$. In this coordinate-free form the results acquire the most symmetric and beautiful form, underlying their physical significance.

9.1 2D elasticity

In this section and the next it will be convenient to list the few links that exist together with the exact relations on which they are defined.

1. Hill's exact relation [10, 11]

$$\mathsf{C} = 2\mu_0 \mathsf{I} + \lambda \boldsymbol{I}_2 \otimes \boldsymbol{I}_2, \qquad \langle \frac{1}{2\mu_0 + \lambda} \rangle. \tag{9.4}$$

   Even though the exact relation itself is UFR, the additional volume average formula in (9.4) is not derivable from uniform field ideas.

2. We recall that an elastic tensor C has a square symmetry in two space dimensions if it is unchanged by 90° rotations. Such a tensor is characterized by the condition $\mathsf{C}\boldsymbol{I}_2 = 2\kappa \boldsymbol{I}_2$, where κ is called the bulk modulus. The remaining two eigenvalues of C are denoted $2\mu_1$ and $2\mu_2$, where μ_1 and μ_2 are called the shear moduli. Suppose that all constituents of a composite have square symmetry and one and the same bulk modulus κ_0. Then the effective elastic tensor of the composite will always possess square symmetry and that same bulk modulus κ_0. This is a uniform field relation (UFR), since it can be described by the equation

$$\mathsf{C}\boldsymbol{I}_2 = 2\kappa_0 \boldsymbol{I}_2, \tag{9.5}$$

   and was first described in [15]. There is also a link associated with this exact relation. The link maps square symmetric elastic tensors C, with the same bulk modulus κ_0 to square symmetric elastic tensors C', that have another constant bulk modulus κ_0', same eigenstrains as C and shear moduli μ_1' and μ_2', given by the equations

$$\left(\frac{1}{\kappa_0} + \frac{1}{\mu_1}\right)\left(\frac{1}{\kappa_0'} + \frac{1}{\mu_2'}\right) = \left(\frac{1}{\kappa_0} + \frac{1}{\mu_2}\right)\left(\frac{1}{\kappa_0'} + \frac{1}{\mu_1'}\right) = c_0, \tag{9.6}$$

   where $c_0 > 1/\kappa_0 \kappa_0'$ is a given constant. This link was obtained in [3] for incompressible materials (i.e., $\kappa_0 = +\infty$) and subsequently generalized to (9.6) in [9].

3. $\mathsf{C}\boldsymbol{I}_2 = 2\kappa_0$ and

$$\left(\frac{1}{\kappa_0} + \frac{1}{\mu_1}\right)\left(\frac{1}{\kappa_0} + \frac{1}{\mu_2}\right) = c_0. \tag{9.7}$$

   This beautiful exact relation was discovered by Lurie and Cherkaev in [12].

4. The exact relation we are about to describe first manifested itself in [1], where upper and lower bounds on the effective shear modulus of a two-dimensional isotropic elastic polycrystal collapsed under a somewhat strange condition on the elastic tensor of a crystallite. Graeme Milton had an insight that this strange condition is equivalent to a particular structure of the elastic tensor. The most convenient description is afforded by its quadratic form

$$(\mathsf{C}\varepsilon, \varepsilon)_F = (\boldsymbol{C}, \varepsilon)_F^2 - 4\mu \det \varepsilon,$$

where μ is a scalar and $\boldsymbol{C}$ is a 2×2 symmetric matrix. We notice that $-2 \det \varepsilon$ is an isotropic quadratic form $(\mathsf{T}\varepsilon, \varepsilon)_F$, where the isotropic elastic tensor T has bulk modulus $-1/2$ and shear modulus 1/2. In [6] Milton's insight was confirmed. It was shown that

$$\mathsf{C} = 2\mu_0 \mathsf{T} + \boldsymbol{C} \otimes \boldsymbol{C}, \quad \boldsymbol{C} \in \mathrm{Sym}(\mathbb{R}^2) \tag{9.8}$$

is an exact relation. Virtually the same result was obtained in [5], where it was also shown to apply to polycrystals made with materials that are rigid in certain directions, specifically, materials whose compliance tensor is rank one.

There is also a link associated with this exact relation.

$$\mathsf{C} = 2\mu_0 \mathsf{T} + \boldsymbol{C} \otimes \boldsymbol{C}, \quad \mathsf{C}' = 2\mu_0 \mathsf{T} + \lambda' \boldsymbol{C} \otimes \boldsymbol{C}. \tag{9.9}$$

This link has a new feature that the new composite $\mathsf{C}'(\boldsymbol{x})$ is not completely determined by the original composite $\mathsf{C}(\boldsymbol{x})$. The scalar multiple $\lambda'(\boldsymbol{x})$ has nothing to do with the original composite. In other words, we only know that matrices $\boldsymbol{C}'(\boldsymbol{x})$ and $\boldsymbol{C}(\boldsymbol{x})$ are scalar multiples of one another. The link guarantees that matrices $\boldsymbol{C}_*$ and $\boldsymbol{C}_*'$ entering the effective tensors C_* and C_*', respectively, are scalar multiples of one another.

5. Finally, there is also a two-parameter family of global links discovered in [4].

$$\mathsf{C}' = \alpha_0(\mathsf{I} - \beta_0 \mathsf{C}\mathsf{T})^{-1}\mathsf{C}, \tag{9.10}$$

where the tensor T is the same as in (9.8). In [4] this link was presented in a simpler and more elegant form $\mathsf{S}' = \alpha_0(\mathsf{S} - \beta_0 \mathsf{T})$, relating the compliance tensors $\mathsf{S} = \mathsf{C}^{-1}$. In fact, if we always keep in mind the homogeneity link $\mathsf{S}' = \alpha_0 \mathsf{S}$, then the link (9.10) can essentially be recorded as $\mathsf{S}' = \mathsf{S} - \beta_0 \mathsf{T}$. The reason for this simple form is that in two space dimension the stress tensor can be given by the Airy stress potential and translating compliances by T does not change the underlying equations, because T is an *isotropic null-Lagrangian* [13, 16]. For this reason we called exact relation (9.8) the 'rank one plus a null-Lagrangian' exact relation.

9.2 3D elasticity

In three dimensions there are a lot fewer exact relations and links. Specifically, exact relation (9.7) and links (9.6) and (9.10) are specific to two space dimensions and do not generalize. The remaining exact relations and links hold in any number of space dimensions. We list them explicitly again for reference purposes.

1. Hill's exact relation [10, 11]

$$\mathsf{C} = 2\mu_0 \mathsf{I} + \lambda \boldsymbol{I}_3 \otimes \boldsymbol{I}_3, \qquad \langle \frac{1}{2\mu_0 + \lambda} \rangle. \tag{9.11}$$

   Even though the exact relation itself is UFR, the additional volume average formula in (9.11) is not derivable from uniform field ideas.

2. The uniform field relation [15] (see also [17])

$$\mathsf{C}\boldsymbol{I}_3 = 3\kappa_0 \boldsymbol{I}_3, \tag{9.12}$$

   is of course still valid in 3D. However, the associated link (9.6) is no longer valid in 3D, since the two-dimensional duality (due to Berdichevsky [3] for elasticity) is behind (9.6).

3. Both 'rank one plus a null-Lagrangian' exact relation (9.8) and related link (9.9) have exactly the same form in 3D as in 2D. It is described by a one-parameter family of manifolds

$$\mathsf{C} = 2\mu_0 \mathsf{T} + \boldsymbol{C} \otimes \boldsymbol{C}\text{: } \boldsymbol{C} \in \mathrm{Sym}(\mathbb{R}^3), \tag{9.13}$$

   where T is a unique (up to a multiple) isotropic null-Lagrangian, given in all space dimensions by

$$\mathsf{T}\varepsilon = \varepsilon - (\mathrm{Tr}\ \varepsilon)\boldsymbol{I}_d. \tag{9.14}$$

   The associated link

$$\mathsf{C} = 2\mu_0 \mathsf{T} + \boldsymbol{C} \otimes \boldsymbol{C}, \quad \mathsf{C}' = 2\mu_0 \mathsf{T} + \lambda' \boldsymbol{C} \otimes \boldsymbol{C},$$

   has the same form in all space dimensions. It is remarkable that the only available proof of (9.13) and (9.14) is via the general theory of exact relations described in this book.

   We note that all elastic tensors of the form (9.13) are orthotropic, since T is isotropic and all symmetric 3×3 matrices possess orthotropic symmetry (180° rotations around eigenvectors leave symmetric matrices invariant). However, not every orthotropic tensor can be written in the form (9.13). Indeed, in a basis in which $\mathsf{C}\boldsymbol{I}_3$ is a diagonal matrix, tensors of the form (9.13) will be described by diagonal matrices $\boldsymbol{C}$, which means that $\mathsf{C}\boldsymbol{\xi} = 2\mu_0 \boldsymbol{\xi}$ for every matrix $\boldsymbol{\xi}$ with all zeros on the diagonal. Conversely, if C has this property, then the space of diagonal matrices must be an invariant subspace of C. In that case C will have the form (9.13) if and only if $\mathsf{C} - 2\mu_0 \mathsf{T}$ is a nonnegative semidefinite rank-one transformation on the three-dimensional space diagonal matrices. In particular, elastic tensors of the form (9.13) respond isotropically to strains lying in the four-dimensional subspace of shear strains orthogonal to $\mathrm{dev}(\boldsymbol{C})$ the trace-free part of $\boldsymbol{C}$,

$$\mathrm{dev}(\boldsymbol{C}) = \boldsymbol{C} - \frac{1}{d}(\mathrm{Tr}\,\boldsymbol{C})\boldsymbol{I}_d, \qquad \boldsymbol{C} \in \mathrm{Sym}(\mathbb{R}^d). \tag{9.15}$$

We emphasize that the list above is *complete*. It lists all microstructure-independent relations with none missing or not yet discovered.

9.3 Fibrous elastic composites

In the context of fibrous elastic composites, the microstructure is the same in any 2D cross-section by a plane perpendicular to the fibers. To fix notation we designate the x_3-direction as the direction of the fibers. The existence of this special direction forces us to represent physical space $\mathbb{R}^3$ as a direct sum of the one dimensional subspace $\|=\{x_3\boldsymbol{e}_3\colon x_3 \in \mathbb{R}\}$ parallel to the fibers and the two dimensional subspace $\top=\{\boldsymbol{x} \in \mathbb{R}^3\colon \boldsymbol{x} \cdot \boldsymbol{e}_3 = 0\}$ transversal (orthogonal) to the fibers.

One of the essential features of polycrystalline exact relations and links is the existence of coordinate-free representations. In the fibrous setting the space direction along the fibers is singled out. The transversal plane, however, will remain coordinate-free. Accordingly, the strain space $\mathrm{Sym}(\mathbb{R}^3)$ acquires the following block-structure

$$\varepsilon = \begin{bmatrix} \varepsilon_\top & \varepsilon_\times \\ \varepsilon_\times & \varepsilon_\| \end{bmatrix}. \tag{9.16}$$

The block $\varepsilon_\top\colon \mathbb{R}^2 \to \mathbb{R}^2$ describes infinitesimal deformations in the transversal plane, $\varepsilon_\| \in \mathbb{R}$ is the strain along the fibers, describing their stretching or shrinking, while $\varepsilon_\times \in \mathbb{R}^2$ describes the 'tilt' of the transversal plane. The direction of $\varepsilon_\times$ is along the projection of the normal to the transversal plane after the deformation onto the transversal plane before the deformation, while the magnitude of $\varepsilon_\times$ indicates the angle of the tilt. The action of operator ε on $\boldsymbol{x} = (x_1, x_2, x_3) = (\boldsymbol{x}_\top, x_3)$, where $\boldsymbol{x}_\top \in \mathbb{R}^2$ is

$$\varepsilon \boldsymbol{x} = \begin{bmatrix} \varepsilon_\top & \varepsilon_\times \\ \varepsilon_\times & \varepsilon_\| \end{bmatrix} \begin{bmatrix} \boldsymbol{x}_\top \\ x_3 \end{bmatrix} = \begin{bmatrix} \varepsilon_\top \boldsymbol{x}_\top + x_3 \varepsilon_\times \\ \varepsilon_\times \cdot \boldsymbol{x}_\top + x_3 \varepsilon_\| \end{bmatrix} \in \mathbb{R}^3.$$

In elasticity the strain space is the underlying vector space on which elasticity tensors act. We will therefore want to regard strains not as 2×2 block matrices but as block-vectors with 'coordinates' $\varepsilon_\top$, $\varepsilon_\times$, $\varepsilon_\|$. Keeping in mind that there are standard inner products on $\mathrm{Sym}(\mathbb{R}^2)$ (Frobenius), $\mathbb{R}^2$ (dot product) and $\mathbb{R}$ (product of numbers), we quickly realize that identifying $\mathrm{Sym}(\mathbb{R}^3)$ with $\mathrm{Sym}(\mathbb{R}^2) \oplus \mathbb{R}^2 \oplus \mathbb{R}$ via the 'natural' coordinatization $\varepsilon \mapsto [\varepsilon_\top, \varepsilon_\times, \varepsilon_\|]$ will lead to a representation of elastic tensors as non-symmetric block-matrices. In order to preserve the symmetry in the block-matrix representation of elastic tensors we write the Frobenius inner product on $\mathrm{Sym}(\mathbb{R}^3)$ in terms of the standard inner products on $\mathrm{Sym}(\mathbb{R}^2)$, $\mathbb{R}^2$ and $\mathbb{R}$:

$$(\boldsymbol{\sigma}, \varepsilon)_F = (\boldsymbol{\sigma}_\top, \varepsilon_\top)_F + 2\boldsymbol{\sigma}_\times \cdot \varepsilon_\times + \sigma_\| \varepsilon_\|.$$

This formula shows that if we keep a standard inner product on $\mathbb{R}^2$, then the inner products on $\mathrm{Sym}(\mathbb{R}^2)$ and $\mathbb{R}$ in which the elastic tensor will be represented as a symmetric block-matrix must be

$$(\boldsymbol{\sigma}_\top \boldsymbol{\varepsilon}_\top)_\top = \frac{1}{2}(\boldsymbol{\sigma}_\top, \boldsymbol{\varepsilon}_\top)_F, \qquad (\sigma_\|, \varepsilon_\|)_\| = \frac{1}{2}\sigma_\|\varepsilon_\|.$$

The inner product $(\cdot, \cdot)_\top$ is very convenient, since $(\boldsymbol{I}_2, \boldsymbol{I}_2)_\top = 1$. However, the inner product $(\cdot, \cdot)_\|$ on $\mathbb{R}$ is neither natural, nor convenient for doing algebra. Thus, in order to simplify the process of computation of exact relations and links we decided to identify 3×3 symmetric matrices $\boldsymbol{\varepsilon}$ with block-vectors $[\boldsymbol{\varepsilon}_\top, \boldsymbol{\varepsilon}_\angle, \varepsilon_\|/\sqrt{2}]$, in which case the elastic tensors will be represented by a symmetric 3×3 block-matrix

$$\mathsf{C} = \begin{bmatrix} \mathsf{C}_\top & \mathsf{C}_{\top\angle} & \mathsf{C}_{\top\|} \\ \mathsf{C}_{\top\angle}^T & \mathsf{C}_\angle & \mathsf{C}_{\angle\|} \\ \mathsf{C}_{\top\|}^T & \mathsf{C}_{\angle\|}^T & \mathsf{C}_\| \end{bmatrix},$$

where $\mathsf{C}_\top\colon \mathrm{Sym}(\mathbb{R}^2) \to \mathrm{Sym}(\mathbb{R}^2)$, $\mathsf{C}_{\top\angle}\colon \mathbb{R}^2 \to \mathrm{Sym}(\mathbb{R}^2)$, $\mathsf{C}_{\top\|}\colon \mathbb{R} \to \mathrm{Sym}(\mathbb{R}^2)$, $\mathsf{C}_\angle\colon \mathbb{R}^2 \to \mathbb{R}^2$, $\mathsf{C}_{\angle\|}\colon \mathbb{R} \to \mathbb{R}^2$, $\mathsf{C}_{\top\angle}^T\colon \mathrm{Sym}(\mathbb{R}^2) \to \mathbb{R}^2$, $\mathsf{C}_{\top\|}^T\colon \mathrm{Sym}(\mathbb{R}^2) \to \mathbb{R}$, $\mathsf{C}_{\angle\|}^T\colon \mathbb{R}^2 \to \mathbb{R}$, $\mathsf{C}_\|\colon \mathbb{R} \to \mathbb{R}$. We first observe that maps $\mathsf{C}_{\top\|}\colon \mathbb{R} \to \mathrm{Sym}(\mathbb{R}^2)$, $\mathsf{C}_{\angle\|}\colon \mathbb{R} \to \mathbb{R}^2$ and $\mathsf{C}_\|\colon \mathbb{R} \to \mathbb{R}$ can be uniquely determined by the image of $1 \in \mathbb{R}$. In this way we identify the map $\mathsf{C}_{\top\|}$ with the matrix $\boldsymbol{C}_{\top\|} \in \mathrm{Sym}(\mathbb{R}^2)$, $\mathsf{C}_{\angle\|}$ with the vector $\boldsymbol{c}_{\angle\|} \in \mathbb{R}^2$, and $\mathsf{C}_\|$ with the number $c_\| \in \mathbb{R}$. The symmetric operator $\mathsf{C}_\angle\colon \mathbb{R}^2 \to \mathbb{R}^2$, is naturally identified with a 2×2 symmetric matrix $\boldsymbol{C}_\angle$. The operators $\mathsf{C}_{\top\angle}\colon \mathbb{R}^2 \to \mathrm{Sym}(\mathbb{R}^2)$ will be denoted by $\mathscr{C}_{\top\angle}$ to distinguish their special nature as third-order tensors. The symmetry of the full elastic tensor C as an operator on $\mathrm{Sym}(\mathbb{R}^3)$ implies that the remaining operators are identified by duality with respect to standard inner products on $\mathbb{R}$, $\mathbb{R}^2$ and $\mathrm{Sym}(\mathbb{R}^2)$:

$$\mathsf{C}_{\top\angle}^T\boldsymbol{\varepsilon}_\top \cdot \boldsymbol{\varepsilon}_\angle = (\mathscr{C}_{\top\angle}\boldsymbol{\varepsilon}_\angle, \boldsymbol{\varepsilon}_\top)_\top, \quad (\mathsf{C}_{\top\|}^T\boldsymbol{\varepsilon}_\top)\varepsilon_\| = \varepsilon_\|(\boldsymbol{C}_{\top\|}\boldsymbol{\varepsilon}_\top)_\top, \quad (\mathsf{C}_{\angle\|}^T\boldsymbol{\varepsilon}_\angle)\varepsilon_\| = \varepsilon_\|(\boldsymbol{c}_{\angle\|} \cdot \boldsymbol{\varepsilon}_\angle).$$

These formulas show that we can encode the dual operator $\mathsf{C}_{\angle\|}^T$ by the same vector $\boldsymbol{c}_{\angle\|}$ that is identified with $\mathsf{C}_{\angle\|}$. Similarly, we encode $\mathsf{C}_{\top\|}^T$ by the symmetric 2×2 matrix $\boldsymbol{C}_{\top\|}$ that is identified with $\mathsf{C}_{\top\|}$. Even though the operator $\mathsf{C}_{\top\angle}^T$ is identified with the third order tensor $\mathscr{C}_{\top\angle}$ we will denote it $\mathscr{C}_{\angle\top}$ for improved readability. In summary, the full elastic tensor C will be written in the block form as

$$\mathsf{C} = \begin{bmatrix} \mathsf{C}_\top & \mathscr{C}_{\top\angle} & \boldsymbol{C}_{\top\|} \\ \mathscr{C}_{\angle\top} & \boldsymbol{C}_\angle & \boldsymbol{c}_{\angle\|} \\ \boldsymbol{C}_{\top\|} & \boldsymbol{c}_{\angle\|} & c_\| \end{bmatrix}, \tag{9.17}$$

and understood to describe the constitutive relation

$$\begin{cases} \sigma_{\top} = \mathsf{C}_{\top}\varepsilon_{\top} + \mathscr{C}_{\top\times}\varepsilon_{\times} + \varepsilon_{\|}\dfrac{\boldsymbol{C}_{\top\|}}{\sqrt{2}}, \\ \boldsymbol{\sigma}_{\times} = \mathscr{C}_{\times\top}\varepsilon_{\top} + \boldsymbol{C}_{\times}\boldsymbol{\varepsilon}_{\times} + \varepsilon_{\|}\dfrac{\boldsymbol{c}_{\times\|}}{\sqrt{2}}, \\ \sigma_{\|} = \left(\dfrac{\boldsymbol{C}_{\top\|}}{\sqrt{2}}, \varepsilon_{\top}\right)_F + \sqrt{2}\boldsymbol{c}_{\times\|}\cdot\boldsymbol{\varepsilon}_{\times} + c_{\|}\varepsilon_{\|}, \end{cases}$$

where

$$\sigma = \begin{bmatrix} \sigma_{\top} & \boldsymbol{\sigma}_{\times} \\ \boldsymbol{\sigma}_{\times} & \sigma_{\|} \end{bmatrix}, \qquad \varepsilon = \begin{bmatrix} \varepsilon_{\top} & \boldsymbol{\varepsilon}_{\times} \\ \boldsymbol{\varepsilon}_{\times} & \varepsilon_{\|} \end{bmatrix}$$

is the block-matrix representation (9.16) of the stress and strain tensors. In our notation, scalars are denoted by lower case normal letters, $\mathbb{R}^2$ vectors, by bold lower case letters, matrices, by bold upper case letters, third order tensors, by script upper case letters and fourth order tensors, i.e., operators on Sym($\mathbb{R}^2$) are denoted by sans serif upper case letters. The subscripts $\|$, $\times$, $\|$ will help the reader understand the physical meaning of the corresponding objects.

9.3.1 Exact relations

In the context of fibrous elastic composites there is a veritable plethora of exact relations and links. This fact made it imperative to eliminate those exact relations and links that could be derived from the remaining set. One very effective way to do this was to derive all global links first. This was done by Meredith Hegg in her PhD dissertation [7] (see also [8]). The fact that a global link would map an exact relation into an exact relation means that all exact relations split into equivalence classes of exact relations related by a global link. Only one exact relation from each equivalence class need to be listed. In general, as one can see from prior sections and chapters, exact relations come in infinite families, parameterized by one or more real parameters. Remarkably, exact relations in each such infinite family could be mapped into one another by global links, leaving only finitely many equivalence classes. While the number of such equivalence classes were into several hundreds, a great many of them could be obtained as intersections of other exact relations. The finiteness of the number of equivalence classes meant that the search for intersections could be automated. This was done during the summer of 2013 by a graduate student Tatyana Nuzhnaya, reducing the number of essential exact relations to a few dozen. During the 2013–14 academic year two undergraduate students Mark Mikida and Andrew Schneider computed the final form of essential exact relations. In the list below each exact consists of all positive definite elastic tensors of the form (9.17), where the block components have either specified form or satisfy specified constraints. For example, the first exact relation in our list is given as $\boldsymbol{C}_{\top\|} = 0$, $\boldsymbol{c}_{\times\|} = 0$, $\langle c_{\|}\rangle$. That means that it consists of tensors C of the form

$$\mathsf{C} = \begin{bmatrix} \mathsf{C}_{\top} & \mathscr{C}_{\top\measuredangle} & 0 \\ \mathscr{C}_{\measuredangle\top} & \boldsymbol{C}_{\measuredangle} & 0 \\ 0 & 0 & c_{\parallel} \end{bmatrix}.$$

So that any fibrous composite made of materials of this form will have an effective tensor of the same form, where the blocks $\mathsf{C}_{\top *}$, $\mathscr{C}_{\top\measuredangle *}$, $\boldsymbol{C}_{\measuredangle *}$, $c_{\parallel *}$ depend on the microstructure, but in such a way that $c_{\parallel *} = \langle c_{\parallel} \rangle$. We list exact relations in the order of increasing complexity.

1. $\boldsymbol{C}_{\top\parallel} = 0$, $\boldsymbol{c}_{\measuredangle\parallel} = 0$, $\langle c_{\parallel} \rangle$
2. $\mathscr{C}_{\top\measuredangle} = 0$
3. $\mathscr{C}_{\top\measuredangle} = 0$, $\boldsymbol{c}_{\measuredangle\parallel} = 0$. This exact relation says that if a fibrous composite is made of materials with monoclinic symmetry, whose two-fold axis coincides with fiber direction, then the composite will always have this monoclinic symmetry. This is the only elasticity symmetry class that is stable under homogenization in the context of fibrous composites.
4. $\mathscr{C}_{\top\measuredangle} = 0$, $\boldsymbol{C}_{\top\parallel} = 0$
5. $\mathscr{C}_{\top\measuredangle} = 0$, $\det \boldsymbol{C}_{\measuredangle} = d_0$
6. $\mathscr{C}_{\top\measuredangle} = 0$, $\boldsymbol{C}_{\measuredangle} = \sigma_0 \boldsymbol{I}_2$, $\langle \boldsymbol{c}_{\measuredangle\parallel} \rangle$
7. $\mathsf{C}_{\top}\boldsymbol{I}_2 = 2\kappa_0 \boldsymbol{I}_2$, $\mathscr{C}_{\measuredangle\top}\boldsymbol{I}_2 = 0$, $\langle \operatorname{Tr} \boldsymbol{C}_{\top\parallel} \rangle$
8. $\mathscr{C}_{\top\measuredangle} = 0$, $\mathsf{C}_{\top}\boldsymbol{I}_2 = 2\kappa_0 \boldsymbol{I}_2$, $\langle \operatorname{Tr} \boldsymbol{C}_{\top\parallel} \rangle$. Let $2\mu_1$ and $2\mu_2$ be the other two eigenvalues of $\mathsf{C}_{\top}$ (κ_0, μ_1 and μ_2 have the physical meaning of the bulk and two shear moduli of a square symmetric 2D elastic material). This exact relation is then described by the constraints above and

$$\left(\frac{1}{\kappa_0} + \frac{1}{\mu_1}\right)\left(\frac{1}{\kappa_0} + \frac{1}{\mu_2}\right) = c_0$$

9. $\mathsf{C}_{\top} = 2\mu_0 \mathsf{T} + \boldsymbol{B} \otimes \boldsymbol{B}$, $\mathscr{C}_{\top\measuredangle} = \boldsymbol{B} \otimes \boldsymbol{b}$, $\boldsymbol{B} \in \operatorname{Sym}(\mathbb{R}^2)$, $\boldsymbol{b} \in \mathbb{R}^2$, where T is a 2D isotropic elasticity tensor with bulk modulus $-1/2$ and shear modulus 1/2, given by (9.14). The expression $\mathscr{C}_{\top\measuredangle} = \boldsymbol{B} \otimes \boldsymbol{b}$ means

$$\mathscr{C}_{\top\measuredangle}\varepsilon_{\measuredangle} = (\boldsymbol{b} \cdot \varepsilon_{\measuredangle})\boldsymbol{B}, \qquad \mathscr{C}_{\measuredangle\top}\varepsilon_{\top} = (\varepsilon_{\top}, \boldsymbol{B})_{\top}\boldsymbol{b}.$$

10. $\mathsf{C}_{\top} = 2\mu_0 \mathsf{T} + \boldsymbol{B} \otimes \boldsymbol{B}$, $\mathscr{C}_{\top\measuredangle} = \boldsymbol{B} \otimes \boldsymbol{b}$, $\boldsymbol{C}_{\top\parallel} = \beta \boldsymbol{B}$
11. $\mathsf{C}_{\top} = 2\mu_0 \mathsf{T} + \boldsymbol{B} \otimes \boldsymbol{B}$, $\mathscr{C}_{\top\measuredangle} = \boldsymbol{B} \otimes \boldsymbol{b}$, $\boldsymbol{C}_{\top\parallel} = \beta \boldsymbol{B}$, $\boldsymbol{c}_{\measuredangle\parallel} = \beta \boldsymbol{b}$, $\langle c_{\parallel} - \beta^2 \rangle$
12. $\mathsf{C}_{\top} = 2\mu_0 \mathsf{T} + \boldsymbol{B} \otimes \boldsymbol{B}$, $\mathscr{C}_{\top\measuredangle} = \boldsymbol{B} \otimes \boldsymbol{b}$, $\boldsymbol{C}_{\measuredangle} = \boldsymbol{\sigma} + \boldsymbol{b} \otimes \boldsymbol{b}$, $\det \boldsymbol{\sigma} = d_0$
13. $\mathsf{C}_{\top} = 2\mu_0 \mathsf{T} + \boldsymbol{B} \otimes \boldsymbol{B}$, $\mathscr{C}_{\top\measuredangle} = \boldsymbol{B} \otimes \boldsymbol{b}$, $\boldsymbol{C}_{\measuredangle} = \sigma_0 \boldsymbol{I}_2 + \boldsymbol{b} \otimes \boldsymbol{b}$
14. $\mathsf{C}_{\top} = 2\mu_0 \mathsf{T} + \boldsymbol{B} \otimes \boldsymbol{B}$, $\mathscr{C}_{\top\measuredangle} = \boldsymbol{B} \otimes \boldsymbol{b}$, $\boldsymbol{C}_{\top\parallel} = \beta \boldsymbol{B}$, $\boldsymbol{C}_{\measuredangle} = \sigma_0 \boldsymbol{I}_2 + \boldsymbol{b} \otimes \boldsymbol{b}$, $\langle \boldsymbol{c}_{\measuredangle\parallel} - \beta \boldsymbol{b} \rangle$
15. $\mathsf{C}_{\top} = 2\mu_0 \mathsf{I} + 2\lambda \boldsymbol{I}_2 \otimes \boldsymbol{I}_2$, $\mathscr{C}_{\top\measuredangle} = \boldsymbol{I}_2 \otimes \boldsymbol{b}$, $\langle \operatorname{dev}(\boldsymbol{C}_{\top\parallel}) \rangle$, where $\operatorname{dev}(\cdot)$ denotes the trance-free part of a matrix, and was defined in (9.15).
16. $\mathsf{C}_{\top} = 2\mu_0 \mathsf{I} + 2\lambda \boldsymbol{I}_2 \otimes \boldsymbol{I}_2$, $\mathscr{C}_{\top\measuredangle} = 0$, $\langle \frac{1}{2\mu_0 + \lambda} \rangle$, $\langle \operatorname{dev}(\boldsymbol{C}_{\top\parallel}) \rangle$

17. $\mathsf{C}_{\top} = 2\mu_0\mathsf{I} + 2\lambda \boldsymbol{I}_2 \otimes \boldsymbol{I}_2$, $\mathscr{C}_{\top\measuredangle} = 0$, $\boldsymbol{C}_{\top\|} = \beta\boldsymbol{I}_2$, $\langle\frac{1}{2\mu_0+\lambda}\rangle$, $\langle\frac{\beta}{2\mu_0+\lambda}\rangle$
18. $\mathsf{C}_{\top} = 2\mu_0\mathsf{I} + 2\lambda \boldsymbol{I}_2 \otimes \boldsymbol{I}_2$, $\mathscr{C}_{\top\measuredangle} = \boldsymbol{I}_2 \otimes \boldsymbol{b}$, $\boldsymbol{C}_{\measuredangle} = \sigma_0\boldsymbol{I}_2 + \frac{\boldsymbol{b}\otimes\boldsymbol{b}}{2(\lambda+2\mu_0)}$, $\langle\frac{\boldsymbol{b}}{\lambda+2\mu_0}\rangle$, $\langle\mathrm{dev}(\boldsymbol{C}_{\top\|})\rangle$
19. $\mathsf{C}_{\top} = 2\mu_0\mathsf{I} + 2\lambda \boldsymbol{I}_2 \otimes \boldsymbol{I}_2$, $\mathscr{C}_{\top\measuredangle} = \boldsymbol{I}_2 \otimes \boldsymbol{b}$, $\boldsymbol{C}_{\top\|} = \beta\boldsymbol{I}_2$, $\boldsymbol{c}_{\measuredangle\|} = \frac{\beta\boldsymbol{b}}{2(\lambda+2\mu_0)}$, $\langle\frac{\beta}{\lambda+2\mu_0}\rangle$, $\langle c_{\|} - \frac{\beta^2}{2(\lambda+2\mu_0)}\rangle$
20. $\mathsf{C}_{\top} = 2\mu_0\mathsf{I} + 2\lambda \boldsymbol{I}_2 \otimes \boldsymbol{I}_2$, $\mathscr{C}_{\top\measuredangle} = 0$, $\boldsymbol{C}_{\top\|} = \beta\boldsymbol{I}_2$, $\boldsymbol{c}_{\measuredangle\|} = 0$, $c_{\|} = c_0 + \frac{\beta^2+\alpha_0\beta+\beta_0}{2(\lambda+2\mu_0)}$ $\langle\frac{1}{2\mu_0+\lambda}\rangle$, $\langle\frac{\beta}{\lambda+2\mu_0}\rangle$
21. $\mathsf{C}_{\top} = 2\mu_0\mathsf{I} + 2\lambda \boldsymbol{I}_2 \otimes \boldsymbol{I}_2$, $\mathscr{C}_{\top\measuredangle} = \boldsymbol{I}_2 \otimes \boldsymbol{b}$, $\boldsymbol{C}_{\top\|} = \beta\boldsymbol{I}_2$, $\boldsymbol{C}_{\measuredangle} = \sigma_0\boldsymbol{I}_2 + \frac{\boldsymbol{b}\otimes\boldsymbol{b}}{2(\lambda+2\mu_0)}$, $\boldsymbol{c}_{\measuredangle\|} = \frac{\beta_0\boldsymbol{b}^{\perp}+\beta\boldsymbol{b}}{2(\lambda+2\mu_0)}$, $\langle\frac{\boldsymbol{b}}{\lambda+2\mu_0}\rangle$, $\langle c_{\|} - \frac{\beta_0^2+\beta^2}{2(\lambda+2\mu_0)} - \frac{4\mu_0^2\,|\,\boldsymbol{b}\,|^2}{\sigma_0(\lambda+2\mu_0)^2}\rangle$, where $\boldsymbol{b}^{\perp} = \boldsymbol{R}_{\perp}\boldsymbol{b}$
22. $\mathsf{C}_{\top} = 2\mu_0\mathsf{I} + 2\lambda \boldsymbol{I}_2 \otimes \boldsymbol{I}_2$, $\mathscr{C}_{\top\measuredangle} = \boldsymbol{I}_2 \otimes \boldsymbol{b}$, $\boldsymbol{C}_{\top\|} = \beta\boldsymbol{I}_2$, $\boldsymbol{c}_{\measuredangle\|} = \frac{\beta\boldsymbol{b}}{2(\lambda+2\mu_0)}$, $c_{\|} = c_0 + \frac{\beta(\beta+\beta_0)}{2(\lambda+2\mu_0)}$, $\langle\frac{\beta}{\lambda+2\mu_0}\rangle$
23. An elasticity tensor C belongs to this exact relation if and only if its blocks satisfy the following constraints.
    (a) $\mathsf{C}_{\top}\boldsymbol{I}_2 = 2\kappa_0\boldsymbol{I}_2$, $\mathscr{C}_{\measuredangle\top}\boldsymbol{I}_2 = 0$, where $\kappa_0 > 0$ is an arbitrary constant.
    (b) Choose arbitrary constants $\sigma_0 > 0$ and $0 < \tau_0 < \kappa_0$, and define Schur complements $\mathsf{S}_{\top}$, $\boldsymbol{S}_{\measuredangle}$ and tensor $\mathscr{S}_{\top\measuredangle}$ by the formula

$$\begin{bmatrix} \mathsf{C}_{\top} + \tau_0\mathsf{I} & \mathscr{C}_{\top\measuredangle} \\ \mathscr{C}_{\measuredangle\top} & \boldsymbol{C}_{\measuredangle} + \sigma_0\boldsymbol{I}_2 \end{bmatrix}^{-1} = \begin{bmatrix} (\mathsf{S}_{\top} + \tau_0\mathsf{I})^{-1} & \mathscr{S}_{\top\measuredangle} \\ \mathscr{S}_{\measuredangle\top} & (\boldsymbol{S}_{\measuredangle} + \sigma_0\boldsymbol{I}_2)^{-1} \end{bmatrix}. \tag{9.18}$$

    (c) $\mathscr{S}_{\top\measuredangle}$: $\mathbb{R}^2 \to \mathrm{Sym}(\mathbb{R}^2)$ is given by its action on an arbitrary vector $\boldsymbol{u} \in \mathbb{R}^2$:

$$\mathscr{S}_{\top\measuredangle}\boldsymbol{u} = \psi(cu), \quad \text{or} \quad \mathscr{S}_{\top\measuredangle}\boldsymbol{u} = \psi(c\bar{u}), \tag{9.19}$$

    where $u \in \mathbb{C}$ is a complex number represented by the vector $\boldsymbol{u}$, $c \in \mathbb{C}$, and the map ψ: $\mathbb{C} \to \mathrm{Sym}(\mathbb{R}^2)$ is defined by

$$\psi(a_1 + ia_2) = \begin{bmatrix} a_1 & a_2 \\ a_2 & -a_1 \end{bmatrix}. \tag{9.20}$$

    (d) We require that $\det(\boldsymbol{S}_{\measuredangle}) = \sigma_0^2$.
    (e) Let μ_1 and μ_2 be the shear moduli of $\mathsf{S}_{\top}$ (its bulk modulus is κ_0). We require that

$$\left(\frac{1}{\kappa_0} + \frac{1}{\mu_1}\right)\left(\frac{1}{\kappa_0} + \frac{1}{\mu_2}\right) = \frac{1}{\rho_0^2}, \qquad \tau_0 = \left(\frac{\kappa_0^{-1} + \rho_0^{-1}}{2}\right)^{-1}$$

    (f) $\mathrm{Tr}\,\boldsymbol{C}_{\top\|*} = \mathrm{Tr}\,\langle\boldsymbol{C}_{\top\|}\rangle$.

In order to explain this exact relation we describe how to construct a tensor that satisfies all the requirements. First, we choose any positive definite two-dimensional elastic tensor $\mathsf{S}_{\top}$ with square symmetry, i.e., $\mathsf{S}_{\top}\boldsymbol{I}_2 = 2\kappa_0\boldsymbol{I}_2$. Then, we choose a complex number c and construct an operator $\mathscr{S}_{\top\angle}$: $\mathbb{R}^2 \to \mathrm{Sym}(\mathbb{R}^2)$ given by one of the two forms in (9.19). Finally, we choose two operators on $\mathbb{R}^2$, denoted $\boldsymbol{S}_{\angle} > 0$ and $\boldsymbol{C}_{\top\|}$, a vector $\boldsymbol{c}_{\angle\|} \in \mathbb{R}^2$, and a scalar $c_{\|} > 0$ arbitrarily. Once these choices have been made, we compute

$$\sigma_0 = \sqrt{\det \boldsymbol{S}_{\angle}}, \qquad \rho_0 = \frac{1}{\sqrt{(\kappa_0^{-1} + \mu_1^{-1})(\kappa_0^{-1} + \mu_2^{-1})}}, \qquad \tau_0 = \left(\frac{\kappa_0^{-1} + \rho_0^{-1}}{2}\right)^{-1}$$

where $2\mu_1$ and $2\mu_2$ are the other two eigenvalues of the operator $\mathsf{S}_{\top}$. We define operators $\mathsf{C}_{\top}$, $\mathscr{C}_{\top\angle}$ and $\boldsymbol{C}_{\angle}$ by (9.18). Explicitly,

$$\tilde{\mathsf{C}}_{\top} = (\tilde{\mathsf{S}}_{\top}^{-1} - \mathscr{S}_{\top\angle}\tilde{\boldsymbol{S}}_{\angle}\mathscr{S}_{\angle\top})^{-1}, \quad \boldsymbol{C}_{\angle} = (\tilde{\boldsymbol{S}}_{\angle}^{-1} - \mathscr{S}_{\angle\top}\tilde{\mathsf{S}}_{\top}\mathscr{S}_{\top\angle})^{-1}, \quad \mathscr{C}_{\top\angle} = -\tilde{\mathsf{S}}_{\top}\mathscr{S}_{\top\angle}\boldsymbol{C}_{\angle},$$

where $\tilde{\mathsf{C}}_{\top} = \mathsf{C}_{\top} + \tau_0\mathsf{I}$, $\tilde{\mathsf{S}}_{\top} = \mathsf{S}_{\top} + \tau_0\mathsf{I}$, and $\tilde{\boldsymbol{S}}_{\angle} = \boldsymbol{S}_{\angle} + \sigma_0\boldsymbol{I}_2$. If we use such a material to make a transversely isotropic polycrystal

$$\mathsf{C}_* = \begin{bmatrix} 2\mu_*(\mathsf{I} - \boldsymbol{I}_2 \otimes \boldsymbol{I}_2) + 2\kappa_*\boldsymbol{I}_2 \otimes \boldsymbol{I}_2 & 0 & m_*\boldsymbol{I}_2 \\ 0 & \sigma_*\boldsymbol{I}_2 & 0 \\ m_*\boldsymbol{I}_2 & 0 & c_{\|*} \end{bmatrix},$$

4 out of its 5 parameters are texture-independent:

$$\kappa_* = \kappa_0, \quad \sigma_* = \sqrt{\det \boldsymbol{S}_{\angle}}, \quad 2m_* = \mathrm{Tr}\,\langle \boldsymbol{C}_{\top\|}\rangle, \quad \frac{1}{\mu_*} = \sqrt{\left(\frac{1}{\kappa_0} + \frac{1}{\mu_1}\right)\left(\frac{1}{\kappa_0} + \frac{1}{\mu_2}\right)} - \frac{1}{\kappa_0}.$$

The fifth parameter $c_{\|*}$ is texture-dependent.

9.3.2 Links

All links (or rather families of links) in the context of fibrous elastic composites could be identified using Nuzhnaya's Matlab code. Their number was staggering: more than 500. However, eliminating logically redundant families of links was a far more difficult task. It became possible only due to the joint work of a graduate student Adam Jacoby and the author during the summer of 2014, whose Maple code reduced the number of non-redundant links to several dozen. The computation of links in its present form was started by the undergraduate student Patrick Wynne, working with the author over the 2014–15 academic year, followed by joint efforts of three undergraduates Abraham Lyle, Hansen Pei and Patrick Wynne over the summer of 2015, followed by the work of Hansen Pei during the fall of 2015.

The links in the list below are grouped into categories determined by the exact relations on which they are defined. We list them in the order of increasing complexity.

1. Global links
   (a) $\mathsf{C}_{\top *}$, $\mathscr{C}_{\top \times *}$, $\boldsymbol{C}_{\times *}$, are independent of the remaining blocks $\boldsymbol{C}_{\top \|}$, $\boldsymbol{c}_{\times \|}$, $c_{\|}$.
   (b) $\mathsf{C}_{\top *}$, $\mathscr{C}_{\top \times *}$, $\boldsymbol{C}_{\times *}$, $\boldsymbol{C}_{\top \| *}$, $\boldsymbol{c}_{\times \| *}$, are independent of $c_{\|}$.
   (c) The derivation of this family of global links constitutes an essential part of Meredith Hegg's PhD dissertation [7]. It can be represented as a composition of a 6-parameter family of affine global links

$$\mathsf{C}' = \mathfrak{F}_{\mathsf{Q},\mathsf{F}}(\mathsf{C}) = \mathsf{QCQ}^T + \mathsf{F}, \tag{9.21}$$

where

$$\mathsf{Q} = \begin{bmatrix} q_1 \mathsf{I} & 0 & 0 \\ 0 & q_2 \boldsymbol{I}_2 & 0 \\ q_4 \boldsymbol{I}_2 & 0 & q_3 \end{bmatrix}, \qquad \mathsf{F} = \begin{bmatrix} 0 & 0 & f_1 \boldsymbol{I}_2 \\ 0 & 0 & 0 \\ f_1 \boldsymbol{I}_2 & 0 & f_2 \end{bmatrix},$$

and a 1-parameter family of nonlinear global links

$$\mathsf{C}' = \Lambda_a(\mathsf{C}) = \mathsf{C} - \mathsf{C} \begin{bmatrix} \Theta_a(\mathsf{C}_\top) & 0 & 0 \\ 0 & 0 & 0 \\ 0 & 0 & 0 \end{bmatrix} \mathsf{C}, \tag{9.22}$$

where

$$\Theta_a(\mathsf{C}_\top) = \left(\frac{1}{a} \mathsf{T} + \mathsf{C}_\top \right)^{-1}.$$

Transformations $\mathfrak{F}_{\mathsf{Q},\mathsf{F}}$ and Λ_a generate a group of global links and have the following composition properties

$$\Lambda_a \circ \Lambda_b = \Lambda_{a+b}, \qquad \lim_{a \to 0} \Lambda_a(\mathsf{C}) = \mathsf{C}, \qquad \Lambda_a \circ \mathfrak{F}_{\mathsf{Q},\mathsf{F}} = \mathfrak{F}_{\mathsf{Q}',\mathsf{F}'} \circ \Lambda_{a'}$$

where $a' = a q_1^2$ and

$$\mathsf{Q}' = \begin{bmatrix} q_1 & 0 & 0 \\ 0 & q_2 & 0 \\ a f_1 q_1 + q_4 & 0 & q_3 \end{bmatrix}, \quad \mathsf{F}' = \begin{bmatrix} 0 & 0 & f_1 \boldsymbol{I}_2 \\ 0 & 0 & 0 \\ f_1 \boldsymbol{I}_2 & 0 & f_2 + a f_1^2 \end{bmatrix}.$$

Explicitly we have

$$\mathfrak{F}_{\mathsf{Q},\mathsf{F}}(\mathsf{C}) = \begin{bmatrix} q_1^2 \mathsf{C}_\top & q_1 q_2 \mathscr{C}_{\top \times} & q_1 q_3 \boldsymbol{C}_{\top \|} + q_1 q_4 \mathsf{C}_\top \boldsymbol{I}_2 + f_1 \boldsymbol{I}_2 \\ * & q_2^2 \boldsymbol{C}_\times & q_2 q_3 \boldsymbol{c}_{\times \|} + q_2 q_4 \mathscr{C}_{\times \top} \boldsymbol{I}_2 \\ * & * & q_3^2 c_\| + 2 q_3 q_4 (\boldsymbol{C}_{\top \|}, \boldsymbol{I}_2)_\top + q_4^2 (\mathsf{C}_\top \boldsymbol{I}_2, \boldsymbol{I}_2)_\top + f_2 \end{bmatrix}$$

and

$$\Lambda_a(\mathsf{C}) = \begin{bmatrix} \frac{1}{a}\mathsf{T}\Theta_a(\mathsf{C}_\top)\mathsf{C}_\top & \frac{1}{a}\mathsf{T}\Theta_a(\mathsf{C}_\top)\mathscr{C}_{\top\curlywedge} & \frac{1}{a}\mathsf{T}\Theta_a(\mathsf{C}_\top)\boldsymbol{C}_{\top\|} \\ * & \boldsymbol{C}_\curlywedge - \mathscr{C}_{\curlywedge\top}\Theta_a(\mathsf{C}_\top)\mathscr{C}_{\top\curlywedge} & \boldsymbol{c}_{\curlywedge\|} - \mathscr{C}_{\curlywedge\top}\Theta_a(\mathsf{C}_\top)\boldsymbol{C}_{\top\|} \\ * & * & c_\| - (\Theta_a(\mathsf{C}_\top)\boldsymbol{C}_{\top\|}, \boldsymbol{C}_{\top\|})_\top \end{bmatrix}.$$

A particular instance of this link (but in a more general context of fibrous piezoelectric composites) was discovered by Benveniste [2].

2. $\mathsf{C} = \begin{bmatrix} \mathsf{C}_\top & 0 & \boldsymbol{C}_{\top\|} \\ 0 & \boldsymbol{C}_\curlywedge & \boldsymbol{c}_{\curlywedge\|} \\ \boldsymbol{C}_{\top\|} & \boldsymbol{c}_{\curlywedge\|} & c_\| \end{bmatrix}.$

   (a) $\mathsf{C}_{\top *}$ depends only on $\mathsf{C}_\top$
   (b) $\boldsymbol{C}_{\curlywedge *}$ depends only on $\boldsymbol{C}_\curlywedge$
   (c) $\boldsymbol{c}_{\curlywedge\| *}$ depends only on $\boldsymbol{C}_{\top\|}$ and $\boldsymbol{c}_{\curlywedge\|}$
   (d) $\boldsymbol{C}_{\top\| *}$ depends only on $\boldsymbol{C}_{\top\|}$ and $\mathsf{C}_\top$
   (e) $\mathsf{C}_{\top *}$, $\boldsymbol{C}_{\curlywedge *}$ and $\boldsymbol{C}_{\top\| *}$ are independent of $\boldsymbol{c}_{\curlywedge\|}$ and $c_\|$.

3. $\mathsf{C} = \begin{bmatrix} \mathsf{C}_\top & 0 & 0 \\ 0 & \boldsymbol{C}_\curlywedge & \boldsymbol{c}_{\curlywedge\|} \\ 0 & \boldsymbol{c}_{\curlywedge\|} & c_\| \end{bmatrix}.$

   (a) $c_{\| *}$ is independent of $\mathsf{C}_\top$.

   (b) $\mathsf{C}' = \begin{bmatrix} \mathsf{C}_\top & 0 & 0 \\ 0 & \frac{\boldsymbol{C}_\curlywedge}{\det \boldsymbol{C}_\curlywedge} & \frac{\boldsymbol{C}_\curlywedge \boldsymbol{c}^\perp_{\curlywedge\|}}{\det \boldsymbol{C}_\curlywedge} \\ 0 & \frac{\boldsymbol{C}_\curlywedge \boldsymbol{c}^\perp_{\curlywedge\|}}{\det \boldsymbol{C}_\curlywedge} & \boldsymbol{C}_\curlywedge^{-1}\boldsymbol{c}_{\curlywedge\|} \cdot \boldsymbol{c}_{\curlywedge\|} - c_\| + c_0 \end{bmatrix}$

4. $\mathsf{C} = \begin{bmatrix} \mathsf{C}_\top & 0 & \boldsymbol{C}_{\top\|} \\ 0 & \boldsymbol{C}_\curlywedge & 0 \\ \boldsymbol{C}_{\top\|} & 0 & c_\| \end{bmatrix}.$

   (a) $c_{\| *}$ is independent of $\boldsymbol{C}_\curlywedge$.

   (b) $\mathsf{C}' = \begin{bmatrix} \mathsf{C}_\top & 0 & \boldsymbol{C}_{\top\|} \\ 0 & \frac{\boldsymbol{C}_\curlywedge}{\det \boldsymbol{C}_\curlywedge} & 0 \\ \boldsymbol{C}_{\top\|} & 0 & c_\| \end{bmatrix}$

5. $\mathsf{C} = \begin{bmatrix} \mathsf{C}_\top & 0 & \boldsymbol{C}_{\top\|} \\ 0 & \boldsymbol{C}_\times & \boldsymbol{c}_{\times\|} \\ \boldsymbol{C}_{\top\|} & \boldsymbol{c}_{\times\|} & c_\| \end{bmatrix}, \quad \mathsf{C}_\top \boldsymbol{I}_2 = 2\kappa_0 \boldsymbol{I}_2, \quad \mathrm{Tr}\, \boldsymbol{C}_{\top\|} = t_0.$

The blocks in C' are constructed according to the following rules. $\mathsf{C}'_\top \boldsymbol{I}_2 = 2\kappa_0 \boldsymbol{I}_2$. Let $\mu_0 \in (0, \kappa_0)$ be a constant. Then the two shear moduli μ'_1, μ'_2 of $\mathsf{C}'_\top$ are defined by

$$\left(\frac{1}{\mu'_1} + \frac{1}{\kappa_0}\right)\left(\frac{1}{\mu_2} + \frac{1}{\kappa_0}\right) = \left(\frac{1}{\mu'_2} + \frac{1}{\kappa_0}\right)\left(\frac{1}{\mu_1} + \frac{1}{\kappa_0}\right) = \frac{1}{\kappa_0}\left(\frac{1}{\kappa_0} + \frac{1}{\mu_0}\right) = \frac{1}{a_0^2}.$$

We also have $\mathrm{Tr}\, \boldsymbol{C}'_{\top\|} = t_0$ and

$$\mathrm{dev}(\boldsymbol{C}'_{\top\|}) = \frac{2\kappa_0\mu_0}{a_0} \boldsymbol{R}_\perp (\mathsf{C}_\top - 2\mu_0 \mathsf{I})^{-1} \mathrm{dev}(\boldsymbol{C}_{\top\|}),$$

where dev(·) was defined in (9.15),

$$\boldsymbol{C}'_\times = \sigma_0^2 \frac{\boldsymbol{C}_\times}{\det \boldsymbol{C}_\times}, \qquad \boldsymbol{c}'_{\times\|} = \sigma_0 \frac{\boldsymbol{C}_\times}{\det \boldsymbol{C}_\times} \boldsymbol{c}^\perp_{\times\|},$$

$$c'_\| = b_0 - c_\| + \boldsymbol{C}_\times^{-1} \boldsymbol{c}_{\times\|} \cdot \boldsymbol{c}_{\times\|} + ((\mathsf{C}_\top - 2a_0 \mathsf{I})^{-1} \boldsymbol{C}_{\top\|}, \boldsymbol{C}_{\top\|})_\top$$

6. $\mathsf{C} = \begin{bmatrix} 2\mu_0 \mathsf{T} + \boldsymbol{B} \otimes \boldsymbol{B} & \boldsymbol{B} \otimes \boldsymbol{b} & \boldsymbol{C}_{\top\|} \\ \boldsymbol{b} \otimes \boldsymbol{B} & \boldsymbol{C}_\times & \boldsymbol{c}_{\times\|} \\ \boldsymbol{C}_{\top\|} & \boldsymbol{c}_{\times\|} & c_\| \end{bmatrix}.$

(a) $\boldsymbol{C}'_\times - \boldsymbol{b}' \otimes \boldsymbol{b}' = \boldsymbol{C}_\times - \boldsymbol{b} \otimes \boldsymbol{b}$
(b) $\boldsymbol{B}' = \Lambda \boldsymbol{B}$, where $\Lambda = \Lambda(\boldsymbol{x})$ is a scalar.
(c) $\boldsymbol{B}' = \Lambda \boldsymbol{B}, \quad \boldsymbol{C}'_\times - \boldsymbol{b}' \otimes \boldsymbol{b}' = \boldsymbol{C}_\times - \boldsymbol{b} \otimes \boldsymbol{b}, \quad \boldsymbol{C}'_{\top\|} = \boldsymbol{C}_{\top\|} + \nu \boldsymbol{B}$, where $\nu = \nu(\boldsymbol{x})$ is a scalar.
(d) $\boldsymbol{B}' = \Lambda \boldsymbol{B}, \ \boldsymbol{C}'_\times - \boldsymbol{b}' \otimes \boldsymbol{b}' = \boldsymbol{C}_\times - \boldsymbol{b} \otimes \boldsymbol{b}, \ \boldsymbol{C}'_{\top\|} = \boldsymbol{C}_{\top\|} + \nu \boldsymbol{B},$

$\boldsymbol{c}'_{\times\|} = \boldsymbol{c}_{\times\|} + \nu \boldsymbol{b}.$

7. $\mathsf{C} = \begin{bmatrix} 2\mu_0 \mathsf{T} + \boldsymbol{B} \otimes \boldsymbol{B} & \boldsymbol{B} \otimes \boldsymbol{b} & \beta \boldsymbol{B} \\ \boldsymbol{b} \otimes \boldsymbol{B} & \boldsymbol{C}_\times & \boldsymbol{c}_{\times\|} \\ \beta \boldsymbol{B} & \boldsymbol{c}_{\times\|} & c_\| \end{bmatrix}.$

(a) $\boldsymbol{C}'_\times - \boldsymbol{b}' \otimes \boldsymbol{b}' = \boldsymbol{C}_\times - \boldsymbol{b} \otimes \boldsymbol{b}, \quad \boldsymbol{c}'_{\times\|} - \beta' \boldsymbol{b}' = \boldsymbol{c}_{\times\|} - \beta \boldsymbol{b}, \quad \boldsymbol{B}' = \Lambda \boldsymbol{B},$ $\beta' = \Lambda \beta,$

(b) $\boldsymbol{C}'_\times - \boldsymbol{b}' \otimes \boldsymbol{b}' = \boldsymbol{C}_\times - \boldsymbol{b} \otimes \boldsymbol{b}, \quad \boldsymbol{c}'_{\times\|} - \beta' \boldsymbol{b}' = \boldsymbol{c}_{\times\|} - \beta \boldsymbol{b},$ $c'_\| - (\beta')^2 = c_\| - \beta^2,$

8. $$\mathsf{C} = \begin{bmatrix} 2\mu_0\mathsf{T} + \boldsymbol{B}\otimes\boldsymbol{B} & \boldsymbol{B}\otimes\boldsymbol{b} & \beta\boldsymbol{B} \\ \boldsymbol{b}\otimes\boldsymbol{B} & \boldsymbol{C}_{\times} & \beta\boldsymbol{b} \\ \beta\boldsymbol{B} & \beta\boldsymbol{b} & c_{\parallel} \end{bmatrix}.$$

(a) $$\mathsf{C}' = \begin{bmatrix} 2\mu_0\mathsf{T} + \boldsymbol{B}\otimes\boldsymbol{B} & \boldsymbol{B}\otimes\boldsymbol{b} & \beta\boldsymbol{B} \\ \boldsymbol{b}\otimes\boldsymbol{B} & \boldsymbol{C}_{\times} & \beta\boldsymbol{b} \\ \beta\boldsymbol{B} & \beta\boldsymbol{b} & a_0c_{\parallel} + (1-a_0)\beta^2 \end{bmatrix}$$

(b) $$\mathsf{C}' = \begin{bmatrix} 2\mu_0\mathsf{T} + \Lambda\boldsymbol{B}\otimes\boldsymbol{B} & \boldsymbol{B}\otimes\boldsymbol{b}' & \Lambda\beta\boldsymbol{B} \\ \boldsymbol{b}'\otimes\boldsymbol{B} & \boldsymbol{C}'_{\times} & \beta\boldsymbol{b}' \\ \Lambda\beta\boldsymbol{B} & \beta\boldsymbol{b}' & c'_{\parallel} \end{bmatrix}$$

9. $$\mathsf{C} = \begin{bmatrix} 2\mu_0\mathsf{T} + \boldsymbol{B}\otimes\boldsymbol{B} & 0 & \beta\boldsymbol{B} \\ 0 & \boldsymbol{C}_{\times} & \boldsymbol{c}_{\times\parallel} \\ \beta\boldsymbol{B} & \boldsymbol{c}_{\times\parallel} & c_{\parallel} \end{bmatrix}.$$

(a) $$\mathsf{C}' = \begin{bmatrix} 2\mu_0\mathsf{T} + \boldsymbol{B}\otimes\boldsymbol{B} & 0 & c_0\beta\boldsymbol{B} \\ 0 & \boldsymbol{C}_{\times} & \boldsymbol{c}_{\times\parallel} \\ c_0\beta\boldsymbol{B} & \boldsymbol{c}_{\times\parallel} & c_{\parallel} + (c_0^2-1)\beta^2 \end{bmatrix},$$

(b) $$\mathsf{C}' = \begin{bmatrix} 2\mu_0\mathsf{T} + \boldsymbol{B}\otimes\boldsymbol{B} & 0 & \beta\boldsymbol{B} \\ 0 & \dfrac{\boldsymbol{C}_{\times}}{\det\boldsymbol{C}_{\times}} & \dfrac{\boldsymbol{C}_{\times}\boldsymbol{c}^{\perp}_{\times\parallel}}{\det\boldsymbol{C}_{\times}} \\ \beta\boldsymbol{B} & \dfrac{\boldsymbol{C}_{\times}\boldsymbol{c}^{\perp}_{\times\parallel}}{\det\boldsymbol{C}_{\times}} & \boldsymbol{C}_{\times}^{-1}\boldsymbol{c}_{\times\parallel}\cdot\boldsymbol{c}_{\times\parallel} + 2\beta^2 - c_{\parallel} + c_0 \end{bmatrix}$$

10. $$\mathsf{C} = \begin{bmatrix} 2\lambda\boldsymbol{I}_2\otimes\boldsymbol{I}_2 + 2\mu_0\mathsf{I} & 0 & \beta\boldsymbol{I}_2 \\ 0 & \boldsymbol{C}_{\times} & \boldsymbol{c}_{\times\parallel} \\ \beta\boldsymbol{I}_2 & \boldsymbol{c}_{\times\parallel} & c_{\parallel} \end{bmatrix}.$$

$$\frac{\lambda'}{\lambda'+2\mu_0} = a_0\frac{\lambda}{\lambda+2\mu_0} + b_0\frac{\beta}{\lambda+2\mu_0}, \qquad \frac{\beta'}{\lambda'+2\mu_0} = \frac{\beta}{\lambda+2\mu_0},$$

$$c'_{\parallel} - \frac{(\beta')^2}{2(\lambda'+2\mu_0)} = c_{\parallel} - \frac{\beta^2}{2(\lambda+2\mu_0)} + \frac{c_0}{\lambda+2\mu_0}.$$

11. $$\mathsf{C} = \begin{bmatrix} 2\lambda\boldsymbol{I}_2\otimes\boldsymbol{I}_2 + 2\mu_0\mathsf{I} & 0 & \boldsymbol{C}_{\top\parallel} \\ 0 & \boldsymbol{C}_{\times} & \boldsymbol{c}_{\times\parallel} \\ \boldsymbol{C}_{\top\parallel} & \boldsymbol{c}_{\times\parallel} & c_{\parallel} \end{bmatrix}, \qquad c'_{\parallel} = c_{\parallel} + \tfrac{a_0}{\lambda+2\mu_0}.$$

12. $$\mathsf{C} = \begin{bmatrix} 2\lambda\boldsymbol{I}_2\otimes\boldsymbol{I}_2 + 2\mu_0\mathsf{I} & \boldsymbol{I}_2\otimes\boldsymbol{b} & \beta\boldsymbol{I}_2 \\ \boldsymbol{b}\otimes\boldsymbol{I}_2 & \boldsymbol{C}_{\times} & \boldsymbol{c}_{\times\parallel} \\ \beta\boldsymbol{I}_2 & \boldsymbol{c}_{\times\parallel} & c_{\parallel} \end{bmatrix}.$$

(a) $\frac{1}{\lambda' + 2\mu_0} = \frac{1}{\lambda + 2\mu_0} + b_0, \qquad \frac{\boldsymbol{b}'}{\lambda' + 2\mu_0} = \frac{\boldsymbol{b}}{\lambda + 2\mu_0},$

$$\frac{\beta'}{\lambda' + 2\mu_0} = \frac{\beta}{\lambda + 2\mu_0}, \qquad \boldsymbol{C}'_{\angle} - \frac{\boldsymbol{b}' \otimes \boldsymbol{b}'}{2(\lambda' + 2\mu_0)} = \boldsymbol{C}_{\angle} - \frac{\boldsymbol{b} \otimes \boldsymbol{b}}{2(\lambda + 2\mu_0)},$$

$$\boldsymbol{c}'_{\angle\|} - \frac{\beta'\boldsymbol{b}'}{2(\lambda' + 2\mu_0)} = \boldsymbol{c}_{\angle\|} - \frac{\beta\boldsymbol{b}}{2(\lambda + 2\mu_0)}, \quad c'_{\|} - \frac{(\beta')^2}{2(\lambda' + 2\mu_0)} = c_{\|} - \frac{\beta^2}{2(\lambda + 2\mu_0)} + c_0,$$

(b) $\frac{2\mu_0^2}{\lambda' + 2\mu_0} = \frac{\beta^2}{2(\lambda + 2\mu_0)} - c_{\|} + b_0, \qquad \frac{\mu_0 \boldsymbol{b}'}{\lambda' + 2\mu_0} = \boldsymbol{c}_{\angle\|} - \frac{\beta\boldsymbol{b}}{2(\lambda + 2\mu_0)},$

$$\boldsymbol{c}'_{\angle\|} - \frac{\beta'\boldsymbol{b}'}{2(\lambda' + 2\mu_0)} = a_0\left(\boldsymbol{c}_{\angle\|} - \frac{(\beta - \beta_0)\boldsymbol{b}}{2(\lambda + 2\mu_0)}\right), \qquad \frac{\mu_0\beta'}{\lambda' + 2\mu_0} = a_0\left(c_{\|} - \frac{\beta(\beta - \beta_0)}{2(\lambda + 2\mu_0)}\right),$$

$$\boldsymbol{C}'_{\angle} - \frac{\boldsymbol{b}' \otimes \boldsymbol{b}'}{2(\lambda' + 2\mu_0)} = \boldsymbol{C}_{\angle} - \frac{\boldsymbol{b} \otimes \boldsymbol{b}}{2(\lambda + 2\mu_0)}, \qquad c'_{\|} - \frac{(\beta')^2}{2(\lambda' + 2\mu_0)} = a_0^2\left(c_{\|} - \frac{(\beta - \beta_0)^2}{2(\lambda + 2\mu_0)}\right) + c_0,$$

(c) This exact relations looks better if we use the bulk modulus $\kappa = \lambda + \mu_0$ instead of the Lamé modulus λ:

$$\frac{\mu_0^2}{\kappa'} = \kappa - \frac{1}{2}\boldsymbol{C}_{\angle}^{-1}\boldsymbol{b} \cdot \boldsymbol{b} \qquad \frac{\boldsymbol{b}'}{2\kappa'} = \frac{\boldsymbol{C}_{\angle}\boldsymbol{b}^{\perp}}{\det \boldsymbol{C}_{\angle}},$$

$$\boldsymbol{C}'_{\angle} - \frac{\boldsymbol{b}' \otimes \boldsymbol{b}'}{2\kappa'} = \frac{4\mu_0^2 \boldsymbol{C}_{\angle}}{\det \boldsymbol{C}_{\angle}}, \qquad \frac{\mu_0\beta'}{\kappa'} = \boldsymbol{C}_{\angle}^{-1}\boldsymbol{b} \cdot \boldsymbol{c}_{\angle\|} - \beta,$$

$$\boldsymbol{c}'_{\angle\|} - \frac{\beta'\boldsymbol{b}'}{2\kappa'} = \frac{2\mu_0 \boldsymbol{C}_{\angle}\boldsymbol{c}_{\angle\|}^{\perp}}{\det \boldsymbol{C}_{\angle}}, \qquad c'_{\|} - \frac{(\beta')^2}{2\kappa'} = c_0 - c_{\|} + \boldsymbol{C}_{\angle}^{-1}\boldsymbol{c}_{\angle\|} \cdot \boldsymbol{c}_{\angle\|}.$$

13. $\mathsf{C} = \begin{bmatrix} 2\mu_0 \mathsf{I} + 2\lambda \boldsymbol{I}_2 \otimes \boldsymbol{I}_2 & \boldsymbol{I}_2 \otimes \boldsymbol{b} & \beta \boldsymbol{I}_2 \\ \boldsymbol{b} \otimes \boldsymbol{I}_2 & \boldsymbol{\sigma} + \frac{\boldsymbol{b} \otimes \boldsymbol{b}}{2(\lambda + 2\mu_0)} & \boldsymbol{c}_{\angle\|} \\ \beta \boldsymbol{I}_2 & \boldsymbol{c}_{\angle\|} & c_{\|} \end{bmatrix}, \quad \det \boldsymbol{\sigma} = d_0^2.$

(a) In C' the parameters λ' and β' are arbitrary and independent of C. Then, for arbitrary constant angles ϕ_0, β_0 we have

$$\frac{2\mu_0 \boldsymbol{b}'}{\lambda' + 2\mu_0} = \frac{1}{\cos\phi_0}\left(\cos\beta_0 \boldsymbol{w}_1 - \frac{\sin\beta_0}{d_0}\boldsymbol{\sigma}\boldsymbol{w}_2^{\perp}\right),$$

$$\boldsymbol{c}'_{\angle\|} \quad \frac{\beta'\boldsymbol{b}'}{2(\lambda' + 2\mu_0)} = \frac{1}{\cos\phi_0}\left(-\cos\beta_0 \boldsymbol{w}_2 + \frac{\sin\beta_0}{d_0}\boldsymbol{\sigma}\boldsymbol{w}_1^{\perp}\right),$$

where

$$\boldsymbol{w}_1 = \frac{2\mu_0 \boldsymbol{b}}{\lambda + 2\mu_0} \mp \left(\boldsymbol{c}_{\angle\|} - \frac{\beta \boldsymbol{b}}{2(\lambda + 2\mu_0)}\right) \sin\phi_0, \quad \boldsymbol{w}_2 = \frac{2\boldsymbol{b}\mu_0 \sin\phi_0}{\lambda + 2\mu_0} \mp \left(\boldsymbol{c}_{\angle\|} - \frac{\beta \boldsymbol{b}}{2(\lambda + 2\mu_0)}\right).$$

$$c_{\|}' - \frac{(\beta')^2}{2(\lambda' + 2\mu_0)} - \frac{4\mu_0 \lambda'}{\lambda' + 2\mu_0} = c_{\|} - \frac{\beta^2}{2(\lambda + 2\mu_0)} - \frac{4\mu_0 \lambda}{\lambda + 2\mu_0} +$$

$$\sin^2 \beta_0 \left(\frac{(\beta^2 - 16\mu_0^2)\boldsymbol{\sigma}^{-1}\boldsymbol{b} \cdot \boldsymbol{b}}{4(\lambda + 2\mu_0)^2} + \boldsymbol{\sigma}^{-1}\boldsymbol{c}_{\angle\|} \cdot \boldsymbol{c}_{\angle\|} - \frac{\beta \boldsymbol{\sigma}^{-1}\boldsymbol{b} \cdot \boldsymbol{c}_{\angle\|}}{\lambda + 2\mu_0}\right) \pm \frac{2 \sin 2\beta_0}{d_0} \boldsymbol{b}^{\perp} \cdot \boldsymbol{c}_{\angle\|}.$$

(b) Moduli λ' and β' are arbitrary and independent of C, while

$$\frac{\boldsymbol{b}'}{\lambda' + 2\mu_0} = \frac{\boldsymbol{b}}{\lambda + 2\mu_0}, \qquad \boldsymbol{c}'_{\angle\|} - \frac{\beta' \boldsymbol{b}'}{\lambda' + 2\mu_0} = \boldsymbol{\sigma}\left(\boldsymbol{c}_{\angle\|} - \frac{\beta \boldsymbol{b}}{\lambda + 2\mu_0}\right)^{\perp}.$$

$$c'_{\|} - \frac{(\beta')^2}{2(\lambda' + 2\mu_0)} - \frac{4\mu_0 \lambda'}{\lambda' + 2\mu_0} = c_0 - \left(c_{\|} - \frac{\beta^2}{2(\lambda + 2\mu_0)}\right) - \frac{4\mu_0 \lambda}{\lambda + 2\mu_0} +$$

$$\boldsymbol{\sigma}\left(\boldsymbol{c}_{\angle\|} - \frac{\beta \boldsymbol{b}}{\lambda + 2\mu_0}\right)^{\perp} \cdot \left(\boldsymbol{c}_{\angle\|} - \frac{\beta \boldsymbol{b}}{\lambda + 2\mu_0}\right)^{\perp}.$$

(c) Lamé modulus λ' is independent of C, $\boldsymbol{\sigma}' = \boldsymbol{\sigma}$, and

$$\frac{\boldsymbol{b}'}{2(\lambda' + 2\mu_0)} = a_0\left(\boldsymbol{c}_{\angle\|} - \frac{\beta \boldsymbol{b}}{2(\lambda + 2\mu_0)}\right) + b_0 \boldsymbol{\sigma}\left(\boldsymbol{c}^{\perp}_{\angle\|} - \frac{(\beta \pm 8\mu_0)\boldsymbol{b}^{\perp}}{2(\lambda + 2\mu_0)}\right),$$

$$\frac{2a_0\beta' + 1}{2(\lambda' + 2\mu_0)} = a_0^2\left(c_{\|} - \frac{\beta^2}{2(\lambda + 2\mu_0)}\right) +$$

$$d_0^2 b_0^2\left(c_{\|} - \frac{(\beta \pm 8\mu_0)^2}{2(\lambda + 2\mu_0)} + \boldsymbol{\sigma}^{-1}\left(\boldsymbol{c}_{\angle\|} - \frac{(\beta \pm 8\mu_0)\boldsymbol{b}}{2(\lambda + 2\mu_0)}\right) \cdot \left(\boldsymbol{c}_{\angle\|} - \frac{(\beta \pm 8\mu_0)\boldsymbol{b}}{2(\lambda + 2\mu_0)}\right)\right).$$

$$\boldsymbol{c}'_{\angle\|} - \frac{\beta' \boldsymbol{b}'}{2(\lambda' + 2\mu_0)} = \boldsymbol{c}_{\angle\|} - \frac{\beta \boldsymbol{b}}{2(\lambda + 2\mu_0)}, \qquad c_{\|}' - \frac{\beta'^2}{2(\lambda' + 2\mu_0)} = c_{\|} - \frac{\beta^2}{2(\lambda + 2\mu_0)} + c_0.$$

14. $$\mathsf{C} = \begin{bmatrix} 2\mu_0 \mathsf{I} + 2\lambda \boldsymbol{I}_2 \otimes \boldsymbol{I}_2 & \boldsymbol{I}_2 \otimes \boldsymbol{b} & \beta \boldsymbol{I}_2 \\ \boldsymbol{b} \otimes \boldsymbol{I}_2 & \sigma_0 \boldsymbol{I}_2 + \frac{\boldsymbol{b} \otimes \boldsymbol{b}}{2(\lambda + 2\mu_0)} & \frac{\beta_0 \boldsymbol{b}^{\perp} + \beta \boldsymbol{b}}{2(\lambda + 2\mu_0)} \\ \beta \boldsymbol{I}_2 & \frac{\beta_0 \boldsymbol{b}^{\perp} + \beta \boldsymbol{b}}{2(\lambda + 2\mu_0)} & c_{\|} \end{bmatrix},$$

Denoting

$$u(\lambda, b) = \frac{b}{\lambda + 2\mu_0}, \qquad \theta(\lambda, b) = \frac{\beta_0^2}{2(\lambda + 2\mu_0)} + \frac{\beta_0^2}{4\sigma_0} \frac{|b|^2}{(\lambda + 2\mu_0)^2}$$

$$\alpha(\lambda, \beta) = \frac{\beta}{\lambda + 2\mu_0}, \qquad \gamma(\lambda, \beta, c_\parallel) = c_\parallel - \frac{\beta^2}{2(\lambda + 2\mu_0)}$$

The link is given by the equations

$$u(\lambda', b') = u(\lambda, b),$$

$$\theta(\lambda', b') = a_0\theta(\lambda, b) + (1 - a_0)\gamma(\lambda, \beta, c_\parallel),$$

$$\alpha(\lambda', \beta') = \alpha(\lambda, \beta) + b_0(\gamma(\lambda, \beta, c_\parallel) - \theta(\lambda, b)),$$

$$\gamma(\lambda', \beta', c_\parallel') = c_0\gamma(\lambda, \beta, c_\parallel) + (1 - c_0)\theta(\lambda, b),$$

where in view of the first equation, the second can be rewritten as a formula for λ':

$$\frac{\beta_0^2}{2(\lambda' + 2\mu_0)} = \frac{a_0\beta_0^2}{2(\lambda + 2\mu_0)} + (1 - a_0)\left(c_\parallel - \frac{\beta^2}{2(\lambda + 2\mu_0)} - \frac{\beta_0^2}{4\sigma_0}\frac{|b|^2}{(\lambda + 2\mu_0)^2}\right)$$

15. Suppose C belongs to the exact relation in item 23 in section 9.3.1. Then there is a link described by the formulas
    - $\mathsf{C}'_\top = \mathsf{C}_\top$, $\mathscr{C}'_{\top\times} = \mathscr{C}_{\top\times}$, $C'_\times = C_\times$.
    - Let $\hat{\mathsf{S}} = (\mathsf{C} + \mathsf{C}_0)^{-1}$ and $\hat{\mathsf{S}}' = (\mathsf{C}' + \mathsf{C}_0)^{-1}$, where $\mathsf{C}_0 = \begin{bmatrix} \tau_0 \mathsf{I} & 0 & 0 \\ 0 & \sigma_0 I_2 & 0 \\ 0 & 0 & 0 \end{bmatrix}$. Then

$$\frac{1}{\hat{s}_\parallel} + \frac{1}{\hat{s}'_\parallel} = \frac{1}{\zeta_0}, \qquad \frac{\hat{s}'_{\times\parallel}}{\hat{s}'_\parallel} = \pm\frac{\hat{s}^\perp_{\times\parallel}}{\hat{s}_\parallel}, \qquad \frac{\mathrm{dev}(\hat{S}'_{\top\parallel})}{\hat{s}'_\parallel} = \pm\frac{R_\perp \mathrm{dev}(\hat{S}_{\top\parallel})}{\hat{s}_\parallel},$$

where the '+' sign corresponds to $\mathscr{S}_{\top\times}u = \psi(c\bar{u})$ and the '−' sign to $\mathscr{S}_{\top\times}u = \psi(cu)$.

References

[1] Avellaneda M, Cherkaev A V, Gibiansky L V, Milton G W and Rudelson M 1996 A complete characterization of the possible bulk and shear moduli of planar polycrystals *J. Mech. Phys. Solids* **44** 1179–218

[2] Benveniste Y 1995 Correspondence relations among equivalent classes of heterogeneous piezoelectric solids under anti-plane mechanical and in-plane electrical fields *J. Mech. Phys. Solids* **43** 553–71

[3] Berdichevsky V L 1983 *Variational Principles of Continuum Mechanics (Russian)* (Moscow: Nauka)
[4] Cherkaev A V, Lurie K A and Milton 1992 Invariant properties of the stress in plane elasticity and equivalence classes of composites *Proc. R. Soc. Lond.* A **438** 519–29
[5] Gibiansky L V 1998 Effective moduli of plane polycrystals with a rank-one local compliance tensor *Proc. R. Soc. Edinburgh* A **128** 1325–54
[6] Grabovsky Y and Milton G W 1998 Rank one plus a null-Lagrangian is an inherited property of two-dimensional compliance tensors under homogenization *Proc. R. Soc. Edinburgh* A **128** 283–99
[7] Hegg M 2012 Exact relations and links for fiber-reinforced elastic composites *PhD Thesis* Temple University, Philadelphia, PA
[8] Hegg M 2013 Links between effective tensors for fiber-reinforced elastic composites *C. R. Méc.* **341** 520–32
[9] Helsing J, Milton G W and Movchan A B 1997 Duality relations, correspondences and numerical results for planar elastic composites *J. Mech. Phys. Solids* **45** 565–90
[10] Hill R 1963 Elastic properties of reinforced solids: some theoretical principles *J. Mech. Phys. Solids* **11** 357–72
[11] Hill R 1964 Theory of mechanical properties of fibre-strengthened materials: I elastic behaviour *J. Mech. Phys. Solids* **12** 199–212
[12] Lurie K A and Cherkaev A V 1984 G-closure of some particular sets of admissible material characteristics for the problem of bending of thin plates *J. Opt. Theor. Appl.* **42** 305–16
[13] Lurie K A and Cherkaev A V 1986 Effective characteristics of composite materials and the optimal design of structural elements (Russian) *Usp. Mekh. = Adv. Mech.* **9** 3–81 English translation in [14]
[14] Lurie K A and Cherkaev A V 1997 Effective characteristics of composite materials and the optimal design of structural elements *Topics in the Mathematical Modeling of Composite Materials* Progress in Nonlinear Differential equations and Their Applications **vol 31** ed A Cherkaev and R Kohn (Boston, MA: Birkhäuser) 175–258
[15] Lurie K A, Cherkaev A V and Fedorov A V 1984 On the existence of solutions to some problems of optimal design for bars and plates *J. Optim. Theor. Appl.* **42** 247–81
[16] Tartar L 1985 Estimation fines des coefficients homogénéisés ed P Kree *E. De Giorgi Colloquium (Paris, 1983)* ed P Kree and E De Giorgi (London: Pitman) 168–87
[17] Walpole L 1985 Evaluation of the elastic moduli of a transversely isotropic aggregate of cubic crystals *J. Mech. Phys. Solids* **33** 623–36

IOP Publishing

Composite Materials (Second Edition)
Mathematical theory and exact relations
Yury Grabovsky

Chapter 10

Piezoelectricity

A piezoelectric material is described by the constitutive relation

$$\begin{bmatrix} \boldsymbol{\sigma}(\boldsymbol{x}) \\ \boldsymbol{e}(\boldsymbol{x}) \end{bmatrix} = \mathsf{L}(\boldsymbol{x}) \begin{bmatrix} \boldsymbol{\varepsilon}(\boldsymbol{x}) \\ \boldsymbol{j}(\boldsymbol{x}) \end{bmatrix}, \qquad \mathsf{L}(\boldsymbol{x}) = \begin{bmatrix} \mathsf{C}(\boldsymbol{x}) & \mathscr{P}(\boldsymbol{x}) \\ \mathscr{P}^T(\boldsymbol{x}) & \boldsymbol{\rho}(\boldsymbol{x}) \end{bmatrix}, \tag{10.1}$$

where symmetric matrix fields $\boldsymbol{\varepsilon}(\boldsymbol{x})$ and $\boldsymbol{\sigma}(\boldsymbol{x})$ denote the elastic strain and stress fields, satisfying differential constraints (9.2)–(9.3), respectively, while vector fields $\boldsymbol{e}(\boldsymbol{x})$ and $\boldsymbol{j}(\boldsymbol{x})$, satisfying (8.1), denote the electric field and the dielectric displacement, respectively. Thus $\boldsymbol{\rho}(\boldsymbol{x})^{-1}$ has the meaning of a dielectric tensor, while $\mathscr{P}(\boldsymbol{x})$ is a piezoelectric coupling tensor.

Most exact relations and links are the same in two and three dimensions, with a few 2D specific results that we list in a separate section. We therefore use d to denote dimension $d = 2$ or 3.

10.1 Exact relations

The complete list of exact relations for three-dimensional piezoelectricity was first published in [4]. There are just five exact relations. Two of them are uniform field relations (UFR), one is related to Hill's exact relation (9.11) and two to the 'rank one plus a null-Lagrangian' exact relation (9.13).

1. (UFR) $\mathsf{C}\boldsymbol{I}_d = d\kappa_0\boldsymbol{I}_d$, $\mathscr{P}^T\boldsymbol{I}_d = 0$
2. $\mathsf{C} = 2\mu_0\mathsf{T} + \boldsymbol{B}\otimes\boldsymbol{B}$, $\mathscr{P} = \boldsymbol{B}\otimes\boldsymbol{b}$
3. (UFR) $\mathsf{C} = 2\mu_0\mathsf{I} + \lambda\boldsymbol{I}_d\otimes\boldsymbol{I}_d$, $\mathscr{P} = \boldsymbol{I}_d\otimes\boldsymbol{b}$
4. $\mathsf{C} = 2\mu_0\mathsf{T} + \boldsymbol{B}\otimes\boldsymbol{B}$, $\mathscr{P} = \boldsymbol{B}\otimes\boldsymbol{b}$, $\boldsymbol{\rho} = \rho_0\boldsymbol{I}_d + \boldsymbol{b}\otimes\boldsymbol{b}$
5. $\mathsf{C} = 2\mu_0\mathsf{I} + \lambda\boldsymbol{I}_d\otimes\boldsymbol{I}_d$, $\mathscr{P} = \boldsymbol{I}_d\otimes\boldsymbol{b}$, $\boldsymbol{\rho} = \rho_0\boldsymbol{I}_d + \frac{\boldsymbol{b}\otimes\boldsymbol{b}}{\lambda+2\mu_0}$, $\left\langle\frac{\boldsymbol{b}}{\lambda+2\mu_0}\right\rangle$

10.2 Links

Links have not been presented anywhere to the best of author's knowledge. Their computation (in three space dimensions) became possible due to the work of an undergraduate student Daniel Lapsley during the summer of 2016, who wrote Maple code for multiplication of SO(3) irreducible representations, based on the theory presented in chapter 17.

1. Global link. $\mathsf{L}' = \begin{bmatrix} \mathsf{C} & -\mathscr{P} \\ -\mathscr{P}^T & \rho \end{bmatrix}$, where $\mathsf{L} = \begin{bmatrix} \mathsf{C} & \mathscr{P} \\ \mathscr{P}^T & \rho \end{bmatrix}$
2. $\mathsf{L} = \begin{bmatrix} 2\mu_0 \mathsf{T} + \boldsymbol{B} \otimes \boldsymbol{B} & \boldsymbol{B} \otimes \boldsymbol{b} \\ \boldsymbol{b} \otimes \boldsymbol{B} & \rho \end{bmatrix}$.
   (a) $\boldsymbol{B}'$ and $\boldsymbol{b}'$ are unrelated to L, while $\rho' - \boldsymbol{b}' \otimes \boldsymbol{b}' = \rho - \boldsymbol{b} \otimes \boldsymbol{b}$
   (b) $\boldsymbol{b}'$ and ρ' are unrelated to L, $\boldsymbol{B}' = \lambda' \boldsymbol{B}$, where λ' is a scalar, unrelated to L
   (c) $\boldsymbol{b}'$ is unrelated to L, $\boldsymbol{B}' = \lambda' \boldsymbol{B}$, $\rho' - \boldsymbol{b}' \otimes \boldsymbol{b}' = \rho - \boldsymbol{b} \otimes \boldsymbol{b}$, where λ' is a scalar, unrelated to L
   (d) $\boldsymbol{b}' = \lambda' \boldsymbol{b}$, $\boldsymbol{B}' = a_0 \lambda'(\boldsymbol{B} + b_0(\mathrm{Tr}\, \boldsymbol{B})\boldsymbol{I}_d)$, $\rho' - \boldsymbol{b}' \otimes \boldsymbol{b}' = \rho - \boldsymbol{b} \otimes \boldsymbol{b}$, where λ' is a scalar, unrelated to L.
3. $\mathsf{L} = \begin{bmatrix} 2\mu_0 \mathsf{I} + \lambda \boldsymbol{I}_d \otimes \boldsymbol{I}_d & \boldsymbol{I}_d \otimes \boldsymbol{b} \\ \boldsymbol{b} \otimes \boldsymbol{I}_d & \rho \end{bmatrix}$.

$$\frac{1}{\lambda' + 2\mu_0'} = \frac{a_0^2}{\lambda + 2\mu_0} + b_0, \qquad \frac{\boldsymbol{b}'}{\lambda' + 2\mu_0'} = a_0 \frac{\boldsymbol{b}}{\lambda + 2\mu_0},$$

$$\rho' - \frac{\boldsymbol{b}' \otimes \boldsymbol{b}'}{\lambda' + 2\mu_0'} = \rho - \frac{\boldsymbol{b} \otimes \boldsymbol{b}}{\lambda + 2\mu_0}.$$

10.3 2D-specific relations and links

We note that the effective behavior of a 2D piezoelectric composite can be determined from the 3D effective elasticity tensor of a fibrous composite with the same microstructure [2, 3, 6]. However, we must note that such relations between effective moduli of composites in different physical contexts have long been known and used (see, e.g., [5]).

In order to match equations of elasticity for a fibrous composite and of two-dimensional piezoelectricity we need to 'interchange' $\boldsymbol{e}$ and $\boldsymbol{j}$ in (10.1). In general this is impossible without destroying either the symmetry of L or its positive definiteness [7], section 2.4. However, in two space dimensions there is another possibility:

$$\varepsilon_{\measuredangle} = \boldsymbol{j}^{\perp}, \qquad \sigma_{\measuredangle} = \boldsymbol{e}^{\perp}. \tag{10.2}$$

The vector fields $\varepsilon_{\measuredangle}$ and $\boldsymbol{\sigma}_{\measuredangle}$ are then curl-free and divergence-free, respectively. In this case

$$\mathsf{C} = \mathsf{C}_{\top}, \qquad \mathscr{P} = \mathscr{C}_{\top\measuredangle} \boldsymbol{R}_{\perp}^T, \qquad \rho = \boldsymbol{R}_{\perp} \boldsymbol{C}_{\measuredangle} \boldsymbol{R}_{\perp}^T. \tag{10.3}$$

Using the conversion rule (10.3) we can read-off all exact relations and links for 2D piezoelectric composites from the complete lists in section 9.3. As we have mentioned before, most exact relations and links hold for both two and three-dimensional piezoelectric polycrystals. There is a 3-parameter family of global links [1, 7], given by

$$\mathsf{L}' = \mathfrak{F}_{a,q_1,q_2}(\mathsf{L}) = \begin{bmatrix} q_1^2\mathsf{T}\Phi_a(\mathsf{C})\mathsf{C} & q_1q_2\mathsf{T}\Phi_a(\mathsf{C})\mathscr{P} \\ q_1q_2\mathscr{P}^T\Phi_a(\mathsf{C})\mathsf{T} & q_2^2(\rho - a\mathscr{P}^T\Phi_a(\mathsf{C})\mathscr{P}) \end{bmatrix}, \tag{10.4}$$

where $\Phi_a(\mathsf{C}) = (\mathsf{T} + a\mathsf{C})^{-1}$.

There are also two special exact relations

1. $\mathsf{L} = \begin{bmatrix} 2\mu_0\mathsf{T} + \boldsymbol{B}\otimes\boldsymbol{B} & \boldsymbol{B}\otimes\boldsymbol{b} \\ \boldsymbol{b}\otimes\boldsymbol{B} & \boldsymbol{\rho} + \boldsymbol{b}\otimes\boldsymbol{b} \end{bmatrix}, \quad \det\boldsymbol{\rho} = \rho_0$
2. A piezoelectric tensor L belongs to this exact relation if and only if its blocks C, $\mathscr{P}$ and $\boldsymbol{\rho}$ satisfy the following constraints.
   - $\mathsf{C}\boldsymbol{I}_2 = 2\kappa_0\boldsymbol{I}_2$, $\mathscr{P}^T\boldsymbol{I}_2 = 0$, where $\kappa_0 > 0$ is an arbitrary constant.
   - Choose arbitrary constants $\sigma_0 > 0$ and $0 < \tau_0 < \kappa_0$, and define tensors S, $\boldsymbol{\alpha}$ and $\mathscr{Q}$ by

$$\begin{bmatrix} \mathsf{C} + \tau_0\mathsf{I} & \mathscr{P} \\ \mathscr{P}^T & \boldsymbol{\rho} + \rho_0\boldsymbol{I}_2 \end{bmatrix}^{-1} = \begin{bmatrix} (\mathsf{S} + \tau_0\mathsf{I})^{-1} & \mathscr{Q} \\ \mathscr{Q} & (\boldsymbol{\alpha} + \rho_0\boldsymbol{I}_2)^{-1} \end{bmatrix}. \tag{10.5}$$

   - The range of the operator $\mathscr{Q}: \mathbb{R}^2 \to \mathrm{Sym}(\mathbb{R}^2)$ must be $\mathrm{Sym}_0(\mathbb{R}^2)$, consisting of all trace-free matrices in $\mathrm{Sym}(\mathbb{R}^2)$, and $\mathscr{Q}^T\mathscr{Q}: \mathbb{R}^2 \to \mathbb{R}^2$ must be a (possibly variable) multiple of $\boldsymbol{I}_2$

$$\mathscr{Q}^T\mathscr{Q} = q\boldsymbol{I}_2. \tag{10.6}$$

   - We require that $\det\boldsymbol{\alpha} = \rho_0^2$.
   - Let μ_1 and μ_2 be the shear moduli of S (its bulk modulus is κ_0). We require that

$$\left(\frac{1}{\kappa_0} + \frac{1}{\mu_1}\right)\left(\frac{1}{\kappa_0} + \frac{1}{\mu_2}\right) = \frac{1}{\alpha_0^2}, \qquad \tau_0 = \left(\frac{\kappa_0^{-1} + \alpha_0^{-1}}{2}\right)^{-1}$$

Of course, instead of the implicit definition (10.5), one can easily find explicit formulas for S, $\boldsymbol{\alpha}$ and $\mathscr{Q}$. However, (10.5) seems like the most elegant way to define these operators.

Let us explain how to construct a material tensor L that belong to this enigmatic exact relation. First, we choose any positive definite two-dimensional elastic tensor S with square symmetry, i.e. $\mathsf{S}\boldsymbol{I}_2 = 2\kappa_0\boldsymbol{I}_2$. Then, we choose a complex number c and construct[1] an operator $\mathcal{Q}: \mathbb{R}^2 \to \mathrm{Sym}_0(\mathbb{R}^2)$, satisfying (10.6). Finally, we choose an arbitrary symmetric operator $\boldsymbol{\alpha} > 0$ on $\mathbb{R}^2$. Once these choices have been made, we compute

$$\rho_0 = \sqrt{\det \boldsymbol{\alpha}}, \quad \alpha_0 = \frac{1}{\sqrt{(\kappa_0^{-1} + \mu_1^{-1})(\kappa_0^{-1} + \mu_2^{-1})}}, \quad \tau_0 = \left(\frac{\kappa_0^{-1} + \alpha_0^{-1}}{2}\right)^{-1}$$

where $2\mu_1$ and $2\mu_2$ are the other two eigenvalues of the operator S. We define operators C, $\mathcal{P}$ and $\boldsymbol{\rho}$ by (10.5). Explicitly,

$$\tilde{\mathsf{C}} = (\tilde{\mathsf{S}}^{-1} - \mathcal{Q}\tilde{\boldsymbol{\alpha}}\mathcal{Q}^T)^{-1}, \quad \boldsymbol{\rho} = (\tilde{\boldsymbol{\alpha}}^{-1} - \mathcal{Q}^T\tilde{\mathsf{S}}\mathcal{Q})^{-1}, \quad \mathcal{P} = -\tilde{\mathsf{S}}\mathcal{Q}\boldsymbol{\rho},$$

where $\tilde{\mathsf{C}} = \mathsf{C} + \tau_0 \mathsf{I}$, $\tilde{\mathsf{S}} = \mathsf{S} + \tau_0 \mathsf{I}$, and $\tilde{\boldsymbol{\alpha}} = \boldsymbol{\alpha} + \rho_0 \boldsymbol{I}_2$. If we use such a material to make a transversely isotropic polycrystal

$$\mathsf{L}_* = \begin{bmatrix} 2\mu_*(\mathsf{I} - \boldsymbol{I}_2 \otimes \boldsymbol{I}_2) + 2\kappa_* \boldsymbol{I}_2 \otimes \boldsymbol{I}_2 & 0 \\ 0 & \rho_* \boldsymbol{I}_2 \end{bmatrix},$$

then all three effective parameters are texture-independent:

$$\kappa_* = \kappa_0, \quad \rho_* = \sqrt{\det \boldsymbol{\alpha}}, \quad \frac{1}{\mu_*} = \sqrt{\left(\frac{1}{\kappa_0} + \frac{1}{\mu_1}\right)\left(\frac{1}{\kappa_0} + \frac{1}{\mu_2}\right)} - \frac{1}{\kappa_0}.$$

References

[1] Benveniste Y 1995 Correspondence relations among equivalent classes of heterogeneous piezoelectric solids under anti-plane mechanical and in-plane electrical fields *J. Mech. Phys. Solids* **43** 553–71

[2] Chen T 1998 Further correspondences between plane piezoelectricity and generalized plane strain in elasticity *Proc. R. Soc. Lond.* A **454** 873–84

[3] Chen T and Lai D 1997 An exact correspondence between plane piezoelectricity and generalized plane strain in elasticity *Proc. R. Soc. Lond.* A **453** 2689–713

[4] Grabovsky Y, Milton G W and Sage D S 2000 Exact relations for effective tensors of polycrystals: necessary conditions and sufficient conditions *Commun. Pure Appl. Math.* **53** 300–53

[5] Hashin Z 1983 Analysis of composite materials–a survey *J. Appl. Mech.* **50** 481–505

[1] The set of admissible operators $\mathcal{Q}$ consists of two components, each parameterized by $c \in \mathbb{C}$. We describe $\mathcal{Q}$ by its action on an arbitrary vector $\boldsymbol{u} \in \mathbb{R}^2$. This is accomplished by identifying $\boldsymbol{u} = (u_1, u_2) \in \mathbb{R}^2$ with a complex number $u = u_1 + iu_2$, and vice versa. Then either $\mathcal{Q}\boldsymbol{u} = \psi(cu)$, or $\mathcal{Q}\boldsymbol{u} = \psi(c\bar{u})$, where $\psi: \mathbb{C} \to \mathrm{Sym}_0(\mathbb{R}^2)$ is defined in (9.20).

[6] Milton G W 1997 Composites: a myriad of microstructure independent relations *Theoretical and Applied Mechanics (Proc. of the XIX Int. Congress of Theoretical and Applied Mechanics, Kyoto, 1996)* ed T Tatsumi, E Watanabe and T Kambe (Amsterdam: Elsevier) 443–59

[7] Milton G W 2002 *The Theory of Composites* Cambridge Monographs on Applied and Computational Mathematics (Cambridge: Cambridge University Press)

IOP Publishing

Composite Materials (Second Edition)
Mathematical theory and exact relations
Yury Grabovsky

Chapter 11

Thermoelasticity

The properties of thermoelastic materials are described by the elastic compliance tensor S, thermal expansion tensor $\boldsymbol{\alpha}$ and the specific heat c_σ per unit volume at constant stress (see [7], section 2.5), so that the constitutive relation has the form

$$\varepsilon = \mathsf{S}\boldsymbol{\sigma} + \boldsymbol{\alpha}\theta, \qquad \varsigma = (\boldsymbol{\alpha}, \boldsymbol{\sigma})_F + c_\sigma \frac{\theta}{T_0}, \tag{11.1}$$

where ε and $\boldsymbol{\sigma}$ are the strain and the stress fields, satisfying (9.2) and (9.3), respectively. The scalar field ς and a constant scalar θ are the change in entropy and temperature, respectively, relative to the stress-free state and temperature T_0. In order to have an easy comparison of our results for thermoelasticity and elasticity it will be convenient to rewrite the constitutive relation (11.1) and rewrite it in the block-matrix form

$$\begin{bmatrix} \boldsymbol{\sigma} \\ -\varsigma \end{bmatrix} = \begin{bmatrix} \mathsf{C} & \boldsymbol{A} \\ \boldsymbol{A} & -c_\varepsilon/T_0 \end{bmatrix} \begin{bmatrix} \varepsilon \\ \theta \end{bmatrix}, \tag{11.2}$$

where C is the elasticity tensor, $\boldsymbol{A}$ is the thermal stress tensor and c_ε is the specific heat per unit volume at constant strain. We have

$$\mathsf{C} = \mathsf{S}^{-1}, \qquad \boldsymbol{A} = -\mathsf{S}^{-1}\boldsymbol{\alpha}, \qquad \frac{c_\varepsilon}{T_0} = \frac{c_\sigma}{T_0} - (\mathsf{S}^{-1}\boldsymbol{\alpha}, \boldsymbol{\alpha})_F. \tag{11.3}$$

We note that we cannot apply our theory to a non-positive definite material tensor in (11.2). Instead we modify the constitutive law (11.3)

$$\begin{bmatrix} \boldsymbol{\sigma} \\ -\varsigma \end{bmatrix} = \begin{bmatrix} \mathsf{C} & \boldsymbol{A} \\ \boldsymbol{A} & (c_0 - c_\varepsilon)/T_0 \end{bmatrix} \begin{bmatrix} \varepsilon \\ \theta \end{bmatrix}, \tag{11.4}$$

doi:10.1088/978-0-7503-6249-8ch11

where the positive constant c_0 is chosen so that $c_0 - c_\varepsilon > 0$ everywhere in the composite. The general theory is now applicable. It is a matter of simple comparison of cell problems for (11.1) and (11.4) to see that

$$\mathsf{C}_* = \mathsf{S}_*^{-1}, \qquad \boldsymbol{A}_* = -\mathsf{S}_*^{-1}\boldsymbol{\alpha}_*, \qquad \frac{c_{\varepsilon *}}{T_0} = \frac{c_{\sigma *}}{T_0} - (\mathsf{S}_*^{-1}\boldsymbol{\alpha}_*, \boldsymbol{\alpha}_*)_F. \tag{11.5}$$

11.1 2D Thermoelasticity

We note that all the exact relations for 2D thermoelasticity are already contained in the exact relations for fibrous elastic composites. Indeed, making the identification

$$\boldsymbol{\sigma}_\top = \boldsymbol{\sigma}, \quad \boldsymbol{\varepsilon}_\top = \boldsymbol{\varepsilon}, \quad \sigma_\| = -\varsigma, \quad \varepsilon_\| = \theta,$$

$$\mathsf{C}_\top = \mathsf{C}, \qquad \boldsymbol{C}_{\top\|} = \sqrt{2}\boldsymbol{A}, \qquad c_\| = \frac{c_0 - c_\varepsilon}{T_0}$$

We reduce (11.4) to the constitutive relation for fibrous elastic composites

$$\boldsymbol{\sigma}_\top = \mathsf{C}_\top \boldsymbol{\varepsilon}_\top + \varepsilon_\| \frac{\boldsymbol{C}_{\top\|}}{\sqrt{2}}, \qquad \sigma_\| = \left(\frac{\boldsymbol{C}_{\top\|}}{\sqrt{2}}, \boldsymbol{\varepsilon}_\top\right)_F + c_\| \varepsilon_\|, \tag{11.6}$$

restricted to the exact relation

$$\mathsf{C} = \begin{bmatrix} \mathsf{C}_\top & 0 & \boldsymbol{C}_{\top\|} \\ 0 & c_\lambda^0 \boldsymbol{I}_2 & 0 \\ \boldsymbol{C}_{\top\|} & 0 & c_\| \end{bmatrix}.$$

This permits us to simply read off all the exact relations and links for 2D thermoelasticity from the ones for fibrous elastic composites.

11.1.1 Exact relations

1. $\mathsf{C}\boldsymbol{I}_2 = 2\kappa_0 \boldsymbol{I}_2$, $\langle \mathrm{Tr}\, \boldsymbol{A} \rangle$
2. $\mathsf{C}\boldsymbol{I}_2 = 2\kappa_0 \boldsymbol{I}_2$, $\left(\frac{1}{\kappa_0} + \frac{1}{\mu_1}\right)\left(\frac{1}{\kappa_0} + \frac{1}{\mu_2}\right) = c_0$, $\langle \mathrm{Tr}\, \boldsymbol{A} \rangle$, where $2\mu_1$ and $2\mu_2$ are the other two eigenvalues of C
3. $\mathsf{C} = 2\mu_0 \mathsf{T} + \boldsymbol{B} \otimes \boldsymbol{B}$
4. $\mathsf{C} = 2\mu_0 \mathsf{T} + \boldsymbol{B} \otimes \boldsymbol{B}$, $\boldsymbol{A} = \beta \boldsymbol{B}$, $\langle c_\varepsilon + 2T_0\beta^2 \rangle$
5. $\mathsf{C} = 2\mu_0 \mathsf{I} + 2\lambda \boldsymbol{I}_2 \otimes \boldsymbol{I}_2$, $\langle \frac{1}{2\mu_0 + \lambda} \rangle$, $\langle \mathrm{dev}(\boldsymbol{A}) \rangle$
6. $\mathsf{C} = 2\mu_0 \mathsf{I} + 2\lambda \boldsymbol{I}_2 \otimes \boldsymbol{I}_2$, $\boldsymbol{A} = \beta \boldsymbol{I}_2$, $\langle \frac{1}{2\mu_0 + \lambda} \rangle$, $\langle \frac{\beta}{2\mu_0 + \lambda} \rangle$, $\langle c_\varepsilon + \frac{T_0\beta^2}{\lambda + 2\mu_0} \rangle$

11.1.2 Links

1. Global links
   (a) The block C_* is independent of the blocks $\boldsymbol{A}$, c_ε.
   (b) Blocks C_*, $\boldsymbol{A}_*$ are independent of c_ε.

(c)

$$\mathsf{L}' = \mathfrak{F}_{a,\mathsf{Q},\mathsf{F}}(\mathsf{L}) = \mathsf{Q}\Lambda_a(\mathsf{L})\mathsf{Q}^T + \mathsf{F}, \tag{11.7}$$

where

$$\mathsf{Q} = \begin{bmatrix} q_1\mathsf{I} & 0 \\ q_4\boldsymbol{I}_2 & q_3 \end{bmatrix}, \qquad \mathsf{F} = \begin{bmatrix} 0 & f_1\boldsymbol{I}_2 \\ f_1\boldsymbol{I}_2 & f_2 \end{bmatrix},$$

and

$$\Lambda_a(\mathsf{L}) = \mathsf{L} - \mathsf{L}\begin{bmatrix} \Theta_a(\mathsf{C}) & 0 \\ 0 & 0 \end{bmatrix}\mathsf{L}, \qquad \Theta_a(\mathsf{C}) = \left(\frac{1}{a}\mathsf{T} + \mathsf{C}\right)^{-1}. \tag{11.8}$$

2. $\mathsf{C}\boldsymbol{I}_2 = 2\kappa_0\boldsymbol{I}_2, \quad \mathrm{Tr}\,\boldsymbol{A} = t_0.$
   The blocks in C' are constructed according to the following rules. $\mathsf{C}'\boldsymbol{I}_2 = 2\kappa_0\boldsymbol{I}_2$. Let $\mu_0 \in (0, \kappa_0)$ be a constant. Then the two shear moduli μ_1', μ_2' of C' are defined by

$$\left(\frac{1}{\mu_1'} + \frac{1}{\kappa_0}\right)\left(\frac{1}{\mu_2} + \frac{1}{\kappa_0}\right) = \left(\frac{1}{\mu_2'} + \frac{1}{\kappa_0}\right)\left(\frac{1}{\mu_1} + \frac{1}{\kappa_0}\right) = \frac{1}{\kappa_0}\left(\frac{1}{\kappa_0} + \frac{1}{\mu_0}\right) = \frac{1}{a_0^2}.$$

   We also have $\mathrm{Tr}\,\boldsymbol{A}' = t_0$ and

$$\mathrm{dev}(\boldsymbol{A}') = \frac{2\kappa_0\mu_0}{a_0}\boldsymbol{R}_\perp(\mathsf{C} - 2\mu_0\mathsf{I})^{-1}\mathrm{dev}(\boldsymbol{A}).$$

$$c_\varepsilon' = b_0 - c_\varepsilon - T_0((\mathsf{C} - 2a_0\mathsf{I})^{-1}\boldsymbol{A}, \boldsymbol{A})_F,$$

   where the constant b_0 is chosen in such a way as to make C' positive definite.
3. $\mathsf{C}\boldsymbol{I}_2 = 2\kappa_0\boldsymbol{I}_2, \quad \left(\frac{1}{\kappa_0} + \frac{1}{\mu_1}\right)\left(\frac{1}{\kappa_0} + \frac{1}{\mu_2}\right) = \frac{1}{\rho_0^2}, \quad \mathrm{Tr}\,\boldsymbol{A} = t_0,$ The link, or rather a family of links parametrized by a constant $\zeta_0 > 0$ can be described in the following way
   - $\mathsf{C}' = \mathsf{C}$, $\mathrm{Tr}\,\boldsymbol{A}' = t_0$
   - Let $$\begin{bmatrix} \mathsf{S} & \boldsymbol{\alpha} \\ \boldsymbol{\alpha} & \xi \end{bmatrix} = \begin{bmatrix} \mathsf{C} + \tau_0\mathsf{I} & \boldsymbol{A} \\ \boldsymbol{A} & -c_\varepsilon/T_0 \end{bmatrix}^{-1}$$ and $$\begin{bmatrix} \mathsf{S}' & \boldsymbol{\alpha}' \\ \boldsymbol{\alpha}' & \xi' \end{bmatrix} = \begin{bmatrix} \mathsf{C} + \tau_0\mathsf{I} & \boldsymbol{A}' \\ \boldsymbol{A}' & -c_\varepsilon'/T_0 \end{bmatrix}^{-1}, \text{ where } \tau_0 = \left(\frac{\kappa_0^{-1} + \rho_0^{-1}}{2}\right)^{-1}.$$
   - Then

$$\frac{1}{\xi} + \frac{1}{\xi'} = \frac{1}{\zeta_0}, \qquad \frac{\mathrm{dev}(\boldsymbol{\alpha}')}{\xi'} = \pm\frac{\boldsymbol{R}_\perp\mathrm{dev}(\boldsymbol{\alpha})}{\xi},$$

4. $\mathsf{C} = 2\mu_0\mathsf{T} + \boldsymbol{B}\otimes\boldsymbol{B}$
   (a) $\boldsymbol{B}' = \lambda'\boldsymbol{B}$, where λ' is independent of L

(b) $\boldsymbol{B}' = \lambda'\boldsymbol{B}$, $\boldsymbol{A}' = \boldsymbol{A} + \nu'\boldsymbol{B}$, where λ' and ν' are independent of L

5. $\mathsf{C} = 2\mu_0\mathsf{T} + \boldsymbol{B} \otimes \boldsymbol{B}$, $\boldsymbol{A} = \beta\boldsymbol{B}$
   (a) $\boldsymbol{B}' = \lambda'\boldsymbol{B}$, $\beta' = \lambda'\beta$,
   (b) $c_\varepsilon' + 2T_0(\beta')^2 = c_\varepsilon + 2T_0\beta^2$,
   (c) $\boldsymbol{B}' = \boldsymbol{B}$, $\beta' = \beta$, $c_\varepsilon' = c_0 + a_0 c_\varepsilon - 2T_0(1 - a_0)\beta^2$
   (d) $\boldsymbol{B}' = \boldsymbol{B}$, $\beta' = c_0\beta$, $c_\varepsilon' + 2T_0(\beta')^2 = c_\varepsilon + 2T_0\beta^2$
6. $\mathsf{C} = 2\mu_0\mathsf{I} + 2\lambda\boldsymbol{I}_2 \otimes \boldsymbol{I}_2$, $\lambda' = \lambda$, $c_\varepsilon' = c_\varepsilon + \frac{a_0}{\lambda + 2\mu_0}$
7. $\mathsf{C} = 2\mu_0\mathsf{I} + 2\lambda\boldsymbol{I}_2 \otimes \boldsymbol{I}_2$, $\boldsymbol{A} = \beta\boldsymbol{I}_2$
   (a) $\frac{1}{\lambda' + 2\mu_0} = \frac{1}{\lambda + 2\mu_0} + b_0, \quad \frac{\beta'}{\lambda' + 2\mu_0} = \frac{\beta}{\lambda + 2\mu_0}$,

$$c_\varepsilon' + \frac{T_0(\beta')^2}{\lambda' + 2\mu_0} = c_\varepsilon + \frac{T_0\beta^2}{\lambda + 2\mu_0} + c_0,$$

   (b) $\frac{2\mu_0^2 T_0}{\lambda' + 2\mu_0} = \frac{T_0\beta^2}{\lambda + 2\mu_0} + c_\varepsilon + b_0, \quad \frac{\mu_0 T_0\beta'}{\lambda' + 2\mu_0} = a_0\left(c_\varepsilon + \frac{T_0\beta(\beta - \beta_0)}{\lambda + 2\mu_0}\right)$,

$$c_\varepsilon' + \frac{T_0(\beta')^2}{\lambda' + 2\mu_0} = 2a_0^2\left(c_\varepsilon + \frac{T_0(\beta - \beta_0)^2}{\lambda + 2\mu_0}\right) + c_0,$$

   (c) This exact relation looks better if we use the bulk modulus $\kappa = \lambda + \mu_0$ instead of the Lamé modulus λ:

$$\frac{\mu_0^2}{\kappa'} = \kappa, \qquad \frac{\mu_0\beta'}{\kappa'} = -\beta, \qquad c_\varepsilon' + \frac{T_0(\beta')^2}{\kappa'} = c_0 - c_\varepsilon.$$

   (d) Here instead of giving rather cumbersome explicit formulas for the moduli λ', β' and c_ε' we determine them from a system of equations that can be written nicely as

$$\begin{cases} \dfrac{T_0\beta_0^2}{\lambda' + 2\mu_0} = a_0\left(c_\varepsilon + \dfrac{T_0\beta^2}{\lambda + 2\mu_0}\right) + (a_0 + b_0)\dfrac{T_0\beta_0^2}{\lambda + 2\mu_0}, \\ \dfrac{\beta'}{\lambda' + 2\mu_0} = \dfrac{b_0\beta}{\lambda + 2\mu_0} + c_0\left(c_\varepsilon + T_0\dfrac{\beta^2 + \beta_0^2}{\lambda + 2\mu_0}\right), \\ c_\varepsilon' + \dfrac{T_0(\beta')^2}{\lambda + 2\mu_0} = d_0\dfrac{T_0\beta_0^2}{\lambda + 2\mu_0} + (b_0 + d_0)\left(c_\varepsilon + \dfrac{T_0\beta^2}{\lambda + 2\mu_0}\right). \end{cases}$$

   (e) Moduli λ' and β' are arbitrary and independent of L, while

$$c_\varepsilon' + \frac{T_0(\beta')^2}{\lambda' + 2\mu_0} = c_\varepsilon + \frac{T_0\beta^2}{\lambda + 2\mu_0}$$

(f) Moduli λ' and β' are arbitrary, while

$$c_\varepsilon' + \frac{T_0(\beta')^2}{\lambda' + 2\mu_0} = c_0 - \left(c_\varepsilon + \frac{T_0\beta^2}{\lambda + 2\mu_0}\right)$$

(g) Lamé modulus λ' is independent of L, and

$$\frac{2a_0^2\beta_0(\beta' + \beta_0)}{\lambda' + 2\mu_0} = d_0 - a_0^2\left(c_\varepsilon + \frac{T_0\beta^2}{\lambda + 2\mu_0}\right) - b_0^2\left(c_\varepsilon + \frac{T_0(\beta \pm \mu_0)^2}{\lambda + 2\mu_0}\right).$$

$$c_\varepsilon' + \frac{T_0\beta'^2}{\lambda' + 2\mu_0} = c_\varepsilon + \frac{T_0\beta^2}{\lambda + 2\mu_0} + c_0.$$

11.2 3D Thermoelasticity

This complete list of exact relations for three-dimensional thermoelasticity is taken from [1].

1. $\mathsf{C} = 2\mu_0\mathsf{T} + \boldsymbol{B} \otimes \boldsymbol{B}$, $\boldsymbol{A} = \alpha_0\boldsymbol{I}_3 + \beta\boldsymbol{B}$
2. Another exact relation is obtained if we replace β in item 1 by ν_0 Tr $\boldsymbol{B}$.
3. $\mathsf{C}\boldsymbol{I}_3 = \kappa_0\boldsymbol{I}_3 + a_0\boldsymbol{A}$. This exact relation was first discovered by Rosen and Hashin [8] and Laws [5] for the case of an isotropic composite of two isotropic phases. It was then generalized further by Hashin [2] to the case of an isotropic polycrystal, and to the general case by Schulgasser [9].
4. $\mathsf{C} = 2\mu_0\mathsf{I} + \lambda\boldsymbol{I}_3 \otimes \boldsymbol{I}_3$, $\boldsymbol{A} = a\boldsymbol{I}_3$. As we have mentioned at the beginning of this section, the effective elastic tensor C^* does not depend on the local thermoelastic coefficient $\boldsymbol{A}$. Therefore, according to Hill's exact relation for elasticity [3, 4], C^* will be isotropic with the same shear modulus μ_0 and the effective Lamé modulus λ^* given by (4.1).

   Our new thermoelastic exact relation says that the effective thermal stress coefficient $\boldsymbol{A}^*$ will always be isotropic and

$$\frac{a^*}{\lambda^* + 2\mu_0} = \left\langle \frac{a}{\lambda + 2\mu_0} \right\rangle. \tag{11.9}$$

5. If in the previous item we assume in addition that there are constants a_0 b_0 and c_0 such that $a_0\alpha + b_0\lambda = c_0$ then $a_0\alpha^* + b_0\lambda^* = c_0$. In particular for a two-phase composite this exact relation can be written as

$$\det\begin{vmatrix} a_1 & a_2 & a^* \\ \lambda_1 & \lambda_2 & \lambda^* \\ 1 & 1 & 1 \end{vmatrix} = 0.$$

   This exact relation is due to Levin [6].

References

[1] Grabovsky Y, Milton G W and Sage D S 2000 Exact relations for effective tensors of polycrystals: necessary conditions and sufficient conditions *Commun. Pure Appl. Math.* **53** 300–53

[2] Hashin Z 1984 Thermal expansion of polycrystalline aggregates: I exact analysis *J. Mech. Phys. Solids* **32** 149–57

[3] Hill R 1963 Elastic properties of reinforced solids: some theoretical principles *J. Mech. Phys. Solids* **11** 357–72

[4] Hill R 1964 Theory of mechanical properties of fibre-strengthened materials: I elastic behaviour *J. Mech. Phys. Solids* **12** 199–212

[5] Laws N 1973 On thermostatics of composite materials *J. Mech. Phys. Solids* **21** 9–17

[6] Levin V M 1967 Thermal Expansion Coefficients of Heterogeneous Materials *MTT* **2** 88–94

[7] Milton G W 2002 *The Theory of Composites Cambridge Monographs on Applied and Computational Mathematics* (Cambridge: Cambridge University Press)

[8] Rosen B W and Hashin Z 1970 Effective thermal expansion coefficients and specific heats of composite materials *Int. J. Eng. Sci.* **8** 157–73

[9] Schulgasser K 1987 Thermal expansion of polycrystalline aggregates with texture *J. Mech. Phys. Solids* **35** 34–42

Chapter 12

Thermoelectricity

12.1 Equations of thermoelectricity

Thermoelectric effects manifest themselves by an electric field produced by a non-uniformly heated material, or by a temperature gradient produced by an electric current flowing through the material. Thermoelectric properties of a material can be described by the relations between the gradient $\nabla\mu$ of an electrochemical potential, temperature gradient ∇T, current density $\boldsymbol{j}_E$ and internal energy flux $\boldsymbol{j}_U$. Naively postulating a linear constitutive relation between the pair $[\nabla\mu, \nabla T]$ and the pair of the divergence-free fluxes $[\boldsymbol{j}_E, \boldsymbol{j}_U]$ by means of a symmetric and positive definite tensor of material properties is physically incorrect. Thus, if we want to place thermoelectricity into our general Hilbert space framework, described in part I, it will be necessary to derive the correct equations of thermoelectricity from thermodynamical principles. As a result, we will also obtain the relations between the components of the tensor of material properties and the physical coupling coefficients used in constitutive laws. This has already been done in [4]. However, our way of explaining the derivation of equations of thermoelectricity from thermodynamic principles is somewhat different, and we believe that the reader may find some aspects of it more enlightening.

Thermodynamics postulates the existence of the energy density u as a smooth function of the entropy density s and charge density q: $u = U(s, q)$. During a thermoelectric process we have conservation of energy and charge

$$\begin{cases} u_t + \nabla \cdot \boldsymbol{j}_U = 0, \\ q_t + \nabla \cdot \boldsymbol{j}_E = 0, \end{cases} \tag{12.1}$$

where $\boldsymbol{j}_U$ and $\boldsymbol{j}_E$ are the energy and the electric charge flux, respectively. These equations say that both energy and electric charge are neither produced nor destroyed. They simply move from place to place. By contrast, the second law of

doi:10.1088/978-0-7503-6249-8ch12

thermodynamics says that entropy (while also moving from place to place) is constantly produced and is never destroyed. We write it as

$$s_t + \nabla \cdot \boldsymbol{j}_S = r > 0, \tag{12.2}$$

where $\boldsymbol{j}_S$ is the entropy flux, and r is the positive entropy production density. In thermodynamics the partial derivatives $T = U_s$ and $\mu = U_q$ are called the absolute temperature and electrochemical potential, respectively. By chain rule

$$u_t = Ts_t + \mu q_t. \tag{12.3}$$

Using the balance equations (12.1) and (12.2) we obtain

$$\nabla \cdot \boldsymbol{j}_U = T\nabla \cdot \boldsymbol{j}_S + \mu\nabla \cdot \boldsymbol{j}_E - Tr$$

Integrating this equation over a test volume Ω we obtain after integrating by parts

$$\int_{\partial\Omega} \boldsymbol{j}_U \cdot \boldsymbol{n} dS = \int_{\partial\Omega} (T\boldsymbol{j}_S + \mu\boldsymbol{j}_E) \cdot \boldsymbol{n} dS - \int_{\Omega} (\boldsymbol{j}_S \cdot \nabla T + \boldsymbol{j}_E \cdot \nabla\mu + Tr) d\boldsymbol{x}. \tag{12.4}$$

This equation suggests the definition for the entropy flux, which does not have an independent definition, unlike energy and charge fluxes. Hence, we *define* entropy flux, so that the surface terms balance on their own:

$$\boldsymbol{j}_S = \frac{1}{T}\boldsymbol{j}_U - \frac{\mu}{T}\boldsymbol{j}_E. \tag{12.5}$$

Similarly, the volume term can serve as the definition of the entropy production density:

$$r = -\frac{1}{T}\boldsymbol{j}_S \cdot \nabla T - \frac{1}{T}\boldsymbol{j}_E \cdot \nabla\mu > 0. \tag{12.6}$$

In equilibrium, energy and charge densities are stationary and therefore conservation laws (12.1) become

$$\nabla \cdot \boldsymbol{j}_E = 0, \qquad \nabla \cdot \boldsymbol{j}_U = 0. \tag{12.7}$$

Balance equations (12.5), (12.6) and conservation laws (12.7) must be augmented by constitutive relations expressing the fact that the inhomogeneities in the thermodynamic potentials T and μ are the causes of corresponding fluxes, except the two effects must be coupled in a thermoelectric medium:

$$\begin{cases} \boldsymbol{j}_E = \boldsymbol{\sigma}\nabla(-\mu) + \boldsymbol{\alpha}\nabla(-T), \\ \boldsymbol{j}_S = \boldsymbol{\alpha}^T\nabla(-\mu) + \boldsymbol{\gamma}\nabla(-T), \end{cases} \qquad \boldsymbol{\sigma}^T = \boldsymbol{\sigma}, \quad \boldsymbol{\gamma}^T = \boldsymbol{\gamma}, \tag{12.8}$$

where the Onsager reciprocity principle (see, e.g., [3, 8]) requiring that the tensor of material properties

$$\mathsf{L}' = \begin{bmatrix} \boldsymbol{\sigma} & \boldsymbol{\alpha} \\ \boldsymbol{\alpha}^T & \boldsymbol{\gamma} \end{bmatrix}$$

be symmetric, is incorporated into the constitutive laws (12.8). Inequality (12.6) implies that the material tensor L' must be positive definite. In the next section we will discuss physical meaning of the block-components of the material tensor L'.

12.2 Seebeck effect

The electrochemical potential μ is a sum of the electrostatic potential and a chemical potential. The latter depends only on the temperature and is therefore constant when the temperature is constant. Thus, when the temperature is constant $\boldsymbol{E} = -\nabla\mu$ is the electric field and the first equation in (12.8) reads $\boldsymbol{j}_E = \boldsymbol{\sigma}\boldsymbol{E}$. Therefore, $\boldsymbol{\sigma}$ has the physical meaning of the isothermal conductivity tensor, which must be a symmetric, positive definite 3×3 matrix.

In the absence of the electrical current ($\boldsymbol{j}_E = 0$) the field $-\nabla\mu$ has the meaning of the electromotive force generated by a temperature gradient. This is called the *Seebeck effect*. From the first equation in (12.8) we obtain

$$\boldsymbol{e}_{\mathrm{emf}} = -\nabla\mu = \boldsymbol{\sigma}^{-1}\boldsymbol{\alpha}\nabla T.$$

The 3×3 matrix $\boldsymbol{S} = \boldsymbol{\sigma}^{-1}\boldsymbol{\alpha}$ is called the Seebeck coefficient (tensor). In the literature the Seebeck coefficient is often assumed to be a scalar. However, we will see that a composite made of such materials can have an anisotropic Seebeck coefficient. Another *a priori* assumption found in the literature is that $\boldsymbol{S}$ be symmetric (see, e.g., [5]). We will again see that symmetry of $\boldsymbol{S}$ is not preserved under homogenization.

The heat flux at zero electric current is characterized by the heat conductivity tensor $\boldsymbol{j}_U = -\boldsymbol{\kappa}\nabla T$, which gives a formula for the tensor $\boldsymbol{\gamma}$ in the constitutive equations in terms of the symmetric, positive definite heat conductivity tensor $\boldsymbol{\kappa}$:

$$\boldsymbol{\kappa} = T(\boldsymbol{\gamma} - \boldsymbol{S}^T\boldsymbol{\sigma}\boldsymbol{S}).$$

We note that our assumptions on the possible values of the tensors $\boldsymbol{\sigma}$, $\boldsymbol{S}$, and $\boldsymbol{\kappa}$, that are expressed as the requirements of symmetry and positive definiteness of the 6×6 matrix

$$\mathsf{L}' = \begin{bmatrix} \boldsymbol{\sigma} & \boldsymbol{\sigma}\boldsymbol{S} \\ \boldsymbol{S}^T\boldsymbol{\sigma} & \boldsymbol{S}^T\boldsymbol{\sigma}\boldsymbol{S} + \boldsymbol{\kappa}/T \end{bmatrix} \tag{12.9}$$

are equivalent to the symmetry and positive definiteness of the tensors $\boldsymbol{\sigma}$ and $\boldsymbol{\kappa}$, without any constraints on the Seebeck tensor $\boldsymbol{S}$.

12.3 The canonical form of equations of thermoelectricity

The canonical equations of thermoelectricity must relate a pair of curl-free fields $(\boldsymbol{e}_1, \boldsymbol{e}_2)$ and a pair of divergence-free fields $(\boldsymbol{j}_1, \boldsymbol{j}_2)$ by means of a symmetric and positive definite tensor L of material properties

$$\begin{bmatrix} \boldsymbol{j}_1 \\ \boldsymbol{j}_2 \end{bmatrix} = \mathsf{L}\begin{bmatrix} \boldsymbol{e}_1 \\ \boldsymbol{e}_2 \end{bmatrix}, \quad \mathsf{L} = \begin{bmatrix} \boldsymbol{L}_{11} & \boldsymbol{L}_{12} \\ \boldsymbol{L}_{12}^T & \boldsymbol{L}_{22} \end{bmatrix}. \tag{12.10}$$

Currently equations (12.5), (12.7), and (12.8) of thermoelectricity relate the pair of fluxes (j_E, j_S) to the pair of gradients of the thermoelectric potentials (μ, T).

$$\begin{bmatrix} j_E \\ j_S \end{bmatrix} = \mathsf{L}' \begin{bmatrix} \nabla(-\mu) \\ \nabla(-T) \end{bmatrix},$$

where the tensor of material properties is symmetric and positive definite. However, since the entropy flux is not divergence free, our equations are not in canonical form. In order to convert our equations of thermoelectricity to the canonical form we observe that equations (12.8) suggest that the fluxes j_E and j_U must play the role of j_1 and j_2, respectively. To identify the appropriate curl-free fields e_1 and e_2 we follow Callen [2] and write the entropy flux j_S in terms of the divergence-free fluxes j_E and j_U. From (12.5) we obtain

$$j_S = -\frac{\mu}{T} j_E + \frac{1}{T} j_U.$$

This gives the appropriate potentials

$$\psi_1 = -\frac{\mu}{T}, \qquad \psi_2 = \frac{1}{T}.$$

This corresponds to redoing our analysis starting from the fundamental function $s = S(u, q)$. Comparing the chain rule $s_t = S_u u_t + S_q q_t$ to (12.3) we see that $S_u = 1/T$ and $S_q = -\mu/T$. Thus, mathematically, the potentials ψ_1 and ψ_2 will play the same role as potentials μ and T in our discussion in section 12.1. Denoting

$$e_1 = \nabla\psi_1, \quad e_2 = \nabla\psi_2, \quad j_1 = j_E, \quad j_2 = j_U,$$

we obtain the desired form (12.10), were

$$L_{11} = T\sigma, \quad L_{12} = T(\mu\sigma + T\sigma S), \quad L_{22} = T[\mu^2\sigma + T^2\gamma + T\mu(\sigma S + S^T\sigma)].$$

Indeed, it is easy to check that the material tensor

$$\mathsf{L} = T\begin{bmatrix} \sigma & \sigma(\mu + TS) \\ (\mu + TS)^T\sigma & T\kappa + (\mu + TS)^T\sigma(\mu + TS) \end{bmatrix} \tag{12.11}$$

is symmetric and positive definite if and only if L', given by (12.9), is symmetric and positive definite, i.e., if and only if σ and κ are symmetric and positive definite.

As formula (12.11) shows, the material moduli L_{ij} must depend on the values of T and μ as a matter of general theory. Therefore, even under the assumptions of physical linearity, represented by the linear constitutive laws (12.8), the equations of thermoelectricity are always nonlinear (semilinear) PDEs. The linear version of thermoelectricity describes situations where the values of thermoelectric potentials T and μ do not change much from point to point. Mathematically, we derive the leading order asymptotics of solutions (μ, T) of the form $\mu = \mu_0 + \epsilon\tilde{\mu}$ and $T = T_0 + \epsilon\tilde{T}$. Before performing the linearization, we observe that if (μ, T) solves equations of thermoelectricity (12.5), (12.7), (12.8), so does $(\mu - \mu_0, T)$ for any constant μ_0. Thus, without loss of generality, we can perform the linearization around the constant temperature state (0, T_0). Therefore, for linearized problems we can write

$$\mathsf{L} = T_0^2 \begin{bmatrix} \sigma/T_0 & \sigma S \\ S^T\sigma & \kappa + T_0 S^T \sigma S \end{bmatrix} = \begin{bmatrix} \sigma_0 & \sigma_0 S_0 \\ S_0^T \sigma_0 & \kappa_0 + S_0^T \sigma_0 S_0 \end{bmatrix}, \tag{12.12}$$

where all physical property tensors $\boldsymbol{\sigma}$, $\boldsymbol{\kappa}$ and $\boldsymbol{S}$ are evaluated at $T = T_0$ and $\mu = 0$, and $\boldsymbol{\sigma}_0 = T_0\boldsymbol{\sigma}$, $\boldsymbol{S}_0 = T_0\boldsymbol{S}$, $\boldsymbol{\kappa}_0 = T_0^2\boldsymbol{\kappa}$.

Now, the vector fields $\boldsymbol{E} = (\boldsymbol{e}_1, \boldsymbol{e}_2)$ and $\boldsymbol{J} = (\boldsymbol{j}_1, \boldsymbol{j}_2)$ take their values in the $2d$-dimensional vector space $\mathscr{T} = \mathbb{R}^d \oplus \mathbb{R}^d$, $d = 2$ or 3. The natural inner product on $\mathscr{T}$ is defined by

$$(\boldsymbol{E}, \boldsymbol{E}') = \boldsymbol{e}_1 \cdot \boldsymbol{e}_1' + \boldsymbol{e}_2 \cdot \boldsymbol{e}_2'.$$

The differential equations satisfied by the fields $\boldsymbol{E}$ and $\boldsymbol{J}$ are

$$\nabla \times \boldsymbol{e}_1 = \nabla \times \boldsymbol{e}_2 = 0, \qquad \nabla \cdot \boldsymbol{j}_1 = \nabla \cdot \boldsymbol{j}_2 = 0, \tag{12.13}$$

which places the linearized thermoelectricity in the general Hilbert space framework of our theory. Formulas (12.12) let us recover the tensors of material properties $\boldsymbol{\sigma}$, $\boldsymbol{S}$ and $\boldsymbol{\kappa}$ from the block-components $\boldsymbol{L}_{ij}$ of L:

$$\boldsymbol{\sigma} = \beta_0 \boldsymbol{L}_{11},\ \boldsymbol{S} = \beta_0 \boldsymbol{L}_{11}^{-1}\boldsymbol{L}_{12},\ \boldsymbol{\kappa} = \beta_0^2(\boldsymbol{L}_{22} - \boldsymbol{L}_{12}^T \boldsymbol{L}_{11}^{-1}\boldsymbol{L}_{12}),\ \beta_0 = \frac{1}{T_0}. \tag{12.14}$$

12.4 2D Thermoelectricity

Before we list all exact relations and links for two-dimensional thermoelectric composites we need to note that in two space dimensions the space of thermoelectric isotropic materials is larger than the three-dimensional space $\{\boldsymbol{\sigma} \otimes \boldsymbol{I}_2 : \boldsymbol{\sigma} \in \mathrm{Sym}(\mathbb{R}^2)\}$. In fact, this space is four-dimensional and is given by

$$\mathsf{L} = \boldsymbol{\sigma} \otimes \boldsymbol{I}_2 + r\boldsymbol{R}_\perp \otimes \boldsymbol{R}_\perp, \qquad \boldsymbol{R}_\perp = \begin{bmatrix} 0 & -1 \\ 1 & 0 \end{bmatrix}. \tag{12.15}$$

The tensor (12.15) is symmetric and positive definite if and only if $\boldsymbol{\sigma}^T = \boldsymbol{\sigma}$, $\boldsymbol{\sigma} > 0$ and $\det \boldsymbol{\sigma} > r^2$.

12.4.1 The global link

Parts of this section have been reprinted with permission from [10]. In the case of two-dimensional thermoelectricity, most exact relations and links come in infinite families whose members are related to each other by special symmetries. Thus, in order to make the task of describing all exact relations and links feasible we start by identifying the group of those symmetries and then proceed to list these equivalence classes by exhibiting only one exact relation or link from each class. The group of fundamental symmetries in each class of physical properties is called the 'global link'. It is a function Ψ: $\mathrm{Sym}^+(\mathscr{T}) \to \mathrm{Sym}^+(\mathscr{T})$, such that

$$\Psi(\mathsf{L}(z))_* = \Psi(\mathsf{L}_*), \tag{12.16}$$

This means that $\Psi(\mathsf{L}_*)$ is the effective tensor of the periodic composite with the local tensor $\Psi(\mathsf{L}(z))$, provided that L_* is the effective tensor of the periodic composite with the local tensor $\mathsf{L}(z)$. For two-dimensional thermoelectricity $\mathscr{T} = \mathbb{R}^2 \oplus \mathbb{R}^2$, and the group of all global links is given by

$$\Psi_{A,B}(\mathsf{L}) = (\boldsymbol{B} \otimes \boldsymbol{I}_2)(a_{11}\mathsf{L} + a_{12}\mathsf{T})(a_{21}\mathsf{L} + a_{22}\mathsf{T})^{-1}\mathsf{T}(\boldsymbol{B}^T \otimes \boldsymbol{I}_2), \tag{12.17}$$

where $\boldsymbol{A} = \begin{bmatrix} a_{11} & a_{12} \\ a_{21} & a_{22} \end{bmatrix}$ and $\boldsymbol{B}$ are 2×2 invertible matrices and $\mathsf{T} = \boldsymbol{R}_\perp \otimes \boldsymbol{R}_\perp$. We remark that since $\Psi_{A,B}$: $\mathrm{Sym}(\mathscr{T}) \to \mathrm{Sym}(\mathscr{T})$ we also have

$$\Psi_{A,B}(\mathsf{L}) = (\boldsymbol{B} \otimes \boldsymbol{I}_2)\mathsf{T}(a_{21}\mathsf{L} + a_{22}\mathsf{T})^{-1}(a_{11}\mathsf{L} + a_{12}\mathsf{T})(\boldsymbol{B}^T \otimes \boldsymbol{I}_2).$$

It is important to note that different pairs of matrices $\{\boldsymbol{A}, \boldsymbol{B}\} \subset \mathrm{GL}(2, \mathbb{R})$ can define the same transformation $\Psi_{A,B}$. Specifically,

$$\Psi_{\lambda A,B} = \Psi_{A,B}, \quad \Psi_{A,\lambda B} = \Psi_{A_\lambda,B}, \quad A_\lambda = \begin{bmatrix} \lambda^2 & 0 \\ 0 & 1 \end{bmatrix} \boldsymbol{A} \tag{12.18}$$

for any nonzero real number λ. Thus, without loss of generality, we may assume that $|\det \boldsymbol{A}| = |\det \boldsymbol{B}| = 1$. Even with this assumption we still have symmetries

$$\Psi_{-A,B} = \Psi_{A,B}, \quad \Psi_{A,-B} = \Psi_{A,B}.$$

We note that transformations (12.18) form a Lie group with the composition law

$$\Psi_{A_1,B_1} \circ \Psi_{A_2,B_2} = \Psi_{A_1^{B_2}A_2,B_1B_2}, \tag{12.19}$$

where

$$\boldsymbol{A}^{\boldsymbol{B}} = \begin{bmatrix} \det \boldsymbol{B} & 0 \\ 0 & 1 \end{bmatrix}^{-1} \boldsymbol{A} \begin{bmatrix} \det \boldsymbol{B} & 0 \\ 0 & 1 \end{bmatrix}.$$

We note that if $\det \boldsymbol{B}_2 = 1$, then $\Psi_{A_1,B_1} \circ \Psi_{A_2,B_2} = \Psi_{A_1A_2,B_1B_2}$. For future reference we observe that the most general transformation $\Psi_{A,B}$, such that $\Psi_{A,B}(\mathsf{I}) = \mathsf{I}$ has

$$\boldsymbol{A} = \begin{bmatrix} a_0 & 1 \\ 1 & a_0 \end{bmatrix}, \quad \boldsymbol{B} \in O(2, \mathbb{R}). \tag{12.20}$$

We note that each of the global links (12.17) maps an exact relation into an exact relation. We therefore have the option of deriving only the exact relations passing though $\mathsf{L}_0 = \mathsf{I}$. Any other exact relations can be obtained as images of the ones passing through L_0 by an application of the link (12.17).

12.4.2 Essential exact relations and links

In addition to lumping together all exact relations permuted by transformation (12.17) into one equivalence class, we also observe that intersection of any two exact relations is an exact relation. Eliminating intersections and choosing a single representative from each $\Psi_{A,B}$-class the following list of exact relations and links have been computed. To describe them it will be convenient to introduce the function $\mathfrak{L}$: $\mathrm{Sym}^+(\mathbb{R}^2) \times \mathrm{End}(\mathbb{R}^2) \to \mathrm{Sym}(\mathscr{T})$

$$\mathfrak{L}(\boldsymbol{L}, \boldsymbol{M}) = \begin{bmatrix} \boldsymbol{L} & \boldsymbol{L}\boldsymbol{M} \\ \boldsymbol{M}^T\boldsymbol{L} & \boldsymbol{M}^T\boldsymbol{L}\boldsymbol{M} \end{bmatrix} + \mathsf{T}, \quad \mathsf{T} = \boldsymbol{R}_\perp \otimes \boldsymbol{R}_\perp. \tag{12.21}$$

Here is the list.

List 12.1.

- $\mathbb{M}_1 = \{\boldsymbol{I}_2 \otimes \boldsymbol{L} + t\mathsf{T}: \boldsymbol{L} \in \mathrm{Sym}^+(\mathbb{R}^2),\ \det \boldsymbol{L} > t^2\}$,
- $\mathbb{M}_2 = \{\mathfrak{L}(\boldsymbol{L}, \boldsymbol{R}_\perp): \boldsymbol{L} > \frac{1}{2}\boldsymbol{I}_2\}, \qquad \boldsymbol{L}_* = \langle \boldsymbol{L}^{-1} \rangle^{-1}$.
- $\mathbb{M}_3 = \{\mathfrak{L}(\boldsymbol{L}, \boldsymbol{M}): \boldsymbol{M}^2 = -\boldsymbol{I}_2, \boldsymbol{L} > 0, \boldsymbol{L}^{-1} < -2\boldsymbol{M}\boldsymbol{R}_\perp\}$.

  This exact relation has two different links, which are not consequences of other relations or links listed here.

  1. This is an infinite family of links that we describe in terms of the function $\mathfrak{L}(\boldsymbol{L}, \boldsymbol{M})$, given by (12.21). The family of links are the maps Φ_{γ_0}: $\mathbb{M}_3 \to \mathbb{M}_3$, given by

     $$\Phi_{\gamma_0}(\mathfrak{L}(\boldsymbol{L}, \boldsymbol{M})) = \mathfrak{L}\left(\boldsymbol{P}_{\gamma_0}^{-1}, \boldsymbol{M}\right), \tag{12.22}$$

     where

     $$\boldsymbol{P}_{\gamma_0} = \gamma_0 \boldsymbol{M}\boldsymbol{L}^{-1}\boldsymbol{M}^T + (1 + \gamma_0)\boldsymbol{L}^{-1} + 2\gamma_0 \boldsymbol{M}\boldsymbol{R}_\perp,$$

     and the parameter γ_0 is constrained to lie in the interval $[-1, 0]$, which is equivalent to the inequalities

     $$\boldsymbol{P}_{\gamma_0} > 0, \qquad \boldsymbol{P}_{\gamma_0} + 2\boldsymbol{M}\boldsymbol{R}_\perp < 0,$$

     that hold for any $\mathfrak{L}(\boldsymbol{L}, \boldsymbol{M}) \in \mathbb{M}_3$. We note that the map Φ_{γ_0} is bijective for all $\gamma_0 \in [-1, 0]$, except for $\gamma_0 = -1/2$, where it is no longer surjective. Nonetheless, it is still a valid link between $\mathbb{M}_3$ and $\mathbb{M}_3' = \Phi_{-1/2}(\mathbb{M}_3)$, where

     $$\mathbb{M}_3' = \{\mathfrak{L}(\boldsymbol{L}, \boldsymbol{M}): \boldsymbol{M}^2 = -\boldsymbol{I}_2, \boldsymbol{M}\boldsymbol{L}^{-1} - \boldsymbol{L}^{-1}\boldsymbol{M}^T = 2\boldsymbol{R}_\perp, \boldsymbol{L} > 0\}.$$

     The exact relation $\mathbb{M}_3'$ is not listed because it is the intersection of $\mathbb{M}_7$, $\mathbb{M}_5$ given below, and $\Psi_{A,B}(\mathbb{M}_5)$ for a certain choice of $\boldsymbol{A}$ and $\boldsymbol{B}$. If we restrict other links listed below to this ER, we will also conclude that $\boldsymbol{M}_*$ is independent of $\boldsymbol{L}(z)$ and can be found as $\boldsymbol{M}_* = \boldsymbol{R}_\perp \boldsymbol{\sigma}_*$, where $\boldsymbol{\sigma}_*$ is the effective conductivity of the 2D conducting periodic composite with local tensor $\boldsymbol{\sigma}(z) = \boldsymbol{R}_\perp^T \boldsymbol{M}(z)$.

2. The second link is between $\mathbb{M}_3$ and

$$\mathbb{M}_3'' = \{\mathfrak{L}(\mu\boldsymbol{\sigma}, \boldsymbol{R}_\perp\boldsymbol{\sigma}): \det\boldsymbol{\sigma} = 1, \boldsymbol{\sigma} > 0, \mu > 1/2\}. \quad (12.23)$$

The exact relation $\mathbb{M}_3''$ is not listed because it is the intersection of $\mathbb{M}_1$, $\mathbb{M}_5$, and $\Psi_{\boldsymbol{A},\boldsymbol{B}}(\mathbb{M}_5)$ for a certain choice of $\boldsymbol{A}$ and $\boldsymbol{B}$. The link between $\mathbb{M}_3$ and $\mathbb{M}_3''$ is given by the formulas

$$\boldsymbol{\sigma} = \boldsymbol{R}_\perp^T\boldsymbol{M}, \qquad \mu = \frac{2}{\mathrm{Tr}\,(\boldsymbol{L}\boldsymbol{\sigma})}, \quad (12.24)$$

so that $\boldsymbol{M}_* = \boldsymbol{R}_\perp\boldsymbol{\sigma}_*$, where $\boldsymbol{\sigma}_*$ is the effective conductivity of the 2D conducting composite with local tensor $\boldsymbol{\sigma}(z) = -\boldsymbol{R}_\perp\boldsymbol{M}(z)$, while additionally we have

$$\mathrm{Tr}\,(\boldsymbol{L}_*\boldsymbol{\sigma}_*) = \frac{2}{\mu_*}. \quad (12.25)$$

- $\mathbb{M}_4 = \{\mathfrak{L}(\boldsymbol{L}, \boldsymbol{M}): \boldsymbol{L} > 0, \boldsymbol{L}^{-1} < -2\boldsymbol{M}\boldsymbol{R}_\perp\}$.

  There is also a link associated with this ER. It says that $\boldsymbol{M}_*$ does not depend on $\boldsymbol{L}(z)$ in the parametrization of $\mathbb{M}_4$ above. The effective tensor $\boldsymbol{M}_*$ can be computed from the exact relation described by

$$\mathbb{M}_4' = \{\mathsf{L} = \boldsymbol{\Lambda} \otimes \boldsymbol{P} > 0: \det\mathsf{L} = \det\boldsymbol{\Lambda}\det\boldsymbol{P} = 1\}. \quad (12.26)$$

The exact relation $\mathbb{M}_4'$ is not listed because it is the intersection of $\mathbb{M}_4$ and $\Psi_{\boldsymbol{A},\boldsymbol{B}}(\mathbb{M}_4)$ for a certain choice of $\boldsymbol{A}$ and $\boldsymbol{B}$. Specifically, $\mathsf{L} = \boldsymbol{\Lambda} \otimes \boldsymbol{P} \in \mathbb{M}_4'$ is uniquely determined by a pair of symmetric, positive definite 2×2 matrices $\boldsymbol{\Lambda}$ and $\boldsymbol{P}$, satisfying $\det\boldsymbol{\Lambda}\det\boldsymbol{P} = 1$, provided we fix $\Lambda_{11} = 1$. We will denote this parametrization by $\mathsf{L} = \mathsf{L}_{4'}(\boldsymbol{\Lambda}, \boldsymbol{P})$. The fact that $\mathbb{M}_4'$ is an exact relation means that $\mathsf{L}_{4'}(\boldsymbol{\Lambda}, \boldsymbol{P})_* = \mathsf{L}_{4'}(\boldsymbol{\Lambda}_*, \boldsymbol{P}_*)$, for some $\boldsymbol{\Lambda}_*$ and $\boldsymbol{P}_*$ that depend on the microstructure of the composite. The link between $\mathbb{M}_4$ and $\mathbb{M}_4'$ is given by a bijective transformation $\boldsymbol{M} \mapsto (\boldsymbol{\Lambda}(\boldsymbol{M}), \boldsymbol{P}(\boldsymbol{M}))$

$$\boldsymbol{\Lambda}(\boldsymbol{M}) = \begin{bmatrix} 1 & \mathrm{Tr}\,\boldsymbol{M}/2 \\ \mathrm{Tr}\,\boldsymbol{M}/2 & \det\boldsymbol{M} \end{bmatrix}, \qquad \boldsymbol{P}(\boldsymbol{M}) = -\boldsymbol{R}_\perp\frac{\boldsymbol{M} - (\mathrm{Tr}\,\boldsymbol{M})\boldsymbol{I}_2/2}{\det\boldsymbol{\Lambda}(\boldsymbol{M})}. \quad (12.27)$$

The link says that $\boldsymbol{M}_*$ is determined via the formula

$$\mathsf{L}_{4'}(\boldsymbol{\Lambda}(\boldsymbol{M}), \boldsymbol{P}(\boldsymbol{M}))_* = \mathsf{L}_{4'}(\boldsymbol{\Lambda}(\boldsymbol{M}_*), \boldsymbol{P}(\boldsymbol{M}_*)), \quad (12.28)$$

so, that

$$\boldsymbol{M}_* = \Lambda_{12*}\boldsymbol{I}_2 + \boldsymbol{R}_\perp\boldsymbol{P}_*\det\boldsymbol{\Lambda}_*. \quad (12.29)$$

- $\mathbb{M}_5 = \{\mathsf{L} > 0: (\mathsf{L} - \mathsf{T})(\boldsymbol{J} \otimes \boldsymbol{R}_\perp)(\mathsf{L} - \mathsf{T}) = 0\}, \quad \boldsymbol{J} = \begin{bmatrix} 1 & 0 \\ 0 & -1 \end{bmatrix}$.

  We can also write this ER in parametric form

$$\mathbb{M}_5 = \{\mathfrak{L}(\boldsymbol{L}, \boldsymbol{M}): \det\boldsymbol{M} = 1, \boldsymbol{L} > 0, \boldsymbol{L}^{-1} < -2\boldsymbol{M}\boldsymbol{R}_\perp\}, \quad (12.30)$$

where inequalities are understood in the sense of quadratic forms.

- $\mathbb{M}_6 = \{\mathsf{L} > 0\colon \mathsf{L}(\boldsymbol{J} \otimes \boldsymbol{R}_\perp)\mathsf{L} = \boldsymbol{J} \otimes \boldsymbol{R}_\perp\}$.
- $\mathbb{M}_7 = \{\mathsf{L} > 0\colon \mathsf{L}(\boldsymbol{I}_2 \otimes \boldsymbol{R}_\perp)\mathsf{L} = \boldsymbol{I}_2 \otimes \boldsymbol{R}_\perp\}$.

12.4.3 Application: isotropic polycrystals

We observe that $\mathbb{M}_7$ from list 12.1 passes through the unique isotropic tensor $\mathsf{L} = \mathsf{I}$. The global automorphisms

$$\Psi_{\alpha,\boldsymbol{B}}(\mathsf{L}) = (\boldsymbol{B} \otimes \boldsymbol{I}_2)(\mathsf{L} - \alpha\mathsf{T})(\boldsymbol{B} \otimes \boldsymbol{I}_2), \qquad \boldsymbol{B}^T = \boldsymbol{B},$$

corresponding to $a_{11} = a_{22} = 1$, $a_{21} = 0$, $a_{12} = -\alpha$ in (12.17), map $\mathsf{L} = \mathsf{I}$ into $\boldsymbol{B}^2 \otimes \boldsymbol{I}_2 - (\alpha \det \boldsymbol{B})\mathsf{T}$. This implies that for every isotropic, symmetric and positive definite tensor L there exists a unique symmetric and positive definite real 2×2 matrix $\boldsymbol{B}$ and real number $|\alpha| < 1$, such that $\mathsf{L} = \Psi_{\alpha,\boldsymbol{B}}(\mathsf{I})$. Each such transformation maps the exact relation $\mathbb{M}_7$ into another exact relation, so that the exact relations $\Psi_{\alpha,\boldsymbol{B}}(\mathbb{M}_7)$ foliate an open neighborhood of the space of isotropic tensors. Let L_0 be an anisotropic tensor. Then for sufficiently small $\epsilon > 0$ the tensor $\epsilon\mathsf{L}_0$ will be in that neighborhood foliated by our family of exact relations. Hence, there exists a unique exact relation $\mathbb{M}_\epsilon$ in our family that passes through $\epsilon\mathsf{L}_0$. But then $\epsilon^{-1}\mathbb{M}_\epsilon$ is the exact relation isomorphic to $\mathbb{M}_7$ that passes through L_0. Thus, regardless of texture, the effective tensor L_* of an isotropic polycrystal made of the single crystallite L_0 will be uniquely determined by L_0, (as is the case in 2D conductivity, where $\sigma_* = \sqrt{\det \boldsymbol{\sigma}_0}$). If α and $\boldsymbol{B}$ are chosen such that $\Psi_{\alpha,\boldsymbol{B}}(\mathsf{L}_*) = \mathsf{I}$, then $\mathsf{L}' = \Psi_{\alpha,\boldsymbol{B}}(\mathsf{L}_0)$ must lie on $\mathbb{M}_7$. This argument gives us equations for L_* in terms of L_0. To describe them we use the standard parametrization of $\mathbb{R}^2$ by complex numbers. In this parametrization vectors $\boldsymbol{e} = [\boldsymbol{u}, \boldsymbol{v}] \in \mathscr{T} = \mathbb{R}^2 \oplus \mathbb{R}^2$ are written as vectors $\boldsymbol{w} = (u, v) \in \mathbb{C}^2$, where the complex numbers u and v correspond to vectors $\boldsymbol{u} \in \mathbb{R}^2$ and $\boldsymbol{v} \in \mathbb{R}^2$, respectively. In this parametrization $\mathsf{L} \in \mathrm{Sym}(\mathscr{T})$ can be described as a real-linear operator on $\mathbb{C}^2$, so that $\mathsf{L}[\boldsymbol{u}, \boldsymbol{v}] \in \mathscr{T}$ is encoded by $\boldsymbol{X}\boldsymbol{w} + \boldsymbol{Y}\bar{\boldsymbol{w}}$, where $\boldsymbol{X}$ and $\boldsymbol{Y}$ are 2×2 complex matrices and $\boldsymbol{w} = (u, v) \in \mathbb{C}^2$ encodes the vector $[\boldsymbol{u}, \boldsymbol{v}] \in \mathscr{T}$. The operator L is symmetric if and only if $\boldsymbol{X}$ is Hermitian and $\boldsymbol{Y}$ is symmetric. Isotropic tensors correspond to $\boldsymbol{Y} = 0$. Hence, the task of computing the isotropic effective tensor L_* of the polycrystal made with the single crystallite L_0 can be formulated as computing a complex Hermitian matrix $\boldsymbol{L}_*$ encoding L_*, if one knows the complex Hermitian matrix $\boldsymbol{X}$ and complex symmetric matrix $\boldsymbol{Y}$ that encode L_0. The equation for $\boldsymbol{L}_*$ is

$$\boldsymbol{X} - \boldsymbol{L}_* = \boldsymbol{Y}(\bar{\boldsymbol{X}} + \boldsymbol{L}_*)^{-1}\bar{\boldsymbol{Y}}. \tag{12.31}$$

If we make a change of variables $\boldsymbol{Z} = \bar{\boldsymbol{X}} + \boldsymbol{L}_*$ then $\boldsymbol{Z}$ is still self-adjoint (and positive definite) and solves

$$\boldsymbol{Z} + \boldsymbol{Y}\boldsymbol{Z}^{-1}\boldsymbol{Y}^* = \boldsymbol{X} + \bar{\boldsymbol{X}}. \tag{12.32}$$

We can first solve four real linear equations with four real unknowns:

$$\boldsymbol{Z} + \theta\boldsymbol{Y}\mathrm{cof}(\boldsymbol{Z})^T\boldsymbol{Y}^* = \boldsymbol{X} + \bar{\boldsymbol{X}}, \tag{12.33}$$

Obtaining a solution $\hat{\boldsymbol{Z}}(\theta)$. We then find $\theta > 0$ from the equation $\theta \det \hat{\boldsymbol{Z}}(\theta) = 1$. Let us analyze the linear equation (12.33), assuming first that $\boldsymbol{Y}$ is invertible. Let $\mathfrak{B}_Y \in \mathrm{End}_{\mathbb{R}}(\mathfrak{H}(\mathbb{C}^2))$ be defined by

$$\mathfrak{B}_Y \boldsymbol{Z} = \boldsymbol{Y} \mathrm{cof}(\boldsymbol{Z})^T \boldsymbol{Y}^*.$$

Let λ be an eigenvalue of $\mathfrak{B}_Y$. Then, taking determinants in the equation $\mathfrak{B}_Y \boldsymbol{Z} = \lambda \boldsymbol{Z}$ we obtain

$$\lambda^2 \det \boldsymbol{Z} = |\det \boldsymbol{Y}|^2 \det \boldsymbol{Z}.$$

Hence, either $\lambda = \pm|\det \boldsymbol{Y}|$ or $\det \boldsymbol{Z} = 0$. In the latter case, $\boldsymbol{Z}$ is a real multiple of $\boldsymbol{a} \otimes \bar{\boldsymbol{a}}$ for some nonzero vector $\boldsymbol{a} \in \mathbb{C}^2$. Then

$$\boldsymbol{Y}\boldsymbol{R}_\perp \bar{\boldsymbol{a}} \otimes \bar{\boldsymbol{Y}}\boldsymbol{R}_\perp \boldsymbol{a} = \lambda \boldsymbol{a} \otimes \bar{\boldsymbol{a}}.$$

Taking traces we obtain

$$\lambda = \frac{|\boldsymbol{Y}\boldsymbol{R}_\perp \bar{\boldsymbol{a}}|^2}{|\boldsymbol{a}|^2} \geqslant 0.$$

Thus, the only possible negative eigenvalue of $\mathfrak{B}_Y$ is $\lambda = -|\det \boldsymbol{Y}|$. In fact, we can show that $\lambda = -|\det \boldsymbol{Y}|$ is always an eigenvalue of $\mathfrak{B}_Y$. Hence, we restrict $\theta > 0$ to be in the interval $(0, 1/|\det \boldsymbol{Y}|)$, so that $\mathsf{I} + \theta\mathfrak{B}_Y$ is invertible and positive definite. In this case we have $\hat{\boldsymbol{Z}}(\theta) = (\mathsf{I} + \theta\mathfrak{B}_Y)^{-1}(\boldsymbol{X} + \bar{\boldsymbol{X}})$ and θ is found as a positive root of

$$\theta \det[(\mathsf{I} + \theta\mathfrak{B}_Y)^{-1}(\boldsymbol{X} + \bar{\boldsymbol{X}})] = 1, \tag{12.34}$$

such that $(\mathsf{I} + \theta\mathfrak{B}_Y)^{-1}(\boldsymbol{X} + \bar{\boldsymbol{X}}) > \bar{\boldsymbol{X}}$. Observe that $\mathfrak{B}_Y$ maps positive definite matrices into positive definite ones. Therefore, if $\theta_1 > \theta_2 > 0$, then, since $\boldsymbol{X} + \bar{\boldsymbol{X}} > 0$,

$$(\mathsf{I} + \theta_1\mathfrak{B}_Y)^{-1}(\boldsymbol{X} + \bar{\boldsymbol{X}}) < (\mathsf{I} + \theta_2\mathfrak{B}_Y)^{-1}(\boldsymbol{X} + \bar{\boldsymbol{X}}).$$

Hence, θ must be the smallest positive root of (12.34), since we know that $\boldsymbol{L}_*$ satisfying all the requirements exists and is unique. It is then given by

$$\boldsymbol{L}_* = (\mathsf{I} + \theta\mathfrak{B}_Y)^{-1}(\boldsymbol{X} + \bar{\boldsymbol{X}}) - \bar{\boldsymbol{X}}. \tag{12.35}$$

If $\det \boldsymbol{Y} = 0$, then $\boldsymbol{Y} = \boldsymbol{a} \otimes \boldsymbol{a}$ for some $\boldsymbol{a} \in \mathbb{C}^2$ and equation (12.33) can be solved explicitly. This explicit solution shows that equation (12.34) is quadratic in θ and has a unique positive root. This nongeneric case can also be recovered from formula (12.35) as a limiting case.

12.4.4 Application: two phase composites with isotropic phases

Parts of this section have been reprinted with permission from [10]. We present these results for the sake of completeness. Suppose, we are given two isotropic thermoelectric tensors

$$\mathsf{L}_1 = \sigma_1 \otimes \boldsymbol{I}_2 + r_1\mathsf{T}, \qquad \mathsf{L}_2 = \sigma_2 \otimes \boldsymbol{I}_2 + r_2\mathsf{T}.$$

Our goal is to say as much as we can about the effective thermoelectric tensor L_* of a two phase composite made with L_1 and L_2, provided we know nothing about the microstructure except for the volume fractions with which the two materials are used in the composite. The most natural approach is to use the global link (12.17) to eliminate the thermoelectric coupling coefficients r_j and $(\boldsymbol{\sigma}_j)_{12}$. For three-dimensional thermoelectric composites this has been done in [9] by means of the linear links (12.17) of the form $\Psi_{I_2,\boldsymbol{B}}$. In two space dimensions the same method applies only when $r_1 = r_2 = 0$. We can eliminate the thermoelectric coupling from L_1 by the transformation

$$\Psi_1(\mathsf{L}) = (\boldsymbol{\sigma}_1^{-1/2} \otimes \boldsymbol{I}_2)(\mathsf{L} - r_1\mathsf{T})(\boldsymbol{\sigma}_1^{-1/2} \otimes \boldsymbol{I}_2), \tag{12.36}$$

which maps L_1 to I, while

$$\mathsf{L}' = \Psi_1(\mathsf{L}_2) = \boldsymbol{\sigma} \otimes \boldsymbol{I}_2 + \rho\mathsf{T},$$

where

$$\boldsymbol{\sigma} = \boldsymbol{\sigma}_1^{-1/2}\boldsymbol{\sigma}_2\boldsymbol{\sigma}_1^{-1/2}, \qquad \rho = \frac{r_2 - r_1}{\sqrt{\det \boldsymbol{\sigma}_1}}. \tag{12.37}$$

Next we apply transformation $\Psi_2 = \Psi_{\boldsymbol{A},\boldsymbol{B}}$, where $\boldsymbol{A}$ is given by (12.20) and $\boldsymbol{B} \in SO(2, \mathbb{R})$. This family of transformations has the property that $\Psi_2(\mathsf{I}) = \mathsf{I}$. A direct calculation gives the formula

$$\Psi_2(\boldsymbol{\sigma} \otimes \boldsymbol{I}_2 + \rho\mathsf{T}) = \mu\boldsymbol{B}\boldsymbol{\sigma}\boldsymbol{B}^T \otimes \boldsymbol{I}_2 + \nu\mathsf{T},$$

where

$$\mu = \frac{1 - a_0^2}{\det \boldsymbol{\sigma} - (\rho + a_0)^2}, \qquad \nu = \frac{a_0 \det \boldsymbol{\sigma} - (\rho + a_0)(a_0\rho + 1)}{\det \boldsymbol{\sigma} - (\rho + a_0)^2}. \tag{12.38}$$

Our goal is to choose the value of a_0 and $\boldsymbol{B} \in SO(2)$ so that $\nu = 0$ in (12.38) and $\boldsymbol{\sigma}$ is diagonal. The latter requirement is obviously achievable. The requirement $\nu = 0$ gives the quadratic equation for a_0:

$$(a_0^2 + 1)\rho = a_0(\det \boldsymbol{\sigma} - (\rho^2 + 1)). \tag{12.39}$$

This equation has distinct real roots if and only if

$$|r_1 - r_2| < |\sqrt{\det \boldsymbol{\sigma}_1} - \sqrt{\det \boldsymbol{\sigma}_2}|. \tag{12.40}$$

We call this a 'weakly coupled regime', because we can eliminate all coupling coefficients by means of the global link. If we have equality in (12.40), then we will say that we are in a 'borderline regime'. If the inequality in (12.40) is reversed, then we will say that we are in a 'strongly coupled regime'.

Let us consider the weakly coupled regime first. If (12.40) holds and a_0 solves (12.39), then

$$\Psi_2(\boldsymbol{\sigma} \otimes \boldsymbol{I}_2 + \rho\boldsymbol{R}_\perp \otimes \boldsymbol{R}_\perp) = (a_0\rho + 1)\frac{\boldsymbol{\sigma}}{\det \boldsymbol{\sigma}} \otimes \boldsymbol{I}_2, \tag{12.41}$$

where for simplicity of notation we use the same letter $\boldsymbol{\sigma}$ to denote the diagonal matrix $\boldsymbol{B}\boldsymbol{\sigma}\boldsymbol{B}^T$. This choice of a_0 (by design) eliminates the thermoelectric coupling coefficients, permitting us to express the effective thermoelectric tensor of the composite in terms of the effective tensors of 2D conducting composites.

When we are in a 'strongly coupled regime' we need to find a global link (12.17) that maps L_1 and L_2 into a different exact relation. The appropriate one is $\mathbb{M}_6$ from list (12.1). A simple calculation shows that

$$\boldsymbol{\sigma} \otimes \boldsymbol{I}_2 + r\mathsf{T} \in \mathbb{M}_6 \iff \sigma_{11} = \sigma_{22} \text{ and } \det \boldsymbol{\sigma} = r^2 + 1. \tag{12.42}$$

We therefore choose a different chain of global links. We first apply

$$\Psi_0(\mathsf{L}) = (\boldsymbol{\sigma}_1^{-1/2} \otimes \boldsymbol{I}_2)\mathsf{L}(\boldsymbol{\sigma}_1^{-1/2} \otimes \boldsymbol{I}_2) \tag{12.43}$$

and compute

$$\Psi_0(\mathsf{L}_1) = \mathsf{I} + \rho_1\mathsf{T} = \mathsf{L}_1', \quad \Psi_0(\mathsf{L}_2) = \boldsymbol{\sigma} \otimes \boldsymbol{I}_2 + \rho_2\mathsf{T} = \mathsf{L}_2', \quad \rho_j = \frac{r_j}{\sqrt{\det \boldsymbol{\sigma}_1}}.$$

It turns out that affine transformations

$$\Psi_3(\mathsf{L}) = (\boldsymbol{B} \otimes \boldsymbol{I}_2)(a\mathsf{L} + b\mathsf{T})(\boldsymbol{B}^T \otimes \boldsymbol{I}_2), \quad \boldsymbol{B} \in SO(2),\ a > 0,\ b \in \mathbb{R} \tag{12.44}$$

are sufficient for our goal to map L_j' into $\mathbb{M}_6$. For $\mathsf{L}''_j = \Psi_3(\mathsf{L}_j')$, $j = 1, 2$, we compute

$$\mathsf{L}''_1 = a\mathsf{I} + (a\rho_1 + b)\mathsf{T}, \qquad \mathsf{L}''_2 = a\boldsymbol{B}\boldsymbol{\sigma}\boldsymbol{B}^T \otimes \boldsymbol{I}_2 + (a\rho_2 + b)\mathsf{T}.$$

It is easy to check that for any symmetric 2×2 matrix $\boldsymbol{\sigma}$ there exists $\boldsymbol{B} \in SO(2)$, such that $(\boldsymbol{B}\boldsymbol{\sigma}\boldsymbol{B}^T)_{11} = (\boldsymbol{B}\boldsymbol{\sigma}\boldsymbol{B}^T)_{22}$. Once, again for simplicity of notation we use $\boldsymbol{\sigma}$ to denote the rotated matrix, so that $\sigma_{11} = \sigma_{22}$. The requirement that L''_j lie on $\mathbb{M}_6$ is then equivalent to

$$a^2 = (a\rho_1 + b)^2 + 1, \qquad a^2 \det \boldsymbol{\sigma} = (a\rho_2 + b)^2 + 1.$$

The values of $a > 0$ and $b \in \mathbb{R}$ with these properties are

$$b = a\frac{\det \boldsymbol{\sigma} - 1 + \rho_1^2 - \rho_2^2}{2\rho}, \quad a^2 = \frac{4\rho^2}{((\rho + 1)^2 - \det \boldsymbol{\sigma})(\det \boldsymbol{\sigma} - (\rho - 1)^2)}, \tag{12.45}$$

where $\rho = \rho_2 - \rho_1$, as in (12.37). Recalling that $|\rho_1| < 1$ and $|\rho_2| < \sqrt{\det \boldsymbol{\sigma}}$, we obtain

$$|\rho| = |\rho_2 - \rho_1| \leqslant |\rho_2| + |\rho_1| < \sqrt{\det \boldsymbol{\sigma}} + 1.$$

The positivity of the right-hand side in the expression for a^2 in (12.45) is then equivalent to

$$(\rho + 1 - \sqrt{\det \boldsymbol{\sigma}})(\sqrt{\det \boldsymbol{\sigma}} - 1 + \rho) > 0,$$

which is also equivalent to

$$\rho^2 > (\sqrt{\det \boldsymbol{\sigma}} - 1)^2 \iff |r_1 - r_2| > |\sqrt{\det \boldsymbol{\sigma}_1} - \sqrt{\det \boldsymbol{\sigma}_2}|.$$

We conclude that the desired parameters $a > 0$ and $b \in \mathbb{R}$ exist if and only if we are in the strongly coupled regime. Hence, we can always use the global link (12.44) with $a > 0$ and b, given by (12.45), to map L_j' onto the exact relation $\mathbb{M}_6$. Applying this exact relation permits us to express the ten effective thermoelectric moduli of the composite in terms of only six microstructure-dependent parameters. In addition to the general considerations above, there are special cases, when L_1 and L_2 could be mapped into smaller exact relation, thereby obtaining more detailed microstructure-independent information about the effective thermoelectric tensor of the composite.

In order to describe our results we need to introduce more notation. We will need to distinguish two cases: when $\boldsymbol{\sigma}_1$ and $\boldsymbol{\sigma}_2$ are scalar multiples of one another, and when they are not. When they are, we will write $\boldsymbol{\sigma}_1 = \theta_1\boldsymbol{\sigma}_0$ and $\boldsymbol{\sigma}_2 = \theta_2\boldsymbol{\sigma}_0$ in order to maintain the index interchange symmetry. When they are not scalar multiples of one another, we denote λ_1 and λ_2 the two distinct (and positive) generalized eigenvalues solving

$$\det(\boldsymbol{\sigma}_2 - \lambda\boldsymbol{\sigma}_1) = 0. \tag{12.46}$$

The ordering of the eigenvalues (as well as of $\boldsymbol{\sigma}_1$ and $\boldsymbol{\sigma}_2$) is unimportant. We also introduce the rank-one matrices

$$\boldsymbol{S}_1 = \frac{\boldsymbol{\sigma}_2 - \lambda_1\boldsymbol{\sigma}_1}{\lambda_2 - \lambda_1}, \qquad \boldsymbol{S}_2 = \frac{\boldsymbol{\sigma}_2 - \lambda_2\boldsymbol{\sigma}_1}{\lambda_1 - \lambda_2}. \tag{12.47}$$

Finally, $\Sigma_*(h)$ denotes the effective conductivity of the 2D conducting composite where the conductivity of material 1 is 1 and the conductivity of material 2 is $h > 0$. We will now give formulas for the effective thermoelectric tensor of a two-phase composite.

Deterministic case

$\boldsymbol{\sigma}_1 = \theta_1\boldsymbol{\sigma}_0$, $\boldsymbol{\sigma}_2 = \theta_2\boldsymbol{\sigma}_0$, $|r_1 - r_2| = |\theta_1 - \theta_2|\sqrt{\det\boldsymbol{\sigma}_0}$. In this case the effective thermoelectric tensor is always isotropic and microstructure-independent. It is given by $\mathsf{L}_* = \theta_*\boldsymbol{\sigma}_0 \otimes \boldsymbol{I}_2 + r_*\mathsf{T}$, where

$$\theta_* = \langle\theta(z)^{-1}\rangle^{-1}, \qquad r_* = \frac{\langle r(z)\theta(z)^{-1}\rangle}{\langle\theta(z)^{-1}\rangle}, \tag{12.48}$$

where $\langle\cdot\rangle$ denotes the volume average over the period cell.

Special borderline case

$\boldsymbol{L}_1 = \boldsymbol{\sigma}_1 + ir_0\boldsymbol{R}_\perp$, $\boldsymbol{L}_2 = \boldsymbol{\sigma}_2 + ir_0\boldsymbol{R}_\perp$, moreover, $\det\boldsymbol{\sigma}_1 = \det\boldsymbol{\sigma}_2$. The formula for the effective thermoelectric tensor is

$$\mathsf{L}_* = r_0\mathsf{T} + \left(\frac{\boldsymbol{S}_1}{\det\boldsymbol{\sigma}_*} + \boldsymbol{S}_2\right) \otimes \boldsymbol{\sigma}_*, \qquad \boldsymbol{\sigma}_* = \Sigma_*(\lambda), \tag{12.49}$$

where $\lambda > 0$ is one of the two roots of (12.46). (The other is then $1/\lambda$.) The result is independent of the choice of the root in (12.46) and the choice of which material has index 1 and which has index 2.

Generic borderline case

$|r_1 - r_2| = |\sqrt{\det\boldsymbol{\sigma}_1} - \sqrt{\det\boldsymbol{\sigma}_2}|$, $r_1 \neq r_2$. The formula for the effective thermoelectric tensor is

$$\mathsf{L}_* = \boldsymbol{S}_1 \otimes \boldsymbol{\sigma}_* \mathrm{cof}(\boldsymbol{L}_*)\boldsymbol{\sigma}_* + \boldsymbol{S}_2 \otimes \boldsymbol{L}_* + (r_1 - \alpha)\mathsf{T} + \frac{\alpha}{\det \boldsymbol{\sigma}_1}(\boldsymbol{S}_1\boldsymbol{R}_\perp\boldsymbol{S}_2 \otimes \boldsymbol{\sigma}_*\boldsymbol{R}_\perp\boldsymbol{L}_* + \boldsymbol{S}_2\boldsymbol{R}_\perp\boldsymbol{S}_1 \otimes \boldsymbol{L}_*\boldsymbol{R}_\perp\boldsymbol{\sigma}_*). \tag{12.50}$$

where

$$\alpha = \frac{\sqrt{\det \boldsymbol{\sigma}_1} - \sqrt{\det \boldsymbol{\sigma}_2}}{r_1 - r_2}\sqrt{\det \boldsymbol{\sigma}_1}, \qquad \boldsymbol{\sigma}_* = \boldsymbol{\Sigma}_*\left(\sqrt{\frac{\lambda_2}{\lambda_1}}\right).$$

Special weakly coupled case

$\boldsymbol{\sigma}_1 = \theta_1\boldsymbol{\sigma}_0$, $\boldsymbol{\sigma}_2 = \theta_2\boldsymbol{\sigma}_0$, $|r_1 - r_2| < |\theta_1 - \theta_2|\sqrt{\det \boldsymbol{\sigma}_0}$. The formula for the effective thermoelectric tensor is

$$\mathsf{L}_* = \frac{(1 - a_0^2)}{\det \boldsymbol{\sigma}_* - a_0^2}\boldsymbol{\sigma}_1 \otimes \boldsymbol{\sigma}_* + \left(r_1 + \frac{a_0(1 - \det \boldsymbol{\sigma}_*)\sqrt{\det \boldsymbol{\sigma}_1}}{\det \boldsymbol{\sigma}_* - a_0^2}\right)\mathsf{T}, \tag{12.51}$$

where a_0 solves (12.39) and

$$\boldsymbol{\sigma}_* = \boldsymbol{\Sigma}_*\left(\frac{(r_2 - r_1)a_0 + \sqrt{\det \boldsymbol{\sigma}_1}}{\sqrt{\det \boldsymbol{\sigma}_2}}\right)$$

A uniform field relation

$|r_1 - r_2|^2 = \det(\boldsymbol{\sigma}_1 - \boldsymbol{\sigma}_2)$. The effective tensor is

$$\mathsf{L}_* = \alpha_*\left(\boldsymbol{S}_1 \otimes \boldsymbol{A}_{1*} + \boldsymbol{S}_2 \otimes \boldsymbol{A}_{2*} + \frac{a_0}{\sqrt{\det \boldsymbol{\sigma}_1}}(\boldsymbol{S}_1\boldsymbol{R}_\perp\boldsymbol{S}_2 \otimes \boldsymbol{\sigma}_*\boldsymbol{R}_\perp + \boldsymbol{S}_2\boldsymbol{R}_\perp\boldsymbol{S}_1 \otimes \boldsymbol{R}_\perp\boldsymbol{\sigma}_*)\right) + \beta_*\mathsf{T}, \tag{12.52}$$

where

$$\boldsymbol{A}_{1*} = \boldsymbol{\sigma}_* - a_0^2\boldsymbol{I}_2, \quad \boldsymbol{A}_{2*} = \boldsymbol{I}_2 \det \boldsymbol{\sigma}_* - a_0^2\boldsymbol{\sigma}_*, \quad \alpha_* = \frac{1 - a_0^2}{\det \boldsymbol{A}_{1*}},$$

$$\beta_* = (r_1 - a_0\sqrt{\det \boldsymbol{\sigma}_1}(1 + a_0^2\alpha_*)), \quad a_0 = \frac{\lambda_1 - 1}{r_2 - r_1}\sqrt{\det \boldsymbol{\sigma}_1}, \quad \boldsymbol{\sigma}_* = \boldsymbol{\Sigma}_*\left(\frac{\lambda_1}{\lambda_2}\right).$$

The case $a_0 = \infty$ corresponding to $r_1 = r_2 = r_0$ and $\det(\boldsymbol{\sigma}_1 - \boldsymbol{\sigma}_2) = 0$ is also included by taking a limit as $a_0 \to \infty$ in (12.52):

$$\mathsf{L}_* = r_0\mathsf{T} + \boldsymbol{S}_1 \otimes \boldsymbol{I}_2 + \boldsymbol{S}_2 \otimes \boldsymbol{\Sigma}_*(\lambda_1), \quad \lambda_1 \neq 1 = \lambda_2.$$

If the composite is isotropic, i.e., if $\boldsymbol{\sigma}_* = x_*\boldsymbol{I}_2$, the result simplifies:

$$\mathsf{L}_* = \left(r_1 + \frac{(1 - x_*)a_0}{x_* - a_0^2}\sqrt{\det \boldsymbol{\sigma}_1}\right)\mathsf{T} + \frac{1 - a_0^2}{x_* - a_0^2}(\boldsymbol{S}_1 + x_*\boldsymbol{S}_2) \otimes \boldsymbol{I}_2. \tag{12.53}$$

Generic weakly coupled case

$|r_1 - r_2| < |\sqrt{\det \boldsymbol{\sigma}_1} - \sqrt{\det \boldsymbol{\sigma}_2}|$, assuming that $\boldsymbol{\sigma}_1 \neq \theta\boldsymbol{\sigma}_2$ and $|r_1 - r_2|^2 \neq \det(\boldsymbol{\sigma}_1 - \boldsymbol{\sigma}_2)$. The formula for the effective thermoelectric tensor is

$$\mathsf{L}_* = \alpha_* \left(S_1 \otimes A_{1*} + S_2 \otimes A_{2*} + \frac{a_0}{\sqrt{\det \sigma_1}} (S_1 R_\perp S_2 \otimes \sigma_{1*} R_\perp \sigma_{2*} + S_2 R_\perp S_1 \otimes \sigma_{2*} R_\perp \sigma_{1*}) \right) \tag{12.54}$$
$$+ (r_1 - a_0 \sqrt{\det \sigma_1}(1 + a_0^2 \alpha_*))\mathsf{T},$$

where a_0 is one of the roots of (12.39), S_1, S_2 are given in (12.47),

$$A_{1*} = \sigma_{1*} \det \sigma_{2*} - a_0^2 \sigma_{2*}, \quad A_{2*} = \sigma_{2*} \det \sigma_{1*} - a_0^2 \sigma_{1*},$$
$$\alpha_* = \frac{(1 - a_0^2)\det \sigma_{2*}}{\det A_{1*}} = \frac{(1 - a_0^2)\det \sigma_{1*}}{\det A_{2*}},$$

and

$$\sigma_{1*} = \Sigma_* \left(\frac{a_0 \rho + 1}{\lambda_2} \right), \qquad \sigma_{2*} = \Sigma_* \left(\frac{a_0 \rho + 1}{\lambda_1} \right),$$

where ρ is defined in (12.37).

Generic strongly coupled case
$|r_1 - r_2| > |\sqrt{\det \sigma_1} - \sqrt{\det \sigma_2}|$. The effective tensor L_* satisfies the following equation:

$$(\mathsf{L}_* + A\mathsf{T})\mathsf{T}(Z_0 \otimes R_\perp)\mathsf{T}(\mathsf{L}_* + A\mathsf{T}) + BZ_0 \otimes R_\perp = 0, \qquad Z_0 = S_2 R_\perp S_1 - S_1 R_\perp S_2,$$

where

$$A = \frac{\det \sigma_2 - \det \sigma_1 + r_1^2 - r_2^2}{2\Delta r}, \qquad \Delta r = r_2 - r_1. \tag{12.55}$$

$$B = \frac{((\Delta r)^2 - (\sqrt{\det \sigma_1} - \sqrt{\det \sigma_2})^2)((\sqrt{\det \sigma_1} + \sqrt{\det \sigma_2})^2 - (\Delta r)^2)}{4(\Delta r)^2}. \tag{12.56}$$

Special strongly coupled case
$\sigma_1 = \theta_1 \sigma_0$, $\sigma_2 = \theta_2 \sigma_0$, $|r_1 - r_2| > |\theta_1 - \theta_2|\sqrt{\det \sigma_0}$. The formula for the effective tensor is

$$\mathsf{L}_* = \sigma_1 \otimes L_* + t_* \mathsf{T}, \qquad \det \sigma_1 \det L_* = (t_* + A)^2 + B,$$

where A and B are given by (12.55) and (12.56), respectively.

12.5 3D Thermoelectricity

If the material is isotropic then all four blocks must be scalar multiples of I_3. We have the following three non-trivial exact relations.

1. $\mathsf{L} = \begin{bmatrix} \sigma & -\nu \\ \nu & \sigma \end{bmatrix}$, where σ is symmetric and ν is a *skew-symmetric* 3×3 matrix, such that L is positive definite. This exact relation can be conveniently described using complex arithmetic. The pairs of fields e_1, e_2, and j_1, j_2 from (12.10), can be combined into two complex-valued fields

$$j = j_1 + ij_2 \in \mathbb{C}^3, \qquad e = e_1 + ie_2 \in \mathbb{C}^3.$$

Then thermoelectric tensors L belonging to this exact relations are characterized by the constitutive relation $\boldsymbol{j} = \mathsf{K}\boldsymbol{e}$, where K is a complex-Hermitian 3×3 matrix. We refer to the review article [1] and references therein for a discussion of the physical interpretation of complex conductivity.

2. (UFR) Let $\boldsymbol{n}_0 = [n_0^1, n_0^2]$ be a fixed unit vector in $\mathbb{R}^2$ and let

$$\mathsf{L} = \begin{bmatrix} \sigma_0 \boldsymbol{I}_3 + (n_0^1)^2 \boldsymbol{\sigma} & \nu_0 \boldsymbol{I}_3 + n_0^1 n_0^2 \boldsymbol{\sigma} \\ \nu_0 \boldsymbol{I}_3 + n_0^1 n_0^2 \boldsymbol{\sigma} & \gamma_0 \boldsymbol{I}_3 + (n_0^2)^2 \boldsymbol{\sigma} \end{bmatrix},$$

or using a tensor product notation

$$\mathsf{L} = \begin{bmatrix} \sigma_0 & \nu_0 \\ \nu_0 & \gamma_0 \end{bmatrix} \otimes \boldsymbol{I}_3 + (\boldsymbol{n}_0 \otimes \boldsymbol{n}_0) \otimes \boldsymbol{\sigma},$$

where $\boldsymbol{\sigma}$ is a 3×3 symmetric matrix.

As an example, consider an isotropic composite made of two isotropic materials L_1 and L_2 such that $\det(\mathsf{L}_1 - \mathsf{L}_2) = 0$. Then L_* must satisfy $\det(\mathsf{L}_* - \mathsf{L}_1) = 0$ and $\det(\mathsf{L}_* - \mathsf{L}_2) = 0$, which is also equivalent to the formula

$$\mathsf{L}_* = \alpha_* \mathsf{L}_1 + (1 - \alpha_*) \mathsf{L}_2$$

for some microstructure-dependent scalar α_*.

3. This exact relation is described by the condition that the four blocks comprising L satisfy a given linear relation, i.e., there are four constants c_{ij}^0, $i, j = 1, 2$ such that

$$c_{11}^0 \boldsymbol{L}_{11} + c_{12}^0 \boldsymbol{L}_{12} + c_{21}^0 \boldsymbol{L}_{12}^T + c_{22}^0 \boldsymbol{L}_{22} = 0.$$

This exact relation is due to Milgrom and Shtrikman [7] (see also Milgrom [6]).

As an example, consider a composite made with two isotropic thermoelectric materials

$$\mathsf{L}_i = \begin{bmatrix} \sigma_i \boldsymbol{I}_3 & \nu_i \boldsymbol{I}_3 \\ \nu_i \boldsymbol{I}_3 & \gamma_i \boldsymbol{I}_3 \end{bmatrix}, \quad i = 1, 2.$$

We can easily find three numbers c_1, c_2 and c_3 such that the vector $\boldsymbol{c} = [c_1, c_2, c_3]$ is orthogonal to the two vectors $\boldsymbol{l}_1 = [\sigma_1, \gamma_1, \nu_1]$ and $\boldsymbol{l}_2 = [\sigma_2, \gamma_2, \nu_2]$. Then our exact relation tells us that the vector $\boldsymbol{l}_* = (\sigma_*, \gamma_*, \nu_*)$ made with components of the isotropic effective tensor L_* is also orthogonal to $\boldsymbol{c}$. In other words, the three vectors $\boldsymbol{l}_1$, $\boldsymbol{l}_2$ and $\boldsymbol{l}_*$ are linearly dependent, i.e.,

$$\det \begin{vmatrix} \sigma_* & \gamma_* & \nu_* \\ \sigma_1 & \gamma_1 & \nu_1 \\ \sigma_2 & \gamma_2 & \nu_2 \end{vmatrix} = 0.$$

References

[1] Bergman D J and Stroud D 1992 Physical properties of macroscopically inhomogeneous media *Solid State Phys. (New York 1955)* **46** 147–269

[2] Callen H B 1960 *Thermodynamics* (New York: Wiley)

[3] Lifshitz E M and Pitaevskii L P 1980 *Statistical Physics* 3rd edn (Oxford: Pergamon)

[4] Liu L 2012 A continuum theory of thermoelectric bodies and effective properties of thermoelectric composites *Int. J. Eng. Sci.* **55** 35–53

[5] Lundgaard C and Sigmund O 2018 A density-based topology optimization methodology for thermoelectric energy conversion problems *Struct. Multidiscip. Optim.* **57** 1427–42

[6] Milgrom M 1997 Some more exact results concerning multifield moduli of two-phase composites *J. Mech. Phys. Solids* **45** 399–404

[7] Milgrom M and Shtrikman S 1989 Linear response of two-phase composites with cross moduli: exact universal relations *Phys. Rev.* **40** 1568–75

[8] Milton G W 2002 *The Theory of Composites Cambridge Monographs on Applied and Computational Mathematics* (Cambridge: Cambridge University Press)

[9] Straley J P 1981 Thermoelectric properties of inhomogeneous materials *J. Phys.* D **14** 2101–5

[10] Childs S 2020 Exact relations satisfied by the effective tensors of two-dimensional two-phase thermoelectric composites, Master's Thesis, Temple University, Philadelphia, PA, https://scholarshare.temple.edu/items/a5005582-3e30-4ee9-b3e4-80253601972d

Part IV

Appendices

Chapter 13

Closedness of $\mathscr{E}(B_1)$ and $\mathscr{J}(B_1)$ for conductivity and elasticity

13.1 Conductivity

$$\mathscr{E} = \left\{ \boldsymbol{E} \in L^2(\mathbb{R}^d; \mathbb{R}^d): \hat{\boldsymbol{E}}(\boldsymbol{\xi}) = \frac{(\hat{\boldsymbol{E}}(\boldsymbol{\xi}) \cdot \boldsymbol{\xi})\boldsymbol{\xi}}{|\boldsymbol{\xi}|^2} \right\}.$$

$$\mathscr{J} = \{ \boldsymbol{J} \in L^2(\mathbb{R}^d; \mathbb{R}^d): \hat{\boldsymbol{J}}(\boldsymbol{\xi}) \cdot \boldsymbol{\xi} = 0 \}.$$

Theorem 13.1. *Let $\Omega \subset \mathbb{R}^d$ be an open, bounded, connected set with Lipschitz boundary. Then*

$$\mathscr{E}(\Omega) = \{\nabla\phi: \phi \in H^1(\Omega)\}. \tag{13.1}$$

Moreover, $\mathscr{E}(\Omega)$ is a closed subspace of $L^2(\Omega; \mathbb{R}^d)$.

Proof. Let $\mathscr{X} = \{\nabla\phi: \phi \in H^1(\Omega)\}$. The closedness of $\mathscr{X}$ in $L^2(\Omega)$ follows from the Poincaré inequality

$$\|\phi - \langle\phi\rangle\|_{H^1(\Omega)} \leqslant C(\Omega)\|\nabla\phi\|_{L^2(\Omega)}, \tag{13.2}$$

where $\langle\phi\rangle$ is the average of ϕ over Ω. Indeed, if $\nabla\phi_n \to \boldsymbol{f}$ in $L^2(\Omega)$, then $\psi_n = \phi_n - \langle\phi_n\rangle$ is bounded in $H^1(\Omega)$ by the Poincaré inequality. Extracting a weakly convergent subsequence, we conclude that there exists $\phi_0 \in H^1(\Omega)$, such that $\boldsymbol{f} = \nabla\phi_0$. Hence, $\boldsymbol{f} \in \mathscr{X}$.

doi:10.1088/978-0-7503-6249-8ch13

By the density theorem 2.22 every $\boldsymbol{E} \in \mathscr{E}$ is a limit in $L^2(\mathbb{R}^d; \mathbb{R}^d)$ of a sequence of functions $\boldsymbol{E}_n$ of the form

$$\boldsymbol{E}_n(\boldsymbol{x}) = \nabla\phi_n$$

for some $\phi_n \in C_0^\infty(\mathbb{R}^d)$. Thus, the restriction of $\boldsymbol{E}_n$ to Ω is in $\mathscr{X}$. Since $\mathscr{X}$ is closed, we conclude that the restriction of any $\boldsymbol{E} \in \mathscr{E}$ to Ω is in $\mathscr{X}$. This proves that $\mathscr{E}(\Omega) \subset \mathscr{X}$. To prove the reverse inclusion we need to show that every $\boldsymbol{E} \in \mathscr{X}$ can be extended to $\tilde{\boldsymbol{E}}\in\mathscr{E}$. Existence of such an extension is guaranteed by a theorem of Calderon [1] (see also Stein [2]), that says that there exists a (bounded) extension operator from all of $H^1(\Omega)$ to $H^1(\mathbb{R}^d)$. Therefore, every function $\nabla\psi \in \mathscr{X}$ is a restriction of some function $\nabla\Psi$ to Ω, where $\Psi \in H^1(\mathbb{R}^d)$. But then $\nabla\Psi \in \mathscr{E}$. Hence, $\mathscr{X} \subset \mathscr{E}(\Omega)$, establishing the theorem. □

The proof that

$$\mathscr{J}(\Omega) = \mathscr{Y}(\Omega) = \{\boldsymbol{J} \in L^2(\Omega; \mathbb{R}^d)\colon \nabla \cdot \boldsymbol{J} = 0\}$$

is done using a completely different idea. It is obvious that $\mathscr{J}(\Omega) \subset \mathscr{Y}(\Omega)$. It is also obvious that $\mathscr{Y}(\Omega)$ is a closed subspace in $L^2(\Omega; \mathbb{R}^d)$. The nontrivial part is the possibility of extension of $\boldsymbol{J} \in \mathscr{Y}(\Omega)$ to $\tilde{\boldsymbol{J}}\in\mathscr{J}$. If Ω is a unit ball B_1, such an extension can be constructed explicitly using the Kelvin transform.

Lemma 13.2. *Let $\boldsymbol{J} \in \mathscr{Y}(B_1)$. Let $\alpha > 0$ and let*

$$\boldsymbol{K}_\alpha(\boldsymbol{x}) = \frac{\boldsymbol{x}}{|\boldsymbol{x}|^{1+\alpha}} \tag{13.3}$$

be the Kelvin transform. Let

$$\tilde{\boldsymbol{J}}(\boldsymbol{x}) = \begin{cases} \boldsymbol{J}(\boldsymbol{x}), & |\boldsymbol{x}| < 1, \\ \operatorname{cof}(\nabla\boldsymbol{K}_\alpha(\boldsymbol{x}))^T\boldsymbol{J}(\boldsymbol{K}_\alpha(\boldsymbol{x})), & |\boldsymbol{x}| > 1. \end{cases}$$

Then $\tilde{\boldsymbol{J}}\in\mathscr{J}$.

Proof. The Kelvin transform maps interior of the unit ball onto the exterior and vice versa, while $\boldsymbol{K}_\alpha(\boldsymbol{x}) = \boldsymbol{x}$, when $|\boldsymbol{x}| = 1$. If $|\boldsymbol{x}| < 1$, then $\nabla \cdot \tilde{\boldsymbol{J}}(\boldsymbol{x}) = \nabla \cdot \boldsymbol{J} = 0$. If $|\boldsymbol{x}| > 1$, then

$$\nabla \cdot \tilde{\boldsymbol{J}}(\boldsymbol{x}) = \det(\nabla\boldsymbol{K}_\alpha)(\nabla \cdot \boldsymbol{J})(\boldsymbol{K}_\alpha(\boldsymbol{x})) = 0.$$

Finally, we compute $\nabla\boldsymbol{K}_\alpha(\boldsymbol{x}) = \boldsymbol{I}_d - (\alpha + 1)\boldsymbol{n} \otimes \boldsymbol{n}$ at $|\boldsymbol{x}| = 1$, where $\boldsymbol{n} = \boldsymbol{x}/|\boldsymbol{x}|$ is the outward unit normal to B_1. Therefore, on $|\boldsymbol{x}| = 1$

$$\operatorname{cof}(\boldsymbol{I}_d - (\alpha + 1)\boldsymbol{n} \otimes \boldsymbol{n})^T\boldsymbol{J}(\boldsymbol{K}_\alpha(\boldsymbol{x})) \cdot \boldsymbol{n} = \boldsymbol{J}(\boldsymbol{x}) \cdot \boldsymbol{n}.$$

Thus, $\nabla \cdot \tilde{\boldsymbol{J}} = 0$ on $\mathbb{R}^d$ in the sense of distributions. It remains to verify that $\tilde{\boldsymbol{J}}\in L^2(B_1^c; \mathbb{R}^d)$. We compute, making the substitution $\boldsymbol{y} = \boldsymbol{K}_\alpha(\boldsymbol{x})$,

$$\int_{B_1^c} |\tilde{J}(x)|^2 dx=$$
$$\int_{B_1} |y|^{(1+1/\alpha)(d-2)}|\mathrm{cof}(I_d - (\alpha + 1)\hat{y} \otimes \hat{y})J(y)|^2 \det(I_d - (\alpha + 1)\hat{y} \otimes \hat{y})dy,$$

where $\hat{y} = y/|y|$. This calculation shows that $\tilde{J} \in L^2(B_1^c; \mathbb{R}^d)$, when $d \geqslant 2$. □

Lemma 13.2 is sufficient to establish the closedness of the subspace $\mathscr{J}(B_1)$ for conductivity. The same method can be used to prove that $\mathscr{J}(\Omega) = \mathscr{Y}(\Omega)$ for domains Ω diffeomorphic to a unit ball (where the diffeomorphism extends up to the boundary).

13.2 Elasticity

The proof of closedness of subspaces $\mathscr{E}(B_1)$ and $\mathscr{J}(B_1)$ is a fairly straightforward extension of the same proof for conductivity.

$$\mathscr{E} = \left\{ E \in L^2(\mathbb{R}^d; \mathrm{Sym}(\mathbb{R}^d)) \colon \hat{E}(\xi) = \frac{2\hat{E}(\xi)\xi \odot \xi}{|\xi|^2} - \frac{(\hat{E}(\xi)\xi \cdot \xi)\xi \otimes \xi}{|\xi|^4} \right\},$$

$$\mathscr{J} = \{\sigma \in L^2(\mathbb{R}^d; \mathrm{Sym}(\mathbb{R}^d)) \colon \hat{\sigma}(\xi)\xi = 0\},$$

where

$$a \odot b = \frac{1}{2}(a \otimes b + b \otimes a).$$

Theorem 13.3. *Let $\Omega \subset \mathbb{R}^d$ be an open, bounded, connected Lipschitz domain. Then*

$$\mathscr{E}(\Omega) = \{e(u) \colon u \in H^1(\Omega; \mathbb{R}^d)\}, \tag{13.4}$$

where

$$e(u) = \frac{1}{2}(\nabla u + (\nabla u)^T).$$

Moreover, $\mathscr{E}(\Omega)$ is a closed subspace of $L^2(\Omega; \mathrm{Sym}(\mathbb{R}^d))$.

Proof. Let $\mathscr{X} = \{e(u) \colon u \in H^1(\Omega; \mathbb{R}^d)\}$. Let us first prove that $\mathscr{X}$ is closed in $L^2(\Omega)$. Suppose, that $e(u_n) \to E$ in $L^2(\Omega)$. By the Korn inequality there exists $C = C(\Omega)$ and skew-symmetric matrices $S_n \in \mathrm{Skew}(\mathbb{R}^d)$, such that

$$\|\nabla u_n - S_n\|_{L^2(\Omega)} \leqslant C(\Omega)\|e(u_n)\|_{L^2(\Omega)},$$

Thus, $\|\nabla v_n\|_{L^2(\Omega)}$ is bounded, where $v_n = u_n - S_n x$. By the Poincaré inequality (13.2), $w_n = v_n - \langle v_n \rangle$ is bounded in $H^1(\Omega; \mathbb{R}^d)$. Extracting a weakly convergent subsequence, we conclude that there exists $w_0 \in H^1(\Omega; \mathbb{R}^d)$, such that $E = e(w_0)$. Hence, $E \in \mathscr{X}$.

By the density theorem 2.22 every $\boldsymbol{E} \in \mathscr{E}$ is a limit in $L^2(\mathbb{R}^d; \mathrm{Sym}(\mathbb{R}^d))$ of a sequence of functions $\boldsymbol{E}_n$ of the form $\boldsymbol{E}_n = e(\boldsymbol{\phi}_n)$, for some $\boldsymbol{\phi}_n \in C_0^\infty(\mathbb{R}^d; \mathbb{R}^d)$. It follows that the restriction of $\boldsymbol{E}_n$ to Ω is in $\mathscr{X}$. Since $\mathscr{X}$ is closed, we conclude that the restriction of any $\boldsymbol{E} \in \mathscr{E}$ to Ω is in $\mathscr{X}$. This proves that $\mathscr{E}(\Omega) \subset \mathscr{X}$. To prove the reverse inclusion we need to show that every $\boldsymbol{E} \in \mathscr{X}$ can be extended to $\tilde{\boldsymbol{E}} \in \mathscr{E}$. Existence of such an extension is guaranteed by a theorem of Calderon [1] (see also Stein [2]), that says that there exists a bounded extension operator from $H^1(\Omega; \mathbb{R}^d)$ to $H^1(\mathbb{R}^d; \mathbb{R}^d)$. Therefore, every function $e(\boldsymbol{u}) \in \mathscr{X}$ is a restriction of some function $e(\boldsymbol{U})$ for some $\boldsymbol{U} \in H^1(\mathbb{R}^d; \mathbb{R}^d)$. Observing that $e(\boldsymbol{U}) \in \mathscr{E}$ for every $\boldsymbol{U} \in H^1(\mathbb{R}^d; \mathbb{R}^d)$, implies $\mathscr{X} \subset \mathscr{E}(\Omega)$, establishing the theorem. □

To prove the closedness of $\mathscr{J}(B_1)$ for elasticity we modify the Kelvin transform method that we used for conductivity. Suppose $\boldsymbol{\sigma} \in L^2(B_1; \mathrm{Sym}(\mathbb{R}^d)$ is such that $\nabla \cdot \boldsymbol{\sigma} = 0$. We first define operators

$$X_\alpha \boldsymbol{\sigma} = \mathrm{cof}(\nabla \boldsymbol{K}_\alpha(\boldsymbol{x}))^T \boldsymbol{\sigma}(\boldsymbol{K}_\alpha(\boldsymbol{x}))\mathrm{cof}(\nabla \boldsymbol{K}_\alpha(\boldsymbol{x})), \qquad \alpha > 0,$$

where $\boldsymbol{K}_\alpha$ is given by (13.3). Then

$$\nabla \cdot (X_\alpha \boldsymbol{\sigma}) = \det(\nabla \boldsymbol{K}_\alpha)\mathrm{cof}(\nabla \boldsymbol{K}_\alpha(\boldsymbol{x}))^T(\nabla \cdot \boldsymbol{\sigma})(\boldsymbol{K}_\alpha(\boldsymbol{x})),$$

proving that $\nabla \cdot (X_\alpha \boldsymbol{\sigma}) = 0$ on B_1^c. We also compute

$$\mathrm{cof}(\nabla \boldsymbol{K}_\alpha) = \boldsymbol{I}_d + (\alpha + 1)(\boldsymbol{n} \otimes \boldsymbol{n} - \boldsymbol{I}_d), \qquad |\boldsymbol{x}| = 1.$$

Therefore,

$$(X_\alpha \boldsymbol{\sigma})\boldsymbol{n} = \boldsymbol{\sigma}\boldsymbol{n} + (\alpha + 1)(\boldsymbol{n} \otimes \boldsymbol{n} - \boldsymbol{I}_d)\boldsymbol{\sigma}\boldsymbol{n}, \qquad |\boldsymbol{x}| = 1.$$

This shows that no X_α can work as an extension operator. The idea is then to define the extension operator as

$$X_{\mathscr{J}}(\Omega) = \sum_{j=1}^{N} \gamma_j X_{\alpha_j}$$

where the real numbers γ_j are chosen in such a way as to satisfy

$$\sum_{j=1}^{N} \gamma_j = 1, \qquad \sum_{j=1}^{N} \gamma_j(\alpha_j + 1) = 0.$$

It is enough to choose $N = 2$, any $\alpha_1 \neq \alpha_2$ and

$$\gamma_1 = \frac{\alpha_2 + 1}{\alpha_2 - \alpha_1}, \qquad \gamma_2 = \frac{\alpha_1 + 1}{\alpha_1 - \alpha_2}.$$

Defining

$$\tilde{\boldsymbol{\sigma}} = \gamma_1 X_{\alpha_1} \boldsymbol{\sigma} + \gamma_2 X_{\alpha_2} \boldsymbol{\sigma}$$

we obtain a divergence-free extension of $\boldsymbol{\sigma}$ to all of $\mathbb{R}^d$. It remains to verify that $\tilde{\boldsymbol{\sigma}} \in L^2(B_1^c)$, which would follow from the estimate

$$\|X_\alpha \boldsymbol{\sigma}\|_{L^2(B_1^c)} \leqslant C\|\boldsymbol{\sigma}\|_{L^2(B_1)}. \qquad (13.5)$$

We prove (13.5) in the same way as in the case of conductivity, by changing variables $y = K_\alpha(x)$ in the integral

$$\|X_\alpha \sigma\|^2_{L^2(B_1^c)} = \int_{B_1^c} |\mathrm{cof}(\nabla K_\alpha(x))^T \sigma(K_\alpha(x))\mathrm{cof}(\nabla K_\alpha(x))|^2 \, dx.$$

Using formulas

$$\nabla K_\alpha(x) = |y|^{1+1/\alpha}(I_d - (\alpha + 1)\hat{y} \otimes \hat{y}),$$

$$dx = \det\left(|y|^{-(1+1/\alpha)}\left(I_d - \left(1 + \frac{1}{\alpha}\right)\hat{y} \otimes \hat{y}\right)\right)dy,$$

We obtain the estimate

$$\|X_\alpha \sigma\|^2_{L^2(B_1^c)} \leqslant C \int_{B_1} |y|^{(3d-4)(1+1/\alpha)} |\sigma(y)|^2 \, dy \leqslant C' \|\sigma\|^2_{L^2(B_1)},$$

whenever $d \geqslant 2$.

References

[1] Calderón A 1961 Lebesgue spaces of differentiable functions *Proc. Symp. Pure Math.* 4 33–49

[2] Stein E M 1970 Singular Integrals and Differentiability Properties of Functions Princeton Mathematical Series, No *30* (Princeton, NJ: Princeton University Press)

IOP Publishing

Composite Materials (Second Edition)
Mathematical theory and exact relations
Yury Grabovsky

Chapter 14

Characterization of all global Jordan isomorphisms of $\mathrm{Sym}(\mathscr{T})$

Let M_1 and M_2 be symmetric $n \times n$ matrices and suppose that there exists a linear isomorphism ϕ: $\mathrm{Sym}(\mathbb{R}^n) \to \mathrm{Sym}(\mathbb{R}^n)$, such that

$$\phi(XM_1X) = \phi(X)M_2\phi(X). \tag{14.1}$$

It was proved in [1], lemma 4.11 that if M_1 and M_2 are invertible, then either M_2 or $-M_2$ has the same signature as M_1, and therefore, at least one of the sets

$$\mathscr{C}_{\pm}(M_1, M_2) = \{C \in \mathrm{GL}(\mathbb{R}^n) : M_1 = \pm CM_2C^T\}$$

is nonempty. Then there exists $C \in \mathscr{C}_{\pm}(M_1, M_2)$, such that $\phi(X) = \pm C^TXC$.

In this section we prove the same statement without assuming that M_1 and M_2 are invertible.

Lemma 14.1. Suppose $X \in \mathrm{Sym}(\mathbb{R}^n)$ is such that $\mathscr{R}(X) \subset \ker(M_1)$. Then $\mathscr{R}(\phi(X)) \subset \ker M_2$.

Proof. Let $\mathscr{X}(M_1) = \{X \in \mathrm{Sym}(\mathbb{R}^n)$: $\mathscr{R}(X) \subset \ker(M_1)\}$. We observe that $M_1X = XM_1 = 0$ for any $X \in \mathscr{X}(M_1)$. Therefore $XM_1Y + YM_1X = 0$ for any $Y \in \mathrm{Sym}(\mathbb{R}^n)$. But then

$$0 = \phi(XM_1Y + YM_1X) = \phi(X)M_2\phi(Y) + \phi(Y)M_2\phi(X).$$

Observe that since ϕ is a linear isomorphism, for every $Z \in \mathrm{Sym}(\mathbb{R}^n)$ there exists $Y \in \mathrm{Sym}(\mathbb{R}^n)$, such that $Z = \phi(Y)$. Let $A = M_2\phi(X)$. Then $A^TZ + ZA = 0$ for every $Z \in \mathrm{Sym}(\mathbb{R}^n)$. Taking $Z = I_n$ we obtain that $A^T = -A$ and that $AZ = ZA$. But the only matrices that commute with all symmetric matrices are multiples of the identity. The only antisymmetric multiple of the identity is zero. Hence, $M_2\phi(X) = 0$, or equivalently, $\mathscr{R}(\phi(X)) \subset \ker(M_2)$. □

doi:10.1088/978-0-7503-6249-8ch14

Corollary 14.2. $\operatorname{rank}(M_1) = \operatorname{rank}(M_2)$ and $\phi(\mathscr{X}(M_1)) = \mathscr{X}(M_2)$.

Proof. Lemma 14.1 says that $\phi: \mathscr{X}(M_1) \to \mathscr{X}(M_2)$. Observe that $\psi = \phi^{-1}$ has the property (14.1), with M_1 and M_2 interchanged. That means that $\psi: \mathscr{X}(M_2) \to \mathscr{X}(M_1)$. In other words, $\phi(\mathscr{X}(M_1)) = \mathscr{X}(M_2)$. If $\operatorname{rank}(M) = r$, then $\dim(\mathscr{X}(M)) = r(r+1)/2$, which is a strictly monotone increasing function of r on $[0, +\infty)$. Since ϕ is an isomorphism, $\dim \phi(\mathscr{X}(M_1)) = \dim \mathscr{X}(M_1)$. Thus, $\operatorname{rank}(M_1) = \operatorname{rank}(M_2)$. □

We will denote the common rank of M_1 and M_2 by r. Let us choose the o.n.b. of $\mathbb{R}^n$ to be the union of the o.n.b. for $\mathscr{R}(M_1)$ and $\ker(M_1)$. In this basis the matrix M_1 will have the block-structure

$$M_1 = \begin{bmatrix} M_1' & 0 \\ 0 & 0 \end{bmatrix},$$

where M_1' is an invertible, symmetric $r \times r$ matrix. It will be convenient to write the output of ϕ in an o.n.b. that is the union of the o.n.b. for $\mathscr{R}(M_2)$ and $\ker(M_2)$, in which

$$M_2 = \begin{bmatrix} M_2' & 0 \\ 0 & 0 \end{bmatrix},$$

where M_2' is an invertible, symmetric $r \times r$ matrix. Moreover, according to corollary 14.2 the sizes of M_1' and M_2' are equal. For any $X = \begin{bmatrix} X' & 0 \\ 0 & 0 \end{bmatrix} \in \operatorname{Sym}(\mathbb{R}^n)$ we can write

$$\phi(X) = \begin{bmatrix} \phi_{11}(X') & \phi_{12}(X') \\ \phi_{12}(X')^T & \phi_{22}(X') \end{bmatrix},$$

where ϕ_{ij} are linear maps into corresponding spaces. We observe that $XM_1X = \begin{bmatrix} X'M_1'X' & 0 \\ 0 & 0 \end{bmatrix}$. Therefore, (14.1) implies

$$\begin{bmatrix} \phi_{11}(X'M'_1X') & \phi_{12}(X'M'_1X') \\ \phi_{12}(X'M'_1X')^T & \phi_{22}(X'M'_1X') \end{bmatrix} = \begin{bmatrix} \phi_{11}(X') & \phi_{12}(X') \\ \phi_{12}(X')^T & \phi_{22}(X') \end{bmatrix} \begin{bmatrix} M'_2 & 0 \\ 0 & 0 \end{bmatrix} \begin{bmatrix} \phi_{11}(X') & \phi_{12}(X') \\ \phi_{12}(X')^T & \phi_{22}(X') \end{bmatrix} = \begin{bmatrix} \phi_{11}(X')M'_2\phi_{11}(X') & \phi_{11}(X')M'_2\phi_{12}(X') \\ \phi_{12}(X')^T M'_2\phi_{11}(X') & \phi_{12}(X')^T M'_2\phi_{12}(X') \end{bmatrix}. \qquad (14.2)$$

We conclude that ϕ_{11} satisfies (14.1), but with nonsingular M_1' and M_2'.

Lemma 14.3. ϕ_{11}: Sym($\mathbb{R}^r$) $\to$ Sym($\mathbb{R}^r$) is a Jordan algebra isomorphism.

Proof. Recall that we have shown that ϕ_{11} satisfies (14.1). Then it is a Jordan algebra homomorphism. Then, ker ϕ_{11} is a Jordan ideal. But Sym($\mathbb{R}^r$) is a simple algebra, since M_1' is invertible. (See lemma 14.4 below.) And therefore, either ϕ_{11} is invertible, or ϕ_{11} is identically zero. In the latter case, we compute on the one hand

$$\phi(X*_{M_1} Y) = \begin{bmatrix} 0 & \phi_{12}(X'*_{M_1'} Y') \\ \phi_{12}(X'*_{M_1'} Y')^T & \phi_{22}(X'*_{M_1'} Y') \end{bmatrix}.$$

On the other,

$$\phi(X*_{M_1} Y) = \phi(X)*_{M_2} \phi(Y) = \begin{bmatrix} 0 & 0 \\ 0 & * \end{bmatrix}.$$

But that implies that $\phi(X*_{M_1} Y) = 0$, which is impossible, since ϕ is invertible. We conclude that ϕ_{11} is a Jordan algebra isomorphism. □

Lemma 14.4. Suppose $M \in$ Sym($\mathbb{R}^r$) is invertible. Then the Jordan algebra with the multiplication $X*_M Y$ is simple.

Proof. Suppose $\mathscr{I}$ is a proper Jordan ideal in Sym($\mathbb{R}^r$). Suppose that $X \in \mathscr{I}$. Then, taking $Y = M^{-2}$, and $Y = M^{-1}X^k M^{-1}$ we conclude that for any polynomial $p(\lambda)$ with $p(0) = 0$ we have $p(X)M^{-1} + M^{-1}p(X) \in \mathscr{I}$. If X is invertible, then there exists a polynomial, $p(\lambda)$ with $p(0) = 0$, such that $p(X) = I_r$. But then $M^{-1} \in \mathscr{I}$. Thus, for every $Y \in$ Sym($\mathbb{R}^r$) we have $Y = M^{-1}*_M Y \in \mathscr{I}$. This shows that if $\mathscr{I}$ is a proper Jordan ideal then all elements of $\mathscr{I}$ must be singular. Let $X \in \mathscr{I}$ be the matrix of maximal rank. Then the same argument as above shows that $P \in \mathscr{I}$, where P is the orthogonal projection onto the range of X. If in the basis in which $P = \begin{bmatrix} I_\rho & 0 \\ 0 & 0 \end{bmatrix}$, the matrix M has the form $M = \begin{bmatrix} M_{11} & M_{12} \\ M_{12}^T & M_{22} \end{bmatrix}$, with $M_{12} \neq 0$, then $2P*_M I_r = \begin{bmatrix} 2M_{11} & M_{12} \\ M_{12}^T & 0 \end{bmatrix} \in \mathscr{I}$. Let us show that when $\lambda > 0$ is so large that $2M_{11} + \lambda I_\rho$ is positive definite, then the rank of $Z = \begin{bmatrix} 2M_{11} + \lambda I_\rho & M_{12} \\ M_{12}^T & 0 \end{bmatrix} \in \mathscr{I}$ is larger than ρ. Let $\begin{bmatrix} u \\ v \end{bmatrix} \in \ker Z$. Then

$$(2M_{11} + \lambda I_\rho)u + M_{12}v = 0, \quad M_{12}^T u = 0.$$

Eliminating u by means of the first equation, we obtain

$$M_{12}^T(2M_{11} + \lambda I_\rho)^{-1}M_{12}v = 0.$$

Taking a dot product with v and using positive definiteness of $2M_{11} + \lambda I_\rho$ implies that $M_{12}v = 0$. We conclude that

$$\ker(Z) = \left\{ \begin{bmatrix} 0 \\ v \end{bmatrix} : v \in \ker M_{12} \right\}.$$

In particular, $\text{rank}(Z) = r - \dim\ker M_{12}$.

Since $M_{12} \neq 0$ is a $\rho \times (r - \rho)$ matrix, $\dim\ker(M_{12}) < r - \rho$. Therefore, $\text{rank}(Z) > \rho$, in contradiction with the assumption that ρ is the maximal rank of matrices in $\mathscr{I}$.

It remains to examine the case when $M_{12} = 0$. In that case both M_{11} and M_{22} must be invertible. Then $Z = P^*{}_M Y \in \mathscr{I}$ for every $Y \in \text{Sym}(\mathbb{R}^r)$. We compute (taking into account that $M_{12} = 0$)

$$2Z = \begin{bmatrix} 2M_{11}Y_{11} & M_{11}Y_{12} \\ Y_{12}^T M_{11} & 0 \end{bmatrix}.$$

It will be convenient to choose $Y_{11} = M_{11}^{-1}/2$, so that $\begin{bmatrix} I_\rho & M_{11}Y_{12} \\ Y_{12}^T M_{11} & 0 \end{bmatrix} \in \mathscr{I}$. If $\text{rank}(Z) = \rho$, then the last $r - \rho$ columns of Z are linear combinations of the first ρ columns of Z, which are obviously linearly independent. That means that there exists a $\rho \times r - \rho$ matrix Λ, such that $I_\rho \Lambda = M_{11}Y_{12}$ and $Y_{12}^T M_{11}\Lambda = 0$. Therefore, $Y_{12}^T M_{11}^2 Y_{12} = 0$. But this implies that $M_{11}Y_{12} = 0$, which, in turn, implies that $Y_{12} = 0$. Choosing $Y \in \text{Sym}(\mathbb{R}^r)$, such that $Y_{12} \neq 0$, therefore, gives us a matrix Z, whose rank is larger than ρ. This finishes the proof of the lemma. □

Now, by lemma 14.3 all conditions of [1, lemma 4.11] are satisfied and therefore, we must have $\phi_{11}(X') = \pm C^T X' C$, where $M_1' = \pm C M_2' C^T$. Applying this formula to (14.2) we see that ψ: $\text{Sym}(\mathbb{R}^r) \to \text{Hom}(\mathbb{R}^m, \mathbb{R}^r)$, defined by $\psi(X) = C^{-T}\phi_{12}(X)$, satisfies

$$\psi(XMX) = XM\psi(X). \tag{14.3}$$

Lemma 14.5. If ψ: $\mathrm{Sym}(\mathbb{R}^r) \to \mathrm{Hom}(\mathbb{R}^m, \mathbb{R}^r)$ satisfies (14.3) with an invertible $M \in \mathrm{Sym}(\mathbb{R}^r)$, then there exists $U \in \mathrm{Hom}(\mathbb{R}^m, \mathbb{R}^r)$, such that $\psi(X) = XU$.

Proof. Let $U = \psi(I_r)$. We note that (14.3) is equivalent to

$$\psi(XMY + YMX) = XM\psi(Y) + YM\psi(X).$$

Then, taking $Y = M^{-1}$ we obtain

$$2\psi(X) = XM\psi(M^{-1}) + \psi(X),$$

which gives us the formula $\psi(X) = XM\psi(M^{-1})$. If we take $X = I_r$ in this formula we obtain $U = M\psi(M^{-1})$, which gives $\psi(M^{-1}) = M^{-1}U$. But then $\psi(X) = XMM^{-1}U = XU$. □

That means that $\phi_{12}(X) = C^T XU$. Thus, we get a formula for ϕ_{22}:

$$\phi_{22}(XMX) = U^T XCM_2 C^T XU = \pm U^T XMXU.$$

We observe that $\mathrm{Span}(\{XMX\colon X \in \mathrm{Sym}(\mathbb{R}^r)\})$ is a Jordan ideal in $\mathrm{Sym}(\mathbb{R}^r)$, which is clearly non-zero. Since $\mathrm{Sym}(\mathbb{R}^r)$ is a simple Jordan algebra, we conclude that for any $Z \in \mathrm{Sym}(\mathbb{R}^r)$ we have

$$Z = \sum_{j=1}^{N} \alpha_j X_j M X_j$$

for some real numbers α_j and matrices $X_j \in \mathrm{Sym}(\mathbb{R}^r)$. Hence,

$$\phi_{22}(Z) = \pm U^T Z U.$$

We conclude that

$$\phi\left(\begin{bmatrix} X' & 0 \\ 0 & 0 \end{bmatrix}\right) = \begin{bmatrix} \pm\, C^T X' C & C^T X' U \\ U^T X' C & \pm\, U^T X' U \end{bmatrix} = \pm \mathscr{C}_\pm^T X \mathscr{C}_\pm, \tag{14.4}$$

where

$$\mathscr{C}_\pm = \begin{bmatrix} C & \pm\, U \\ * & * \end{bmatrix}$$

We now need to compute the map

$$\phi\left(\begin{bmatrix} 0 & Y \\ Y^T & 0 \end{bmatrix}\right) = \begin{bmatrix} \phi_{11}(Y) & \phi_{12}(Y) \\ \phi_{12}^T(Y) & \phi_{22}(Y) \end{bmatrix},$$

where we for simplicity of notation have repurposed symbols ϕ_{ij}. As before, we apply (14.1) to the product

$$\begin{bmatrix} 0 & Y \\ Y^T & 0 \end{bmatrix} *_{M_1} \begin{bmatrix} X & 0 \\ 0 & 0 \end{bmatrix} = \frac{1}{2} \begin{bmatrix} 0 & XM_1' Y \\ Y^T M_1' X & 0 \end{bmatrix}.$$

We obtain, using formula (14.4)

$$\phi_{11}(XM_1' Y) = \phi_{11}(Y) C^{-1} M_1' XC + C^T XM_1' C^{-T} \phi_{11}(Y), \tag{14.5}$$

$$\phi_{12}(XM_1' Y) = \pm\phi_{11}(Y) C^{-1} M_1' XU + C^T XM_1' C^{-T} \phi_{12}(Y), \tag{14.6}$$

$$\phi_{22}(XM_1' Y) = \pm\phi_{12}(Y)^T C^{-1} M_1' XU \pm U^T XM_1' C^{-T} \phi_{12}(Y). \tag{14.7}$$

For $\psi_{11}(Y) = \phi_{11}(Y) C^{-1}$ we obtain from (14.5)

$$\psi_{11}(XM_1' Y) = \psi_{11}(Y) M_1' X + C^T XM_1' C^{-T} \psi_{11}(Y).$$

Taking $X = (M_1')^{-1}$ we obtain $\psi_{11}(Y) = 0$. Using this result in (14.6) we obtain

$$\phi_{12}(XM_1' Y) = C^T XM_1' C^{-T} \phi_{12}(Y).$$

Denoting $\psi_{12}(Y) = C^{-T} \phi_{12}(Y)$ we get

$$\psi_{12}(XM_1' Y) = XM_1' \psi_{12}(Y) \tag{14.8}$$

Let us first fix a vector $\boldsymbol{u}_0 \in \mathbb{R}^m$ and an index $1 \leqslant \alpha \leqslant m$. Then the function $f_{\boldsymbol{u}_0,\alpha}(\boldsymbol{y}) = \psi_{12}(\boldsymbol{y} \otimes \boldsymbol{e}_\alpha)\boldsymbol{u}_0$ is a linear operator on $\mathbb{R}^r$. Equation (14.8) can then be written as

$$f_{\boldsymbol{u}_0,\alpha}(XM_1' \boldsymbol{y}) = XM_1' f_{\boldsymbol{u}_0,\alpha}(\boldsymbol{y}).$$

By linearity of $f_{\boldsymbol{u}_0,\alpha}$ there exists an $r \times r$ matrix, $F_\alpha(\boldsymbol{u}_0)$ such that $f_{\boldsymbol{u}_0,\alpha}(\boldsymbol{y}) = F_\alpha(\boldsymbol{u}_0)\boldsymbol{y}$. It follows that

$$F_\alpha(\boldsymbol{u}_0) XM_1' \boldsymbol{y} = XM_1' F_\alpha(\boldsymbol{u}_0)\boldsymbol{y}, \quad \forall \boldsymbol{y} \in \mathbb{R}^r, \forall X \in \mathrm{Sym}(\mathbb{R}^r).$$

Hence,

$$F_\alpha(\boldsymbol{u}_0) XM_1' = XM_1' F_\alpha(\boldsymbol{u}_0), \quad \forall X \in \mathrm{Sym}(\mathbb{R}^r). \tag{14.9}$$

Taking $X = I_r$ we see that the matrices M_1' and $F_\alpha(\boldsymbol{u}_0)$ commute. It follows that

$$F_\alpha(\boldsymbol{u}_0) XM_1' = XF_\alpha(\boldsymbol{u}_0) M_1', \quad \forall X \in \mathrm{Sym}(\mathbb{R}^r).$$

By invertibility of M_1' we conclude that $F_\alpha(\boldsymbol{u}_0)$ commutes with all symmetric matrices. Therefore, $F_\alpha(\boldsymbol{u}_0)$ must be a multiple of the identity. Moreover, this multiple of the identity is a linear function of $\boldsymbol{u}_0$. Thus, there exist vectors $\{\boldsymbol{v}_1, \dots \boldsymbol{v}_m\} \subset \mathbb{R}^m$, such that $F_\alpha(\boldsymbol{u}_0) = (\boldsymbol{v}_\alpha \cdot \boldsymbol{u}_0) I_r$. We conclude that

$$\psi_{12}(\boldsymbol{y} \otimes \boldsymbol{e}_\alpha)\boldsymbol{u}_0 = (\boldsymbol{v}_\alpha \cdot \boldsymbol{u}_0)\boldsymbol{y}.$$

This is equivalent to

$$\psi_{12}(\boldsymbol{y} \otimes \boldsymbol{e}_\alpha) = \boldsymbol{y} \otimes \boldsymbol{v}_\alpha.$$

It follows that if V is the $m \times m$ matrix, whose m rows are v_α, $\alpha = 1, 2, \dots, m$. Then, $\psi_{12}(Y) = YV$. Finally, from (14.7) we get

$$\phi_{22}(XM_1'Y) = \pm\psi_{12}(Y)^T M_1' XU \pm U^T XM_1' \psi_{12}(Y) = \pm V^T Y^T M_1' XU \pm U^T XM_1' YV.$$

Taking $X = (M_1')^{-1}$ we get

$$\phi_{22}(Y) = \pm VY^T U \pm U^T YV^T.$$

Putting everything together we obtain

$$\phi\left(\begin{bmatrix} 0 & Y \\ Y^T & 0 \end{bmatrix}\right) = \begin{bmatrix} 0 & C^T YV \\ V^T Y^T C & \pm V^T Y^T U \pm U^T YV \end{bmatrix},$$

or equivalently,

$$\phi\left(\begin{bmatrix} 0 & Y \\ Y^T & 0 \end{bmatrix}\right) = \pm \mathscr{C}^T \begin{bmatrix} 0 & Y \\ Y^T & 0 \end{bmatrix} \mathscr{C}, \quad \mathscr{C} = \begin{bmatrix} C & \pm U \\ 0 & \pm V \end{bmatrix}. \tag{14.10}$$

It remains to compute $\phi\left(\begin{bmatrix} 0 & 0 \\ 0 & Z \end{bmatrix}\right)$. We observe that

$$\begin{bmatrix} 0 & Y \\ Y^T & 0 \end{bmatrix} *_{M_1} \begin{bmatrix} 0 & Y \\ Y^T & 0 \end{bmatrix} = \begin{bmatrix} 0 & 0 \\ 0 & Y^T M_1' Y \end{bmatrix}.$$

Hence, by (14.1) and (14.10)

$$\phi\left(\begin{bmatrix} 0 & 0 \\ 0 & Y^T M_1' Y \end{bmatrix}\right) = \pm \mathscr{C}^T \begin{bmatrix} 0 & 0 \\ 0 & Y^T M_1' Y \end{bmatrix} \mathscr{C}.$$

But that shows that

$$\phi(X) = \pm \begin{bmatrix} C & \pm U \\ 0 & \pm V \end{bmatrix}^T X \begin{bmatrix} C & \pm U \\ 0 & \pm V \end{bmatrix}, \tag{14.11}$$

where $M_1' = \pm CM_2' C^T$. In fact, if M_1 and M_2 or M_1 and $-M_2$ have the same signature, then the set of all invertible matrices C, such that $M_1 = CM_2 C^T$ or $M_1 = -CM_2 C^T$ is exactly the one given in (14.11). This proves theorem 6.1.

Reference

[1] Grabovsky Y, Milton G W and Sage D S 2000 Exact relations for effective tensors of polycrystals: necessary conditions and sufficient conditions *Commun. Pure Appl. Math.* **53** 300–53

Chapter 15

Jordan subalgebras of real symmetric matrices

15.1 Questions about real $n \times n$ matrices

Consider the space of $n \times n$ real symmetric matrices denoted by $\mathrm{Sym}(\mathbb{R}^n)$. In this section we characterize explicitly the canonical forms of subspaces $\Pi \subset \mathrm{Sym}(\mathbb{R}^n)$ satisfying

$$AB + BA \in \Pi, \quad \forall \{A, B\} \subset \Pi. \tag{15.1}$$

Subspaces Π satisfying (15.1) are the Jordan subalgebras of $\mathrm{Sym}(\mathbb{R}^n)$, since the commutative, non-associative multiplication

$$A^*B = \frac{1}{2}(AB + BA) \tag{15.2}$$

satisfies all the axioms of a Jordan algebra (see, e.g., [4, 5]). The canonical form refers to the choice of an orthonormal basis in $\mathbb{R}^n$ in which the structure of Π is simplest.

For each Jordan subalgebra Π we consider the smallest associative algebra Π' containing Π. In other words, Π' is the smallest subspace of real $n \times n$ matrices, which will be denoted by $\mathrm{End}(\mathbb{R}^n)$, such that it contains Π and satisfies $AB \in \Pi'$ for all $\{A, B\} \subset \Pi'$. We observe that if $\mathscr{A}$ is an associative subalgebra of $\mathrm{End}(\mathbb{R}^n)$ containing Π, then so is $\mathscr{A}^T$, thus, so is $\mathscr{A} \cap \mathscr{A}^T$. It follows that $(\Pi')^T = \Pi'$. For this reason the set of all symmetric matrices contained in Π' also coincides with the set of symmetric parts of all matrices in Π':

$$\Pi'_{\mathrm{sym}} := \Pi' \bigcap \mathrm{Sym}(\mathbb{R}^n) = \left\{ \frac{A + A^T}{2} : A \in \Pi' \right\}.$$

Definition 15.1. *We say that a Jordan subalgebra* Π *is reflexive if* $\Pi = \Pi'_{\mathrm{sym}}$, *where* Π' *is the smallest associative algebra containing* Π.

doi:10.1088/978-0-7503-6249-8ch15

Our main goal in this chapter is to characterize explicitly the reflexive and nonreflexive Jordan subalgebras of $\mathrm{Sym}(\mathbb{R}^n)$.

The main observation is that each Jordan subalgebra $\Pi \subset \mathrm{Sym}(\mathbb{R}^n)$ is *formally real*, i.e.,

$$\sum_{j=1}^{N} A_j^2 = 0 \quad \Rightarrow \quad A_1 = \cdots = A_N = 0.$$

All such algebras are direct sums of simple algebras (algebras containing no nontrivial ideals). All simple formally real Jordan algebras have been characterized up to an isomorphism in [3]. Thus, what we are after is the explicit characterization of all representations of formally real Jordan algebras.

15.2 Structure of Jordan subalgebras of Sym($\mathbb{R}^n$)

Examining the invariant subspaces of a Jordan subalgebra Π, it is not too difficult to prove the following basic structure theorem.

Theorem 15.2. (Jordan subalgebras of $\mathrm{Sym}(\mathbb{R}^n)$). *Let* Π *be a Jordan subalgebra of* $\mathrm{Sym}(\mathbb{R}^n)$. *Then there exists an o.n.b. of* $\mathbb{R}^n$ *in which*

$$\Pi = \left\{ \begin{bmatrix} A_1 & & & 0 \\ & \ddots & & \\ & & A_m & \\ 0 & & & 0 \end{bmatrix} : A_1 \in \Pi_1, \ldots, A_m \in \Pi_m \right\}, \tag{15.3}$$

where each $\Pi_\alpha \subset \mathrm{Sym}(\mathbb{R}^{d_\alpha})$ *is isomorphic to a simple formally real Jordan algebra. Moreover, each subalgebra* Π_α *is nonsingular, in the sense that*

$$\{x \in \mathbb{R}^{d_\alpha}: Ax = 0, \forall A \in \Pi_\alpha\} = \{0\}. \tag{15.4}$$

Theorem 15.2 reduces the characterization of all Jordan subalgebras $\Pi \subset \mathrm{Sym}(\mathbb{R}^n)$ to the ones that are nonsingular, in the sense of (15.4), and isomorphic to simple ones. The isomorphism between Π and a corresponding simple formally real Jordan algebra is called the *representation* of the Jordan algebra. Representations of groups and algebras are of great interest because they establish a correspondence between an abstract algebraic object and a specific set of matrices that is closed with respect to corresponding matrix algebra operations.

If Π is a representation of a Jordan algebra in $\mathrm{Sym}(\mathbb{R}^n)$, then we look for subspaces of $\mathbb{R}^n$ that are invariant with respect to all matrices in Π. Obviously, an orthogonal complement to such a subspace will also be invariant. It follows that $\mathbb{R}^n$ is an orthogonal sum of invariant subspaces, on each of which the algebra Π has no invariant subspaces.

Definition 15.3. We say that the Jordan subalgebra $\Pi \subset \mathrm{Sym}(\mathbb{R}^n)$ is irreducible, if it has no invariant subspaces.

We thus obtain the following structure of nonsingular Jordan subalgebras isomorphic to simple ones.

Theorem 15.4. (Simple, nonsingular Jordan subalgebras of $\mathrm{Sym}(\mathbb{R}^n)$). *Let $\Pi \subset \mathrm{Sym}(\mathbb{R}^n)$ be a nonsingular, in the sense of (15.4), Jordan subalgebra isomorphic to a simple formally real Jordan algebra. Then there exists an o.n.b. of $\mathbb{R}^n$ in which*

$$\Pi = \left\{ \begin{bmatrix} A & & & 0 \\ 0 & T_1 A & & 0 \\ 0 & 0 & \ddots & \\ 0 & 0 & \cdots & T_k A \end{bmatrix} : A \in \Pi_0 \right\},$$

where Π_0 is an irreducible Jordan subalgebra of $\subset \mathrm{Sym}(\mathbb{R}^{d_0})$, isomorphic to a simple formally real Jordan algebra, and $T_1, \dots, T_k$ are Jordan isomorphisms of Π_0, i.e, linear maps T_j: $\Pi_0 \to \mathrm{Sym}(\mathbb{R}^{d_j})$, satisfying

$$(T_j X)(T_j Y) + (T_j Y)(T_j X) = T_j(XY + YX), \tag{15.5}$$

and such that $T_j \Pi_0$ is also an irreducible Jordan subalgebra of $\mathrm{Sym}(\mathbb{R}^{d_j})$.

We thus arrive at the problem of characterizing all irreducible representations of simple formally real Jordan algebras. As we have mentioned, all simple formally real Jordan algebras have been characterized, up to an isomorphism, by Jordan, von Neumann and Wigner in [3]. All irreducible representations of such algebras in $\mathrm{Sym}(\mathbb{R}^n)$ have been described in [2]. We summarize their results in the next section.

15.3 Simple, irreducible Jordan subalgebras of Sym($\mathbb{R}^n$)

Theorem 15.5. *Let $\mathfrak{J}$ be a formally real simple Jordan algebra. Then it is isomorphic to one of the following algebras*

- $\mathrm{Sym}(\mathbb{R}^r)$, $r \geqslant 1$;
- $\mathfrak{H}(\mathbb{C}^r)$, $r \geqslant 2$, *denoting complex Hermitian* $r \times r$ matrices;
- $\mathfrak{H}(\mathbb{H}^r)$, $r \geqslant 3$, *denoting quaternionic Hermitian $r \times r$ matrices;*
- *'spin factors'* $\mathfrak{S}_n$, $n \geqslant 5$ *defined by* $\mathfrak{S}_n = \mathrm{Span}\{I_{\mathfrak{S}_n}, s_1, \dots, s_{n-1}\}$, *where* $I_{\mathfrak{S}_n} {}^* I_{\mathfrak{S}_n} = I_{\mathfrak{S}_n}$, $I_{\mathfrak{S}_n} {}^* s_j = s_j$, $s_i^* s_j = \delta_{ij} I_{\mathfrak{S}_n}$. *Here * denotes the Jordan product in* $\mathfrak{J}$;
- *27-dimensional exceptional Albert algebra $\mathfrak{H}(\mathfrak{O}^3)$, where $\mathfrak{O}$ stands for Cayley's octonions.*

Several remarks are in order.

Remark 15.6.

1. *The Jordan multiplication in the first three items in the theorem is given by (15.2);*
2. *If $r = 1$, then* $\mathfrak{H}(\mathbb{C}^r) = \mathrm{Sym}(\mathbb{R}^r) = \mathbb{R}$;
3. *If $r = 2$, then* $\mathfrak{H}(\mathbb{H}^r) = \mathfrak{S}_6$;

4. $\mathfrak{S}_1 = \mathbb{R} = Sym(\mathbb{R}^1)$, $\mathfrak{S}_2 = \mathbb{R} \oplus \mathbb{R}$ *is not simple*, $\mathfrak{S}_3 = \mathrm{Sym}(\mathbb{R}^2)$, $\mathfrak{S}_4 = \mathfrak{H}(\mathbb{C}^2)$;
5. *It was shown by Albert* [1] *that* $\mathfrak{H}(\mathfrak{O}^3)$ *cannot be identified as a subspace of an associative algebra that is closed under the multiplication (15.2). For this reason we did not describe the algebraic structure of the exceptional Albert algebra* $\mathfrak{H}(\mathfrak{O}^3)$.

Of course, algebras $\mathrm{Sym}(\mathbb{R}^r)$, $\mathfrak{H}(\mathbb{C}^r)$ and $\mathfrak{H}(\mathbb{H}^r)$ could have been anticipated from the beginning. The Albert algebra and spin factors were a surprise. Even though the algebra $\mathfrak{H}(\mathfrak{O}^3)$ is the most curious member of the list, it is completely irrelevant for our purposes, since it does not have any faithful representations in $\mathrm{Sym}(\mathbb{R}^n)$.

Let us now describe all irreducible representation of each of the simple formally real Jordan algebras in theorem 15.5. Obviously, the algebra $\mathrm{Sym}(\mathbb{R}^r)$ is trivially represented by itself. To describe the irreducible representations of $\mathfrak{H}(\mathbb{C}^r)$ and $\mathfrak{H}(\mathbb{H}^r)$ we note that complex numbers and quaternions are naturally identified with $\mathbb{R}^2$ and $\mathbb{R}^4$, respectively. Complex and quaternionic multiplication can then be viewed as linear transformations on $\mathbb{R}^2$ and $\mathbb{R}^4$, respectively, whereby a vector $\boldsymbol{x}$ is written as a complex number or a quaternion x, gets multiplied by a complex number of a quaternion h, and the resulting product hx is then reinterpreted as a vector. The corresponding linear maps will be denoted by $\varphi(h)$, for complex numbers and $Q(h)$ for quaternions, respectively. It is easy to calculate that the matrices of these transformations are given by

$$Q(q_0 + iq_1 + jq_2 + kq_3) = \begin{bmatrix} \varphi(q_0 + iq_1) & -\psi(q_2 + iq_3) \\ \psi(q_2 + iq_3) & \varphi(q_0 + iq_1) \end{bmatrix}, \tag{15.6}$$

where

$$\varphi(a + ib) = \begin{bmatrix} a & -b \\ b & a \end{bmatrix}, \quad \psi(x + iy) = \begin{bmatrix} x & y \\ y & -x \end{bmatrix}. \tag{15.7}$$

These functions have the properties

$$\varphi(a)^T = \varphi(a), \quad Q(q)^T = Q(\bar{q}).$$

Then a matrix $A \in \mathfrak{H}(\mathbb{C}^r)$ will be represented as a $2r \times 2r$ block-matrix

$$\begin{bmatrix} \varphi(A_{11}) & \dots & \varphi(A_{1r}) \\ & \ddots & \\ \varphi(A_{1r})^T & \dots & \varphi(A_{rr}) \end{bmatrix}$$

Similarly, a matrix $A \in \mathfrak{H}(\mathbb{H}^r)$ will be represented as a $4r \times 4r$ block-matrix

$$\begin{bmatrix} Q(A_{11}) & \dots & Q(A_{1r}) \\ & \ddots & \\ Q(A_{1r})^T & \dots & Q(A_{rr}) \end{bmatrix}.$$

It turns out that these representations are the only irreducible ones, up to the choice of an orthonormal basis.

Theorem 15.7. *Up to an orthogonal conjugation the following representations* $\Phi: \mathfrak{J} \to \mathrm{Sym}(\mathbb{R}^n)$ *are the unique irreducible representations of the indicated simple formally real Jordan algebras* $\mathfrak{J}$.

(a) $\mathfrak{J} = \mathrm{Sym}(\mathbb{R}^r)$, $r \geqslant 1$. *Then* $n = r$ *and* $\Phi(J) = J$, $J \in \mathrm{Sym}(\mathbb{R}^r)$;
(b) $\mathfrak{J} = \mathfrak{H}(\mathbb{C}^r)$, $r \geqslant 2$. *Then* $n = 2r$ *and* $\mathfrak{H}(\mathbb{C}^r) \ni J \mapsto \Phi(J) = (\varphi(J_{\alpha\beta}))$;
(c) $\mathfrak{J} = \mathfrak{H}(\mathbb{H}^r)$, $r \geqslant 3$. *Then* $n = 4r$ *and* $\mathfrak{H}(\mathbb{H}^r) \ni J \mapsto \Phi(J) = (Q(J_{\alpha\beta}))$.

It remains to describe the irreducible representations of the spin factors $\mathfrak{S}_N$, $N \geqslant 5$.

Theorem 15.8. *Suppose* $\mathfrak{J} = \mathfrak{S}_N$, $N \geqslant 3$. *Then it has a unique (up to an orthogonal conjugation) irreducible representation* $\Phi: \mathfrak{J} \to \mathrm{Sym}(\mathbb{R}^{2d(N-2)})$, *unless* $N = 2 \bmod 4$, *where the numbers* $d(p)$ *are defined in (15.13) below. If* $N = 2 \bmod 4$, *then* $\mathfrak{J}$ *has exactly two distinct (up to an orthogonal conjugation) irreducible representations, related to one other by an involutory Jordan algebra automorphism* $T: \Pi_{\mathfrak{S}_N} \to \Pi_{\mathfrak{S}_N}$. *Specifically,*

$$\Pi_{\mathfrak{S}_N} = \left\{ \begin{bmatrix} \lambda I_{d(N-2)} & A \\ A^T & \mu I_{d(N-2)} \end{bmatrix} : \{\lambda, \mu\} \subset \mathbb{R}, A \in \mathscr{U}_{N-2} \right\}, \tag{15.8}$$

and

$$T\begin{bmatrix} \lambda I_{d(N-2)} & A \\ A^T & \mu I_{d(N-2)} \end{bmatrix} = \begin{bmatrix} \lambda I_{d(N-2)} & A^T \\ A & \mu I_{d(N-2)} \end{bmatrix}. \tag{15.9}$$

The p-dimensional subspaces $\mathscr{U}_p \subset \mathrm{End}(\mathbb{R}^{d(p)})$ *have the following structure*

$$\mathscr{U}_p = \mathbb{R} I_{d(p)} \oplus \mathscr{W}_p,$$

where the subspaces $\mathscr{W}_p \subset \mathrm{Skew}(\mathbb{R}^{d(p)})$, $p \geqslant 9$ *are defined recursively, as follows*

$$\mathscr{W}_p = \left\{ \begin{bmatrix} I_8 \otimes A & \mathbb{O}(q, h) \otimes I_{d(p-8)} \\ -\mathbb{O}(q, h)^T \otimes I_{d(p-8)} & -I_8 \otimes A \end{bmatrix} : A \in \mathscr{W}_{p-8}, \{q, h\} \subset \mathbb{H} \right\}, \tag{15.10}$$

where the spaces $\mathscr{W}_1, \ldots \mathscr{W}_8$ *are given in list 15.9 and* $\mathbb{O}(q, h)$ *is defined in (15.11) below.*

For $m_1 \times n_1$ matrix A and $m_2 \times n_2$ matrix B the tensor product notation $A \otimes B$ denotes $m_1 m_2 \times n_1 n_2$ matrix written in block-form as

$$A \otimes B = \begin{bmatrix} A_{11}B & \ldots & A_{1n_1}B \\ & \ddots & \\ A_{m_1 1}B & \ldots & A_{m_1 n_1}B \end{bmatrix}.$$

For $\{q, h\} \subset \mathbb{H}$ we define

$$\mathbb{O}(q, h) = \begin{bmatrix} Q(q) & \hat{Q}(h) \\ -\hat{Q}(\bar{h}) & Q(\bar{q}) \end{bmatrix}, \tag{15.11}$$

where the map Q was defined in (15.6) and

$$\hat{Q}(q_0 + iq_1 + jq_2 + kq_3) = \begin{bmatrix} \varphi(q_0 + iq_1) & \varphi(q_2 + iq_3) \\ -\varphi(q_2 - iq_3) & \varphi(q_0 - iq_1) \end{bmatrix}, \tag{15.12}$$

We note that the maps $\hat{Q}$ and $\mathbb{O}$ satisfy

$$\hat{Q}(q_1)\hat{Q}(q_2) = \hat{Q}(q_1 q_2), \quad \hat{Q}(q)^T = \hat{Q}(\bar{q}), \quad \hat{Q}(q_1)Q(q_2) = Q(q_2)\hat{Q}(q_1),$$

$$\mathbb{O}(q, h)^T = \mathbb{O}(\bar{q}, -h), \quad \mathbb{O}(q, h)\mathbb{O}(q, h)^T = (|q|^2 + |h|^2)I_8.$$

With this notation we can describe the first 8 subspaces $\mathscr{W}_p$ and $\mathscr{U}_p$:

List 15.9.

- $p = 1$, $\mathscr{W}_1 = \{0\} \subset \mathbb{R}$, $\mathscr{U}_1 = \mathbb{R}$
- $p = 2$,

$$\mathscr{W}_2 = \mathbb{R}\varphi(i) = \mathrm{Skew}(\mathbb{R}^2),$$

$$\mathscr{U}_2 = \{\varphi(z)\colon z \in \mathbb{C}\}.$$

- $p = 3$,

$$\mathscr{W}_3 = \{Q(q)\colon q \in \mathbb{H},\ q = q_1 i + q_2 j\}.$$

$$\mathscr{U}_3 = \{Q(q)\colon q \in \mathbb{H},\ q = q_0 + q_1 i + q_2 j\}.$$

- $p = 4$,

$$\mathscr{W}_4 = \{Q(q)\colon q \in \mathbb{H},\ \mathfrak{R}(q) = 0\},$$

$$\mathscr{U}_4 = \{Q(q)\colon q \in \mathbb{H}\}.$$

- $p = 5$,

$$\mathscr{W}_5 = \{\mathbb{O}(0, h)\colon h \in \mathbb{H}\},$$

$$\mathscr{U}_5 = \{\mathbb{O}(\lambda, h)\colon h \in \mathbb{H},\ \lambda \in \mathbb{R}\}.$$

- $p = 6$,

$$\mathscr{W}_6 = \{\mathbb{O}(q, h)\colon \{q, h\} \subset \mathbb{H},\ \mathfrak{R}(h) = 0,\ q = q_1 i + q_2 j\}.$$

$$\mathscr{U}_6 = \{\mathbb{O}(q, h)\colon \{q, h\} \subset \mathbb{H},\ \mathfrak{R}(h) = 0,\ q = q_0 + q_1 i + q_2 j\}.$$

- $p = 7$,

$$\mathscr{W}_7 = \{\mathbb{O}(q, h) \colon \{q, h\} \subset \mathbb{H}, \mathfrak{R}(q) = 0, \mathfrak{R}(h) = 0\},$$

$$\mathscr{U}_7 = \{\mathbb{O}(q, h) \colon \{q, h\} \subset \mathbb{H}, \mathfrak{R}(h) = 0\}.$$

- $p = 8$,

$$\mathscr{W}_8 = \{\mathbb{O}(q, h) \colon \{q, h\} \subset \mathbb{H}, \mathfrak{R}(q) = 0\},$$

$$\mathscr{U}_8 = \{\mathbb{O}(q, h) \colon \{q, h\} \subset \mathbb{H}\}.$$

The recursive formula (15.10) and the list above show that the dimensions $d(p)$ used in theorem 15.8 can also be defined recursively by

$$d(p + 8) = 16d(p), \tag{15.13}$$

while

$$d(1) = 1,\ d(2) = 2,\ d(3) = d(4) = 4,\ d(5) = d(6) = d(7) = d(8) = 8.$$

15.4 Reflexivity of Jordan subalgebras of Sym($\mathbb{R}^n$)

By theorem 15.2, the reflexivity of a Jordan subalgebras of Sym($\mathbb{R}^n$) depends on the reflexivity of each independent block in (15.3), i.e., on the reflexivity of the representation of a simple formally real Jordan algebra. The reflexivity of the unique representations of Sym($\mathbb{R}^r$), $\mathfrak{H}(\mathbb{C}^r)$ and $\mathfrak{H}(\mathbb{H}^r)$ from theorem 15.7 are obvious, since the images of these Jordan algebras are exactly the set of all symmetric matrices in the images of their minimal associative algebras $\mathrm{End}_{\mathbb{R}}(\mathbb{R}^r)$, $\mathrm{End}_{\mathbb{C}}(\mathbb{C}^r)$ and $\mathrm{End}_{\mathbb{H}}(\mathbb{H}^r)$ in $\mathrm{End}_{\mathbb{R}}(\mathbb{R}^r)$, $\mathrm{End}_{\mathbb{R}}(\mathbb{R}^{2r})$, and $\mathrm{End}_{\mathbb{R}}(\mathbb{R}^{4r})$, respectively. One can also check directly that none of the representations of $\mathfrak{S}_5$ and $\mathfrak{S}_N$, $n \geqslant 7$ are reflexive. The minimal associative algebras containing the images of $\mathfrak{S}_N$ are Clifford algebras, represented by $2^N \times 2^N$ matrices, that contain a much larger subset of symmetric matrices than the images of the generators s_j of $\mathfrak{S}_N$. The algebra $\mathfrak{S}_6 = \mathfrak{H}(\mathbb{H}^2)$ is special. Both irreducible representations of $\mathfrak{S}_6$ map into one and the same reflexive subalgebra $\mathfrak{H}(\mathbb{H}^2)$ of Sym($\mathbb{R}^8$). However, the subalgebra

$$\Pi = \left\{ \begin{bmatrix} A & 0 \\ 0 & TA \end{bmatrix} : A \in \mathfrak{H}(\mathbb{H}^2) \right\} \subset \mathrm{Sym}(\mathbb{R}^{16})$$

isomorphic to $\mathfrak{S}_6$ is not reflexive, because T does not satisfy the 4-chain property

$$\begin{aligned} T(A_1A_2A_3A_4 + A_4A_3A_2A_1) = {} & (TA_1)(TA_2)(TA_3)(TA_4) \\ & + (TA_4)(TA_3)(TA_2)(TA_1). \end{aligned} \tag{15.14}$$

for all $\{A_1, A_2, A_3, A_4\} \subset \Pi_{\mathfrak{S}_N}$, $N \geqslant 5$. We conclude that the smallest nonreflexive Jordan subalgebra of Sym($\mathbb{R}^n$) is $\Pi_{\mathfrak{S}_5} \subset \mathrm{Sym}(\mathbb{R}^8)$. According to (15.8) and list 15.9,

$$\Pi_{\mathfrak{S}_5} = \left\{ \begin{bmatrix} \lambda I_4 & Q(q) \\ Q(\bar{q}) & \mu I_4 \end{bmatrix} : \{\lambda, \mu\} \subset \mathbb{R},\ q = q_0 + q_1 i + q_2 j \right\}.$$

The smallest associative algebra containing $\Pi_{\mathfrak{S}_5}$ is $\mathrm{End}_{\mathbb{H}}(\mathbb{H}^2) \subset \mathrm{End}(\mathbb{R}^8)$, while $\mathrm{End}_{\mathbb{H}}(\mathbb{H}^2) \cap \mathrm{Sym}(\mathbb{R}^8) = \mathfrak{H}(\mathbb{H}^2)$, which contains $\Pi_{\mathfrak{S}_5}$ but is 6-dimensional, while $\dim \Pi_{\mathfrak{S}_5} = 5$.

References

[1] Albert A A 1934 On a certain algebra of quantum mechanics *Ann. Math.* **35** 65–73

[2] Clerc J-L 1992 Représentation d'une algèbre de Jordan, polynômes invariants et harmoniques de Stiefel *J. Angew. Math.* **423** 47–72

[3] Jordan P, Neumann J V and Wigner E 1934 On an algebraic generalization of the quantum mechanical formalism *Ann. Math.* **35** 29–64

[4] Koecher M, Krieg A and Walcher S 1999 *The Minnesota Notes on Jordan Algebras and Their Applications* (Berlin: Springer)

[5] McCrimmon K 2004 *A Taste of Jordan Algebras* (Berlin: Springer)

Chapter 16

A polycrystalline L-relation that is not exact

In this section we describe the construction of a polycrystalline L-relation that is not exact from [1]. This example has important implications in Calculus of Variations. However, this discussion is not directly relevant to the subject of the book, and we omit it here.

Let us recall that for two-dimensional composites theorem 4.13 states that 3-chain property (see Definition 4.7) must be satisfied by all Jordan $\mathscr{A}$-multialgebras corresponding to exact relations. In this chapter we construct a rotationally invariant Jordan $\mathscr{A}$-multialgebra that fails 3-chain property, and hence leads to an explicit example of an L-relation that is not exact. The example is found in the context of two-dimensional multifield response composite materials [3–5], coupling four curl-free fields $\boldsymbol{E} = (\nabla\phi_1, \dots, \nabla\phi_4)$ and four divergence-free fields $\boldsymbol{J} = (\boldsymbol{j}_1, \dots, \boldsymbol{j}_4)$. In this case

$$\mathscr{T} = \underbrace{\mathbb{R}^2 \oplus \cdots \oplus \mathbb{R}^2}_{4} = \mathbb{R}^4 \otimes \mathbb{R}^2$$

can be interpreted in two very different ways. On the one hand we can identify $\mathbb{R}^2$ with the set of complex numbers, so that $\mathscr{T} \cong \mathbb{C}^4$. This representation of the space $\mathscr{T}$ makes it convenient to describe the action of rotations. If $\boldsymbol{R}_\theta$ denotes the counterclockwise rotation of the plane $\mathbb{R}^2$ through the angle θ, then

$$\boldsymbol{R}_\theta \cdot \boldsymbol{u} = e^{i\theta}\boldsymbol{u}, \qquad \boldsymbol{u} \in \mathscr{T} \cong \mathbb{C}^4.$$

Every $\mathsf{K} \in \mathrm{Sym}(\mathscr{T})$ is uniquely determined by a complex Hermitian 4×4 matrix X and a complex symmetric 4×4 matrix Y by the rule

$$\mathsf{K}\boldsymbol{u} = X\boldsymbol{u} + Y\bar{\boldsymbol{u}}, \qquad \boldsymbol{u} \in \mathbb{C}^4.$$

Henceforth we will write $\mathsf{K}(X, Y)$ to indicate this parameterization of $\mathrm{Sym}(\mathscr{T})$. We easily compute the action of rotations $\boldsymbol{R}_\theta$ on $\mathsf{K}(X, Y)$:

$$\boldsymbol{R}_\theta \cdot \mathsf{K}(X, Y) = \mathsf{K}(X, e^{2i\theta}Y). \tag{16.1}$$

doi:10.1088/978-0-7503-6249-8ch16

Therefore, if Π is an SO(2) invariant subspace of $\mathrm{Sym}(\mathscr{T})$ then

$$\Pi = \Pi_{V,W} = \{(X,\, Y)\colon X \in V \subset \mathscr{H}(\mathbb{C}^4),\ Y \in W \subset \mathrm{Sym}(\mathbb{C}^4)\},$$

where V can be any real subspace of $\mathscr{H}(\mathbb{C}^4)$ the set of all complex Hermitian 4×4 matrices, and W can be any complex subspace of $\mathrm{Sym}(\mathbb{C}^4)$ the set of all complex symmetric 4×4 matrices. In this notation, the subspace $\mathscr{A}$ is described by $V = \{0\}$ and $W = \mathbb{C}\boldsymbol{I}_4$. The condition that the subspace $\Pi_{V,W}$ is an SO(2)-invariant Jordan $\mathscr{A}$-multialgebra is equivalent to the conditions

$$Y^2 + XX^T \in W, \quad YX + XY^* \in V \text{ for all } X \in V,\ Y \in W, \tag{16.2}$$

where $Y^* = \bar{Y}^T$ denotes Hermitian conjugation. Finally, the 3-chain condition is equivalent to

$$iX_1X_2^TX_3 + (iX_1X_2^TX_3)^* \in V, \quad \forall\{X_1,\, X_2,\, X_3\} \subset V, \tag{16.3}$$

provided (16.2) holds as well.

There is an alternative interpretation of the space $\mathscr{T} = \mathbb{R}^4 \times \mathbb{R}^2$, whereby we identify $\mathbb{R}^4$ with the set of quaternions $\mathbb{H}$, so that $\mathscr{T} \cong \mathbb{H}^2$. The second, quaternionic representation of $\mathscr{T}$ permits us to think of quaternionic multiplication as a real linear transformation on $\mathbb{R}^4$. For $q \in \mathbb{H}$ we denote by $Q(q)$ a linear transformation

$$\mathbb{R}^4 \ni \boldsymbol{h} \mapsto h \in \mathbb{H} \mapsto qh \in \mathbb{H} \mapsto Q(q)\boldsymbol{h} \in \mathbb{R}^4.$$

In this interpretation $Q(q)$ can be written as a 4×4 real matrix. If we now interpret this matrix as a *complex* 4×4 matrix, whose entries just happen to be real, we will obtain the map $\mathfrak{Q}\colon \mathbb{H} \to \mathrm{End}_{\mathbb{C}}(\mathbb{C}^4) \subset \mathrm{End}_{\mathbb{R}}(\mathscr{T})$ that allows us to think of quaternions as real linear maps on $\mathscr{T}$. It is easy to see that $\mathfrak{Q}(q) = \mathsf{K}(Q(q),\, 0)$. Inspired by the nonreflexive representations of spin factors $\mathfrak{S}_N$ discussed in chapter 15, specifically the cases $p = 3$ and 4 in the list 15.9, we define

$$V = \{i\mathfrak{Q}(q)\colon q \in \mathbb{H},\ \mathfrak{R}(q) = 0\} \subset \mathfrak{H}(\mathbb{C}^4), \quad W = \mathbb{C}\boldsymbol{I}_4 \subset \mathrm{Sym}(\mathbb{C}^4) \tag{16.4}$$

It is not hard to verify that $\Pi_{V,W}$ is an SO(2)-invariant Jordan $\mathscr{A}$-multialgebra that does not satisfy (16.3). The corresponding submanifold $\mathbb{M}$, given by (4.4), consists of 'quaternionic materials' with constant determinant:

$$\mathbb{M} = \left\{\mathsf{L} = \begin{bmatrix} \lambda & h \\ \bar{h} & \mu \end{bmatrix} : \lambda > 0,\ \mu > 0,\ h \in \mathbb{H},\ \det_{\mathbb{H}}(\mathsf{L}) = \lambda\mu - |h|^2 = 1\right\}. \tag{16.5}$$

Here we use the quaternionic representation of $\mathscr{T}$ to describe operators in $\mathrm{Sym}(\mathscr{T})$. Of course, $\det_{\mathbb{H}}(\mathsf{L})$ is set to 1 in (16.5) only for simplicity. It can be any positive constant.

Remark 16.1. *Curiously, in three space dimensions we have a positive result, proved in* [2], *that all L-relations for 3D multifield response polycrystalline composites are exact no matter how many fields are coupled.*

References

[1] Grabovsky Y 2018 From microstructure-independent formulas for composite materials to rank-one convex, non-quasiconvex functions *Arch. Ration. Mech. Anal.* **227** 607–36

[2] Grabovsky Y, Milton G W and Sage D S 2000 Exact relations for effective tensors of polycrystals: necessary conditions and sufficient conditions *Commun. Pure Appl. Math.* **53** 300–53

[3] Milgrom M 1997 Some more exact results concerning multifield moduli of two-phase composites *J. Mech. Phys. Solids* **45** 399–404

[4] Milgrom M and Shtrikman S 1989 Linear response of polycrystals to coupled fields: exact relations among the coefficients *Phys. Rev.* **40** 5991–4

[5] Milgrom M and Shtrikman S 1989 Linear response of two-phase composites with cross moduli: exact universal relations *Phys. Rev.* **40** 1568–75

IOP Publishing

Composite Materials (Second Edition)
Mathematical theory and exact relations
Yury Grabovsky

Chapter 17

Multiplication of SO(3) irreps in endomorphism algebras

17.1 Irreducible representations (irreps) of SO(3)

Here we give an explicit construction of all irreducible representations of $SO(3)$. We follow the exposition in [2] but provide more details. Since the theory is classical and very well known in quantum mechanics as quantization of angular momentum, we limit ourselves to statements of fact without proof or discussion. These can be found in numerous textbooks on both group representations (for mathematicians) and quantum mechanics (for physicists).

The main idea to deal with complicated Lie groups like $SO(3)$ is to replace them with geometrically simpler groups, called universal covers. For $SO(3)$, which is topologically a three-dimensional projective space, the universal cover is $SU(2)$

$$SU(2) = \left\{ \begin{bmatrix} a & b \\ -\bar{b} & \bar{a} \end{bmatrix} : \{a, b\} \subset \mathbb{C}, |a|^2 + |b|^2 = 1 \right\},$$

which is isomorphic to a three-dimensional unit sphere. The complex SU(2) irreps are spaces $\mathrm{Sym}^m(\mathbb{C}^2)$, where $g \in SU(2)$ acts on the commutative and associative formal product of any m vectors $\boldsymbol{u}_1 \dots \boldsymbol{u}_m$ by

$$g \cdot \boldsymbol{u}_1 \dots \boldsymbol{u}_m = (g\boldsymbol{u}_1) \dots (g\boldsymbol{u}_m).$$

A basis of $\mathrm{Sym}^m(\mathbb{C}^2)$ is $\{\boldsymbol{e}_1^j \boldsymbol{e}_2^{m-j} : j = 0, \dots, m\}$. For $\boldsymbol{x} \in \mathbb{R}^3$ we define

$$\rho(\boldsymbol{x}) = \begin{bmatrix} ix_1 & x_2 + ix_3 \\ -x_2 + ix_3 & -ix_1 \end{bmatrix}.$$

We define $\pi(g) \in SO(3)$ by its action on $\boldsymbol{x} \in \mathbb{R}^3$:

$$\pi(g)\boldsymbol{x} = \rho^{-1}(g\rho(\boldsymbol{x})g^{-1}).$$

doi:10.1088/978-0-7503-6249-8ch17

We note that $\pi : SU(2) \to SO(3)$ is a double cover[1], whose kernel is $\pm \boldsymbol{I}_2$. Thus, any irrep of SU(2) where $-\boldsymbol{I}_2$ acts trivially will also be an irrep of SO(3). Hence SO(3) irreps are $\mathrm{Sym}^m(\mathbb{C}^2)$ for even m.

The subgroup

$$T = \left\{ \begin{bmatrix} e^{i\theta} & 0 \\ 0 & e^{-i\theta} \end{bmatrix} : \theta \in \mathbb{R} \bmod 2\pi \right\}$$

is a maximal commutative subgroup of SU(2). It is called a maximal torus. Observe that the 'standard' basis vectors $\boldsymbol{e}_1^j \boldsymbol{e}_2^{m-j}$ are eigenvectors of all $g \in T$. If

$$g_\theta = \begin{bmatrix} e^{i\theta} & 0 \\ 0 & e^{-i\theta} \end{bmatrix}$$

then $g_\theta \boldsymbol{e}_1 = e^{i\theta}\boldsymbol{e}_1$ and $g_\theta \boldsymbol{e}_2 = e^{-i\theta}\boldsymbol{e}_2$. Therefore,

$$g_\theta \cdot \boldsymbol{e}_1^j \boldsymbol{e}_2^{m-j} = e^{(2j-m)i\theta} \boldsymbol{e}_1^j \boldsymbol{e}_2^{m-j}.$$

The common eigenvectors of a maximal torus are called the weight vectors. We define the weight of $\boldsymbol{e}_1^j \boldsymbol{e}_2^{m-j}$ to be $s = m/2 - j$, $j = 0, \ldots, m$. In other words, we denote

$$v_s = \boldsymbol{e}_1^{m/2+s} \boldsymbol{e}_2^{m/2-s}, \quad s = -m/2, -m/2+1, \ldots, m/2-1, m/2$$

the weight vectors of weight s. When m is even the weights are integers. From now on we will consider only even m, i.e., we will only deal with SO(3) irreps. It turns out that all SO(3) irreps are of real type. That means that there exists a basis of $\mathrm{Sym}^m(\mathbb{C}^2)$ in which all matrices of operators $\pi(g)$ are real. We denote the real SO(3) irreps by W_k, $k = 0, 1, 2, \ldots$ so that $W_k \otimes \mathbb{C} \cong \mathrm{Sym}^{2k}(\mathbb{C}^2)$. Our goal will be to build a basis of W_k and $W_k \otimes \mathbb{C}$ that is independent of the specific model of irreps. In other words, we assume that the action of SO(3) on W_k is known and that a specific maximal torus $T \subset SO(3)$ has been chosen. To fix notation we choose this maximal torus to be the group of all rotations around x_1 axis. A natural choice of basis of $W_k \otimes \mathbb{C}$ would be the eigenvectors of T (weight vectors). However, the eigenvectors are defined up to complex multiples, which will also have to be chosen. We start by considering the zero weight vector w_0: a fixed point of T. This vector can be chosen to lie in W_k, i.e., to be real, since its eigenvalue is 1. Of course, w_0 is still defined up to a real scalar multiple. However, once a specific choice of w_0 is made, the bases of W_k and $W_k \otimes \mathbb{C}$ will be constructed in a canonical way.

17.2 Raising and lowering operators

We recall that every representation of a Lie group[2] defines an action of its Lie algebra as a differentiation. Recall that elements of a Lie algebra are tangent vectors

[1] A continuous map, for which the preimage of any point consists of exactly two points.

[2] A Lie group is a group and a manifold at the same time, where we require that group multiplication be continuous.

$A = \dot{g}(0)$ to curves $g(t) \in SU(2)$, such that $g(0) = I_2$. A tangent vector A can then be regarded as an operator on W_k defined by its action

$$A \cdot w = \left. \frac{d(\pi(g(t))w)}{dt} \right|_{t=0}.$$

We consider the following special curves in SU(2):

$$g_0(t) = \begin{bmatrix} e^{it} & 0 \\ 0 & e^{-it} \end{bmatrix}, \quad g_1(t) = \frac{1}{\sqrt{1+t^2}} \begin{bmatrix} 1 & t \\ -t & 1 \end{bmatrix}, \quad g_2(t) = \frac{1}{\sqrt{1+t^2}} \begin{bmatrix} 1 & it \\ it & 1 \end{bmatrix}.$$

The corresponding operators A will be called Z_0, Z_1 and Z_2, respectively. We easily compute

$$Z_0 = \dot{g}_0(0) = \begin{bmatrix} i & 0 \\ 0 & -i \end{bmatrix}, \quad Z_1 = \dot{g}_1(0) = \begin{bmatrix} 0 & 1 \\ -1 & 0 \end{bmatrix}, \quad Z_2 = \dot{g}_2(0) = \begin{bmatrix} 0 & i \\ i & 0 \end{bmatrix}$$

Returning to our model of $W_k \otimes \mathbb{C}$ we compute (for $A = \dot{g}(0)$)

$$A \cdot v_s = \frac{d}{dt}[(g(t)e_1)^{k+s}(g(t)e_2)^{k-s}] = (k+s)e_1^{k+s-1}e_2^{k-s}(Ae_1) + (k-s)e_1^{k+s}e_2^{k-s-1}(Ae_2)$$

The result is linear in A, and hence this formula will also define the action of any complex linear combination of Z_0, Z_1 and Z_2. We define

$$H = -iZ_0 = \begin{bmatrix} 1 & 0 \\ 0 & -1 \end{bmatrix}, \quad X_+ = \frac{Z_1 - iZ_2}{2} = \begin{bmatrix} 0 & 1 \\ 0 & 0 \end{bmatrix}, \quad X_- = \frac{Z_1 + iZ_2}{2} = \begin{bmatrix} 0 & 0 \\ -1 & 0 \end{bmatrix}.$$

Using the relations

$$He_1 = e_1, \quad He_2 = -e_1, \quad X_+e_1 = 0, \quad X_+e_2 = e_1, \quad X_-e_1 = -e_2, \quad X_-e_2 = 0$$

we obtain

$$H \cdot v_s = (k+s)e_1^{k-s}e_2^{k+s} - (k-s)e_1^{k-s}e_2^{k+s} = 2sv_s.$$

$$X_+ \cdot v_s = (k-s)e_1^{k+s+1}e_2^{k-s-1} = (k-s)v_{s+1}$$

$$X_- \cdot v_s = -(k+s)e_1^{k+s-1}e_2^{k-s+1} = -(k+s)v_{s-1}$$

For this reason operators X_+ and X_- are called the raising and lowering operators, respectively. Thus, if $w_0 \in W_k$ is the zero weight vector then we can generate the canonical basis $\{v_j: -k \leq j \leq k\}$ of $W_k \otimes \mathbb{C}$ defining recursively

$$v_{j+1} = \frac{X_+ \cdot v_j}{k-j}, \quad v_{-j-1} = -\frac{X_- \cdot v_{-j}}{k-j}, \quad j = 0, 1, \ldots, k-1,$$

where $v_0 = w_0$. In other words,

$$v_j = \frac{(k-j)!}{k!} X_+^j \cdot w_0, \quad v_{-j} = (-1)^j \frac{(k-j)!}{k!} X_-^j \cdot w_0, \quad j = 1, \ldots, k.$$

Taking complex conjugates we obtain

$$\overline{v_j} = (-1)^j v_{-j}.$$

It follows that $w_j = \Re(v_j)$ and $w_{-j} = \Im(v_j)$, $j = 1, \dots, k$, together with w_0 forms the basis of the real vector space W_k.

It is known that there exists a unique, up to a positive scalar multiple, inner product on $W_k \otimes \mathbb{C}$. Let $\{u, v\} \subset W_k \otimes \mathbb{C}$ and let $g(t)$ be a smooth curve in SU(2) with $g(0) = I_2$. Differentiating $(g(t) \cdot u, g(t) \cdot v) = (u, v)$ at $t = 0$ we obtain

$$(A \cdot u, v) + (u, A \cdot v) = 0$$

(This means that A is a skew-adjoint operator with respect to such an inner product.) In particular, for $A = Z_0$ we have

$$0 = (Z_0 v_j, v_s) + (v_j, Z_0 v_s) = (2ijv_j, v_s) + (v_j, 2isv_s) = 2i(j - s)(v_j, v_s).$$

It follows that the weight basis $\{v_j\colon -k \leq j \leq k\}$ is orthogonal. We compute

$$Z_1 \cdot v_j = X_+ \cdot v_j + X_- \cdot v_j = (k - j)v_{j+1} - (k + j)v_{j-1}.$$

Similarly,

$$Z_2 \cdot v_j = -i(X_- - X_+) \cdot v_j = i(k + j)v_{j-1} + i(k - j)v_{j+1}.$$

Using these formulas and the skew-adjointness of Z_1 and Z_2 we obtain

$$\left\|v_j\right\|^2 = \frac{n + j}{n - j + 1}\left\|v_{j-1}\right\|^2,$$

from which we obtain the values of the norms of v_j:

$$\|v_j\|^2 = \frac{\binom{2k}{k}}{\binom{2k}{k + j}}\|w_0\|^2, \ -k \leqslant j \leqslant k. \tag{17.1}$$

This formula shows, in particular, that $\|v_j\| = \|v_{-j}\|$. It is then easy to check (via formulas)

$$w_j = \frac{v_j + (-1)^j v_{-j}}{2}, \qquad w_{-j} = \frac{v_j - (-1)^j v_{-j}}{2i}, \quad j = 1, \dots, k,$$

that $\{w_j\}$ is an orthogonal system. We compute

$$\|w_j\|^2 = \|w_{-j}\|^2 = \frac{\|v_j\|^2}{2}, \quad j = 1, \dots, k.$$

17.3 Homogeneous coordinates of irreps in endomorphism algebras

From now on we follow the exposition in [1], while giving more details on topics that do not occur in the classical theory of quantization of angular momentum.

Let

$$V = \mathbb{R}^p \otimes (W_0 \oplus W_1 \oplus \cdots \oplus W_n).$$

Then

$$\mathrm{End}(V) = \mathrm{End}(\mathbb{R}^p) \otimes \bigoplus_{j,k=0}^{n} \mathrm{Hom}(W_j, W_k).$$

Recall that $\mathrm{Hom}(W_j, W_k) \cong W_k \otimes W_j$. The decomposition of a tensor product of irreps into a direct sum of irreps is called the Clebsch–Gordan decomposition in representation theory. For SO(3)

$$W_k \otimes W_j \cong \bigoplus_{\alpha=|j-k|}^{j+k} W_\alpha. \tag{17.2}$$

Let $0 \le \alpha \le 2n$. Formula (17.2) shows that for each ordered pair (j, k), such that $|j-k| \le \alpha \le j+k$ there is a unique copy of W_α in $\mathrm{Hom}(W_j, W_k)$, which contains a unique, up to a scalar multiple zero weight vector. Suppose that we have made a choice of a real zero weight vector $\mathsf{E}^{jk\alpha}$ for every such $W_\alpha \subset \mathrm{Hom}(W_j, W_k)$. Then any zero weight vector of a copy of $W_\alpha \subset \mathrm{End}(V)$ can be written as

$$E^\alpha(\mathbb{X}) = \sum_{j,k} \boldsymbol{X}_{jk} \otimes \mathsf{E}^{jk\alpha}$$

for some $p \times p$ matrices $\boldsymbol{X}_{jk} \in \mathrm{End}(\mathbb{R}^p)$. Clearly, $E^\alpha(\mathbb{X})$ and $E^\alpha(\mathbb{Y})$ determine the same $W_\alpha \subset \mathrm{End}(V)$ if and only if $\mathbb{Y}$ is a scalar multiple of $\mathbb{X}$. For this reason the block-matrix $\mathbb{X}$ will be called the matrix of homogeneous coordinates of $W_\alpha \subset \mathrm{End}(V)$. We will denote this copy of W_α by $W_\alpha(\mathbb{X})$.

17.4 Clebsch–Gordan coefficients

In this section we apply the classical theory of preceding sections and derive the classical decomposition of the tensor product of two irreps into a direct sum of irreps. We give here all the details because we will be applying the same calculation to the matrix product of two irreps.

Let $\{v_A^j: -j \le A \le j\}$ be the canonical basis in $W_j \otimes \mathbb{C}$. It is clear that operators $v_{B-A}^k \otimes v_A^j$ are weight vectors of weight B. Then there are explicitly given real numbers $C_{A,\,B}^{jk\alpha}$ such that

$$e_B(jk\alpha) = \sum_A C_{A,\,B}^{jk\alpha} v_{B-A}^k \otimes v_A^j$$

form a canonical basis of $W_\alpha \otimes \mathbb{C} \subset \mathrm{Hom}(W_j, W_k) \otimes \mathbb{C}$.

In order to find the coefficients $C_{A,\,B}^{jk\alpha}$ (called the Clebsch–Gordan coefficients) we can first consider the highest weight vector

$$e_{\alpha}(jk\alpha) = \sum_{A=\alpha-k}^{j} C^{jk\alpha}_{A,\,\alpha} v^{k}_{\alpha-A} \otimes v^{j}_{A}.$$

It has the property that it will be killed by X_+:

$$0 = X_+ \cdot e_{\alpha}(jk\alpha) = \sum_{A=\alpha-k}^{j} C^{jk\alpha}_{A,\,\alpha}\{(k-\alpha+A)v^{k}_{\alpha-A+1} \otimes v^{j}_{A} + (j-A)v^{k}_{\alpha-A} \otimes v^{j}_{A+1}\}$$

Therefore,

$$\sum_{A=\alpha-k+1}^{j} \{C^{jk\alpha}_{A-1,\,\alpha}(j-A+1) + (k-\alpha+A)C^{jk\alpha}_{A,\,\alpha}\}v^{k}_{\alpha-A+1} \otimes v^{j}_{A} = 0.$$

This implies that

$$C^{jk\alpha}_{A,\,\alpha} = -\frac{j-A+1}{k-\alpha+A} C^{jk\alpha}_{A-1,\,\alpha}.$$

Hence,

$$C^{jk\alpha}_{A,\,\alpha} = (-1)^{A+k-\alpha}\frac{(j-A+1)\ldots(j+k-\alpha)}{(k-\alpha+A)!}C^{jk\alpha}_{\alpha-k,\,\alpha} = (-1)^{A+k-\alpha}\binom{j+k-\alpha}{j-A}C^{jk\alpha}_{\alpha-k,\,\alpha},$$

and

$$e_{\alpha}(jk\alpha) = C^{jk\alpha}_{\alpha-k,\,\alpha} \sum_{A=\alpha-k}^{j} (-1)^{A+k-\alpha}\binom{j+k-\alpha}{j-A} v^{k}_{\alpha-A} \otimes v^{j}_{A}.$$

The highest weight vector is defined up to a scalar multiple (dependent on j, k and α), hence, changing index of summation $a = A - \alpha + k$ we obtain

$$e_{\alpha}(jk\alpha) = \sum_{a=0}^{j+k-\alpha} (-1)^{a}\binom{j+k-\alpha}{a} v^{k}_{k-a} \otimes v^{j}_{\alpha-k+a}.$$

We now apply the lowering operator X_- s times, using the formula

$$X^{s}_{-} \cdot v^{k}_{A} = (-1)^{s}\binom{k+A}{s} s!\, v^{k}_{A-s}.$$

$$X^{s}_{-} \cdot e_{\alpha}(jk\alpha) = \sum_{a=0}^{j+k-\alpha} (-1)^{a}\binom{j+k-\alpha}{a}\sum_{\nu=0}^{s}\binom{s}{\nu} X^{\nu}_{-} \cdot v^{k}_{k-a} \otimes X^{s-\nu}_{-} \cdot v^{j}_{\alpha-k+a}.$$

Thus,

$$\binom{2\alpha}{s} e_{\alpha-s}(jk\alpha) = \sum_{a=0}^{j+k-\alpha} (-1)^{a}\binom{j+k-\alpha}{a}\sum_{\nu=0}^{s}\binom{2k-a}{\nu}\binom{\alpha+j-k+a}{s-\nu} v^{k}_{k-a-\nu} \otimes v^{j}_{\alpha-k+a-s+\nu}.$$

Introducing notation $B = \alpha - s$ and changing index of summation from ν to $A = k - a - \nu$ we obtain

$$\binom{2\alpha}{\alpha+B}e_B(jk\alpha) = \sum_{a=0}^{j+k-\alpha}(-1)^a\binom{j+k-\alpha}{a}\sum_{A=k-a+B-\alpha}^{k-a}\binom{2k-a}{k+A}\binom{\alpha+j-k+a}{j+B-A}v_A^k\otimes v_{B-A}^j.$$

We observe that if $A > k - a$, then $k + A > 2k - a$, so that $\binom{2k-a}{k+A} = 0$, while, if $A < k - a + B - \alpha$ then $j + B - A > j - k + a + \alpha$, so that $\binom{\alpha+j-k+a}{j+B-A} = 0$. In that case we can write

$$e_B(jk\alpha) = \frac{1}{\binom{2\alpha}{\alpha+B}}\sum_{a=0}^{j+k-\alpha}(-1)^a\binom{j+k-\alpha}{a}\sum_{A=B-j}^{k}\binom{2k-a}{k+A}\binom{\alpha+j-k+a}{j+B-A}v_A^k\otimes v_{B-A}^j,$$

since $0 \leq a \leq j + k - \alpha$. Hence,

$$e_B(jk\alpha) = \sum_{A=B-j}^{k} C_{A,B}^{jk\alpha}v_A^k\otimes v_{B-A}^j, \quad |B| \leqslant \alpha, \tag{17.3}$$

where

$$C_{A,B}^{jk\alpha} = \frac{1}{\binom{2\alpha}{\alpha+B}}\sum_{a=0}^{j+k-\alpha}(-1)^a\binom{j+k-\alpha}{a}\binom{2k-a}{k+A}\binom{\alpha+j-k+a}{j+B-A} \tag{17.4}$$

are the Clebsch–Gordan coefficients[3]. From (17.3) we obtain

$$e_0(jk\alpha) = \sum_{A=-\min(j,k)}^{\min(j,k)} C_{A,0}^{jk\alpha}v_A^k\otimes v_{-A}^j.$$

Taking complex conjugate and changing index of summation from A to $-A$ we obtain

$$\overline{e_0(jk\alpha)} = \sum_{A=-\min(j,k)}^{\min(j,k)} C_{-A,0}^{jk\alpha}v_A^k\otimes v_{-A}^j$$

Changing index of the other summation from $a' = j + k - \alpha - a$ in (17.4) we obtain

$$C_{-A,0}^{jk\alpha} = (-1)^{j+k-\alpha}C_{A,0}^{kj\alpha}.$$

Finally, we use a symmetry of Clebsch–Gordan coefficients $C_{A,0}^{jk\alpha} = C_{A,0}^{kj\alpha}$, and conclude that zero weight vectors $i^{j+k-\alpha}e_0(jk\alpha)$ have to be real.

[3] The Clebsch–Gordan coefficients used in quantum mechanics have different normalization. In a slight abuse of terminology we will call numbers (17.4) the Clebsch–Gordan coefficients.

17.5 Choice structure coefficients

Formula (17.3) gives a canonical choice of real zero weight vectors

$$i^{j+k-\alpha}e_0(jk\alpha) \in W_\alpha \subset \mathrm{Hom}(W_j, W_k).$$

However, in applications we usually describe such 'coordinate irreps' in terms natural for that specific application. For this reason, we usually have a natural choice of zero weight vectors within these irreps. We denoted these chosen zero weight vectors $\mathsf{E}^{jk\alpha}$ and used them to define the homogeneous coordinates for specifying an irrep $W_\alpha \subset \mathrm{End}(V)$. Thus, there exist *real* nonzero constants $K_{jk\alpha}$, such that

$$\mathsf{E}^{jk\alpha} = K_{jk\alpha} i^{j+k-\alpha}e_0(jk\alpha).$$

These constants depend not only on our choice of zero weight vectors $\mathsf{E}^{jk\alpha}$, but also on the choice of real zero weight vectors $w_0^k \in W_k \subset \mathscr{W}$, where

$$\mathscr{W} = W_0 \oplus W_1 \oplus \cdots \oplus W_n,$$

as well as on the specific isomorphisms $\mathrm{Hom}(W_j, W_k) \cong W_k \otimes W_j$. These isomorphisms are determined by a choice of SO(3)-invariant inner products on all $W_k \subset \mathscr{W}$ via the formula

$$(a \otimes b)u = (b, u)_{W_j}a, \quad a \in W_k, \{b, u\} \subset W_j.$$

We call constants $K_{jk\alpha}$ *choice structure coefficients*, since they depend on our choices of zero weight vectors and inner products. In order to compute them we compare the action of our zero weight vectors $\mathsf{E}^{jk\alpha}$ to the action of canonical zero weight vectors $i^{j+k-\alpha}e_0(jk\alpha)$, given by (17.3). In the case when $j + k - \alpha$ is even the numbers $C_{0,0}^{jk\alpha}$ are nonzero and hence,

$$\mathsf{E}^{jk\alpha}w_0^j = K_{jk\alpha} i^{j+k-\alpha} C_{0,0}^{jk\alpha} \|w_0^j\|^2 w_0^k.$$

Taking an inner product with w_0^k, we obtain

$$(\mathsf{E}^{jk\alpha}w_0^j, w_0^k)_{W_k} = K_{jk\alpha} i^{j+k-\alpha} C_{0,0}^{jk\alpha} \|w_0^j\|^2 \|w_0^k\|^2.$$

If $j + k - \alpha$ is odd then we need to apply the raising (or the lowering) operator to w_0^j, obtaining $v_1^j = (X_+ \cdot w_0^j)/j$ and $v_1^k = (X_+ \cdot w_0^k)/k$. Then, using (17.3),

$$e_0(jk\alpha)v_1^j = -C_{1,0}^{jk\alpha}\|v_1^j\|^2 v_1^k = -\frac{j+1}{j} C_{1,0}^{jk\alpha}\|w_0^j\|^2 v_1^k.$$

Hence,

$$\mathsf{E}^{jk\alpha}(X_+ \cdot w_0^j) = -(j+1)K_{jk\alpha} i^{j+k-\alpha}\|w_0^j\|^2 C_{1,0}^{jk\alpha} v_1^k.$$

Taking an inner product with v_1^k we obtain the formula

$$(\mathsf{E}^{jk\alpha}(X_+ \cdot w_0^j), (X_+ \cdot w_0^k))_{W_k} = -(j+1)(k+1)K_{jk\alpha} i^{j+k-\alpha}\|w_0^j\|^2 \|w_0^k\|^2 C_{1,0}^{jk\alpha}.$$

Thus, we have the following formulas for computing the choice structure coefficients:

$$K_{jk\alpha} = \begin{cases} \dfrac{(\mathsf{E}^{jk\alpha} w_0^j, w_0^k)}{i^{j+k-\alpha} C_{0,0}^{jk\alpha} \|w_0^j\|^2 \|w_0^k\|^2}, & j+k+\alpha \text{ is even,} \\ -\dfrac{(\mathsf{E}^{jk\alpha}(X_+ \cdot w_0^j), (X_+ \cdot w_0^k))}{(j+1)(k+1) i^{j+k-\alpha} \|w_0^j\|^2 \|w_0^k\|^2 C_{1,0}^{jk\alpha}}, & j+k+\alpha \text{ is odd.} \end{cases}$$

These formulas are valid only when $|j-k| \le \alpha \le j+k$. Set $K_{jk\alpha} = 0$ when the inequality $|j-k| \le \alpha \le j+k$ is violated. If we agree to always choose $\mathsf{E}^{jk\alpha}$ to be dual to $\mathsf{E}^{kj\alpha}$, when $k \neq j$, then $K_{jk\alpha} = (-1)^{j+k-\alpha} K_{kj\alpha}$. We note that for computing $K_{jk\alpha}$ we do not need to compute $\{e_B(jk\alpha) : |B| \le \alpha\}$.

For the maximal torus consisting of all rotations around the x_1-axis, the action of the raising operator on $\mathbb{C}^3$ is given by the matrix

$$\boldsymbol{X}_+ = \begin{bmatrix} 0 & i & 1 \\ -i & 0 & 0 \\ -1 & 0 & 0 \end{bmatrix}, \quad X_+ \cdot \boldsymbol{x} = \boldsymbol{X}_+ \boldsymbol{x}. \tag{17.5}$$

The action of X_+ on 3×3 matrices is given by the commutator

$$X_+ \cdot \boldsymbol{A} = [\boldsymbol{X}_+, \boldsymbol{A}] = \boldsymbol{X}_+ \boldsymbol{A} - \boldsymbol{A} \boldsymbol{X}_+. \tag{17.6}$$

These formulas are sufficient for all physical applications, where we always have $n \le 2$.

17.6 The Racah coefficients

Let us now take $W_\alpha \subset \mathrm{Hom}(W_j, W_k)$ and $W_\beta \subset \mathrm{Hom}(W_k, W_r)$ and observe that $W_\beta W_\alpha \subset \mathrm{Hom}(W_j, W_r)$. Next we also observe that the map

$$\Phi\colon \mathrm{End}(\mathscr{W}) \otimes \mathrm{End}(\mathscr{W}) \to \mathrm{End}(\mathscr{W}), \qquad \Phi(\mathsf{A} \otimes \mathsf{B}) = \mathsf{AB}$$

is SO(3)-invariant. Then

$$h_0 = i^{\alpha+\beta-\gamma} \sum_{A=-\min(\alpha,\beta)}^{\min(\alpha,\beta)} C_{A,0}^{\alpha\beta\gamma} i^{k+r-\beta} e_A(kr\beta) i^{j+k-\alpha} e_{-A}(jk\alpha)$$

is a real zero weight vector in $W_\gamma \subset \mathrm{Hom}(W_j, W_r)$. Hence, it has to be a real scalar multiple of $i^{j+r-\gamma} e_0(jr\gamma)$. Simplifying out the powers of i we conclude that

$$\hat{h}_0 = \sum_{A=-\min(\alpha,\beta)}^{\min(\alpha,\beta)} C_{A,0}^{\alpha\beta\gamma} e_A(kr\beta) e_{-A}(jk\alpha)$$

must be a real scalar multiple of $e_0(jr\gamma)$. We compute

$$e_A(kr\beta) e_{-A}(jk\alpha) = \left(\sum_b C_{b,A}^{kr\beta} v_b^r \otimes v_{A-b}^k \right) \left(\sum_c C_{c,-A}^{jk\alpha} v_c^k \otimes v_{-A-c}^j \right).$$

Using orthogonality of v_ν^k and (17.1) we obtain

$$e_A(kr\beta)e_{-A}(jk\alpha) = \|w_0^k\|^2 \sum_c (-1)^c \frac{\binom{2k}{k}}{\binom{2k}{k+c}} C^{kr\beta}_{A+c,\,A} C^{jk\alpha}_{c,\,-A} v^r_{A+c} \otimes v^j_{-A-c}.$$

Hence,

$$\hat{h}_0 = \|w_0^k\|^2 \sum_d \sum_c (-1)^c \frac{\binom{2k}{k}}{\binom{2k}{k+c}} C^{\alpha\beta\gamma}_{d-c,\,0} C^{kr\beta}_{d,\,d-c} C^{jk\alpha}_{c,\,c-d} v^r_d \otimes v^j_{-d} = \|w_0^k\|^2 R^{\alpha\beta\gamma}_{jkr} e_0(jr\gamma).$$

We conclude that there are real numbers $R^{\alpha\beta\gamma}_{jkr}$, such that

$$R^{\alpha\beta\gamma}_{jkr} C^{jr\gamma}_{d,\,0} = \sum_c (-1)^c \frac{\binom{2k}{k}}{\binom{2k}{k+c}} C^{\alpha\beta\gamma}_{d-c,\,0} C^{kr\beta}_{d,\,d-c} C^{jk\alpha}_{c,\,c-d}.$$

The absolute constants $R^{\alpha\beta\gamma}_{jkr}$ are called the Racah coefficients in representation theory of SO(3). If W_γ is absent in $W_\beta W_\alpha$ then the Racah coefficient $R^{\alpha\beta\gamma}_{jkr}$ is zero. We can compute the Racah coefficients by means of the formula

$$R^{\alpha\beta\gamma}_{jkr} = \frac{1}{C^{jr\gamma}_{d,\,0}} \sum_{c=-k}^{k} (-1)^c \frac{\binom{2k}{k}}{\binom{2k}{k+c}} C^{\alpha\beta\gamma}_{d-c,\,0} C^{kr\beta}_{d,\,d-c} C^{jk\alpha}_{c,\,c-d}, \tag{17.7}$$

taking $d=0$ if $j+r+\gamma$ is even and $d=1$, if $j+r+\gamma$ is odd. Formula (17.7), as all the subsequent formulas involving division, has the property that the numerator always evaluates to 0 whenever the denominator is 0, in which case we set the value of the fraction to 0. In all of our applications $n \leq 2$, so that formulas (17.4), (17.7) can be used directly to compute all the required Clebsch–Gordan and Racah coefficients.

17.7 Multiplication of irreps in End(V)

If we start with our chosen zero weight vectors $\mathsf{E}^{jk\alpha}$ and construct a canonical basis of weight vectors $\mathsf{E}^{jk\alpha}_A$, $|A| \leq \alpha$ of $W_\alpha \subset \mathrm{Hom}(W_j, W_k)$, then

$$\mathsf{E}^{jk\alpha}_A = i^{j+k-\alpha} K_{jk\alpha} e_A(jk\alpha).$$

The canonical basis for $W_\alpha(\mathbb{X})$ is

$$E^\alpha_B(\mathbb{X}) = \sum_{jk} \boldsymbol{X}_{jk} \otimes \mathsf{E}^{jk\alpha}_B,$$

where $\boldsymbol{X}_{jk} = 0$, if there is no $W_\alpha \subset \mathrm{Hom}(W_j, W_k)$. Similarly, the canonical basis for $W_\beta(\mathbb{Y})$ is

$$E_B^\beta(\mathbb{Y}) = \sum_{sr} \boldsymbol{Y}_{sr} \otimes \mathsf{E}_B^{sr\beta},$$

Then a real zero weight vector $h_0 \in W_\gamma \subset W_\beta(\mathbb{Y}) W_\alpha(\mathbb{X})$ is given by

$$h_0 = i^{\alpha+\beta-\gamma} \sum_A C_{A,\,0}^{\alpha\beta\gamma} E_A^\beta(\mathbb{Y}) E_{-A}^\alpha(\mathbb{X})$$

Putting together the formulas for $E_{-A}^\alpha(\mathbb{X})$, $\mathsf{E}_A^{sr\beta}$ and $e_A(jk\alpha)$ we calculate

$$h_0 = \sum_{jr} \boldsymbol{Z}_{jr} \otimes \mathsf{E}^{jr\gamma},$$

where

$$\boldsymbol{Z}_{jr}^\gamma = \frac{1}{K_{jr\gamma}} \sum_{k=0}^n (-1)^k \|w_0^k\|^2\, \boldsymbol{Y}_{kr} \boldsymbol{X}_{jk} R_{jkr}^{\alpha\beta\gamma} K_{kr\beta} K_{jk\alpha}. \tag{17.8}$$

Using our rule about formulas with division (setting $\boldsymbol{Z}_{jr}^\gamma = 0$, whenever $K_{jr\gamma} = 0$), we can extend the range of all Latin indices to $0, \ldots, n$ and all Greek indices to $0, \ldots, 2n$. Looking at formula (17.8) we see that it would be more natural to work with choice structure constants

$$\hat{K}_{jk\alpha} = \|w_0^k\|^2 K_{jk\alpha} = \begin{cases} \dfrac{(\mathsf{E}^{jk\alpha} w_0^j,\, w_0^k)}{i^{j+k-\alpha} C_{0,\,0}^{jk\alpha} \|w_0^j\|^2}, & j+k+\alpha \text{ is even}, \\[2ex] -\dfrac{(\mathsf{E}^{jk\alpha}(X_+ \cdot w_0^j),\, (X_+ \cdot w_0^k))}{(j+1)(k+1) i^{j+k-\alpha} \|w_0^j\|^2 C_{1,\,0}^{jk\alpha}}, & j+k+\alpha \text{ is odd}. \end{cases} \tag{17.9}$$

while redefining the Racah coefficients

$$\hat{R}_{jkr}^{\alpha\beta\gamma} = (-1)^k R_{jkr}^{\alpha\beta\gamma} = \frac{(-1)^k}{C_{d,\,0}^{jr\gamma}} \sum_{c=-k}^{k} (-1)^c \frac{\binom{2k}{k}}{\binom{2k}{k+c}} C_{d-c,\,0}^{\alpha\beta\gamma} C_{d,\,d-c}^{kr\beta} C_{c,\,c-d}^{jk\alpha}, \tag{17.10}$$

In this case formula (17.8) simplifies to

$$\boldsymbol{Z}_{jr}^\gamma = \frac{1}{\hat{K}_{jr\gamma}} \sum_{k=0}^n \boldsymbol{Y}_{kr} \boldsymbol{X}_{jk} \hat{R}_{jkr}^{\alpha\beta\gamma} \hat{K}_{kr\beta} \hat{K}_{jk\alpha}. \tag{17.11}$$

If we choose $\lambda_j w_0^j$ instead of w_0^j for some real non−zero scalars λ_j, while redefining the inner product on $W_j \subset \mathscr{W}$ by $\mu_j(\cdot, \cdot)_{W_j}$ for some positive scalars μ_j, $j = 1, \ldots, n$, then the constants $\hat{K}_{jk\alpha}$ will get multiplied by $\lambda_k \mu_k / \lambda_j \mu_j$, which obviously will not affect the left-hand side in (17.11).

The analysis of products of irreps in endomorphism algebras appears to be a new question in representation theory. We refer to papers of Sage [3–6], who showed that a similar analysis goes through for other groups and in more general situations.

17.8 Example: choice structure constants for piezoelectricity

For piezoelectricity

$$V = \mathbb{R}^3 \oplus \mathrm{Sym}(\mathbb{R}^3) \cong W_0 \oplus W_1 \oplus W_2.$$

In this example $p = 1$ and $n = 2$. In order to make our choices we identify

$$W_0 = \{c\boldsymbol{I}_3\colon c \in \mathbb{R}\}, \qquad W_1 = \mathbb{R}^3, \qquad W_2 = \{\boldsymbol{A} \in \mathrm{Sym}(\mathbb{R}^3)\colon \mathrm{Tr}\, \boldsymbol{A} = 0\}.$$

We define the following SO(3)-invariant inner products on V:

$$((\boldsymbol{a}, \boldsymbol{A}), (\boldsymbol{b}, \boldsymbol{B})) = \boldsymbol{a} \cdot \boldsymbol{b} + \mathrm{Tr}\,(\boldsymbol{A}\boldsymbol{B}), \qquad \{\boldsymbol{a}, \boldsymbol{b}\} \subset \mathbb{R}^3,\ \{\boldsymbol{A}, \boldsymbol{B}\} \subset \mathrm{Sym}(\mathbb{R}^3).$$

We select the following zero weight vectors in $W_j \subset \mathscr{W} = V$:

$$w_0^0 = \boldsymbol{I}_3 \in W_0, \quad w_0^1 = \boldsymbol{e}_1 \in W_1, \quad w_0^2 = \begin{bmatrix} -2 & 0 & 0 \\ 0 & 1 & 0 \\ 0 & 0 & 1 \end{bmatrix}.$$

These vectors are left invariant by rotations around the x_1 axis. We easily compute

$$\|w_0^0\|^2 = \mathrm{Tr}\,(\boldsymbol{I}_3^2) = 3, \quad \|w_0^1\|^2 = 1, \quad \|w_0^2\|^2 = (-2)^2 + 1^2 + 1^2 = 6.$$

Next we choose zero weight vectors $\mathsf{E}^{jk\alpha}$.

1. $W_0 \cong \mathrm{Hom}(W_0, W_0)$ is spanned by the map (which we will call E^{000}) that sends $\boldsymbol{I}_3$ into $\boldsymbol{I}_3$, while sending $W_1 \oplus W_2 \subset V$ to zero. We observe that

$$\mathsf{E}^{000} - K_0'\left(\frac{1}{3}\right),$$

where $K'(\kappa)$ is defined in (6.12).

2. The map $\mathsf{E}^{011} \in W_1 \simeq \mathrm{Hom}(W_0, W_1)$, sending $w_0^0 - \boldsymbol{I}_3$ to $w_0^1 - \boldsymbol{e}_1$ is obviously a fixed point of rotations around $\boldsymbol{e}_1$ axis, i.e. a zero weight vector. It will be convenient to define $\mathsf{E}^{101} \in W_1 \cong \mathrm{Hom}(W_1, W_0)$ to be the dual of E^{011}. All maps in $\mathrm{Hom}(W_1, W_0)$ have the form $A\boldsymbol{x} = (\boldsymbol{a} \cdot \boldsymbol{x})\boldsymbol{I}_3$ for some $\boldsymbol{a} \in \mathbb{R}^3$. By definition $(\mathsf{E}^{101}\boldsymbol{x}, \boldsymbol{I}_3)_V = (\boldsymbol{x}, \mathsf{E}^{011}\boldsymbol{I}_3)_V = x_1$. Hence, $((\boldsymbol{a} \cdot \boldsymbol{x})\boldsymbol{I}_3, \boldsymbol{I}_3) = x_1$, which implies that $\boldsymbol{a} = \boldsymbol{e}_1/3$. So, that $\mathsf{E}^{101}\boldsymbol{x} = x_1\boldsymbol{I}_3/3$, while sending $W_0 \oplus W_2 \subset V$ to 0.
3. Identifying $\mathsf{E}^{022} \in W_2 \cong \mathrm{Hom}(W_0, W_2)$ is just as easy: $\mathsf{E}^{022}w_0^0 = w_0^2$. The map $\mathsf{E}^{202} \in \mathrm{Hom}(W_2, W_0)$ is defined by duality: $\mathsf{E}^{202}\boldsymbol{X} = \mathrm{Tr}\,(\boldsymbol{A}\boldsymbol{X})\boldsymbol{I}_3$ and

$$3\,\mathrm{Tr}\,(\boldsymbol{A}\boldsymbol{X}) = (\mathsf{E}^{202}\boldsymbol{X}, \boldsymbol{I}_3) = (\boldsymbol{X}, \mathsf{E}^{022}\boldsymbol{I}_3) = (\boldsymbol{X}, w_0^2) = \mathrm{Tr}\,(\boldsymbol{X}w_0^2).$$

Hence, $\mathsf{E}^{202}\boldsymbol{X} = \mathrm{Tr}\,(w_0^2\boldsymbol{X})\boldsymbol{I}_3/3$. We observe that

$$\mathsf{E}^{022} + \mathsf{E}^{202} = \frac{1}{3}K_2'(w_0^2),$$

where $K_2'(\boldsymbol{A})$ is defined in (6.13).

4. $\mathrm{Hom}(W_1, W_1) = \mathrm{End}(W_1)$ splits into multiples of the identity map, skew-symmetric operators and symmetric, trace free operators. In the standard basis of $\mathbb{R}^3$ the maps $\mathsf{E}^{11\alpha}$ will be

$$\mathsf{E}^{110}\boldsymbol{x} = \boldsymbol{x}, \qquad \mathsf{E}^{111}\boldsymbol{x} = \boldsymbol{e}_1 \times \boldsymbol{x}, \qquad \mathsf{E}^{112}\boldsymbol{x} = w_0^2\boldsymbol{x} = (-2x_1, x_2, x_3).$$

These maps also send $W_0 \oplus W_2 \subset V$ to 0.

5. Understanding irreps W_1 and W_2 in $\mathrm{Hom}(W_1, W_2)$ is also easy.

$$W_1 = \{\boldsymbol{x} \mapsto \boldsymbol{x} \odot \boldsymbol{a} - (1/3)(\boldsymbol{a} \cdot \boldsymbol{x})\boldsymbol{I}_3 \colon \boldsymbol{a} \in \mathbb{R}^3\},$$

$$W_2 = \{\boldsymbol{x} \mapsto [\boldsymbol{A}, \boldsymbol{x} \times\,] \colon \boldsymbol{A} \in W_2 \subset V\},$$

where $[\boldsymbol{A}, \boldsymbol{B}] = \boldsymbol{AB} - \boldsymbol{BA}$ is the commutator of two matrices.

$$\mathsf{E}^{121}\boldsymbol{x} = \boldsymbol{x} \odot \boldsymbol{e}_1 - (1/3)x_1\boldsymbol{I}_3, \qquad \mathsf{E}^{122}\boldsymbol{x} = [w_0^2, \boldsymbol{x} \times\,],$$

where $(\boldsymbol{x} \times\,)\boldsymbol{u} = \boldsymbol{x} \times \boldsymbol{u}$. We define E^{211} and E^{212} by duality: $\mathsf{E}^{211}\boldsymbol{X} = \boldsymbol{X}\boldsymbol{e}_1$, $\mathsf{E}^{212}\boldsymbol{X} = 6\boldsymbol{e}_1 \times \boldsymbol{X}\boldsymbol{e}_1$. It is much harder to describe explicitly the irrep $W_3 \subset \mathrm{Hom}(W_1, W_2)$. This, however, is not necessary. We simply declare that we have chosen the zero weight vector E^{123} and its dual $\mathsf{E}^{213} \in W_3 \subset \mathrm{Hom}(W_2, W_1)$. The reason for this is that there is a unique irrep of weight 3 in $\mathrm{Sym}(V)$, and hence, the problem of identifying which $W_3 \subset \mathrm{Sym}(V)$ occurs in $(\Pi_1\mathscr{A}\Pi_2)_{\mathrm{sym}}$ does not arise. For that same reason we just set $K_{123} = K_{213} = 1$.

6. It is easy to describe $\mathsf{E}^{22\alpha} \in W_\alpha \subset \mathrm{End}(W_2)$ for $\alpha = 0, 1, 2$. $\mathsf{E}^{220}\boldsymbol{X} = \boldsymbol{X}$, $\mathsf{E}^{221}\boldsymbol{X} = [\boldsymbol{X}, \boldsymbol{e}_1 \times\,]$, $\mathsf{E}^{222}\boldsymbol{X} = w_0^2\boldsymbol{X} + \boldsymbol{X}w_0^2 - 2\,\mathrm{Tr}\,(w_0^2\boldsymbol{X})\boldsymbol{I}_3/3$. We observe that

$$\mathsf{E}^{220} = K_0\left(\frac{1}{2}\right), \qquad \mathsf{E}^{222} = K_2(w_0^2),$$

where $K_0(\mu)$ and $K_2(\boldsymbol{A})$ are defined in (6.16) and (6.17), respectively. $W_3 \subset \mathrm{End}(W_2)$ consists of antisymmetric operators on V and does not enter into $\mathrm{Sym}(V)$, while there is a unique irrep of weight 4 in $\mathrm{Sym}(V)$. That means that we do not need to describe the maps $\mathsf{E}^{22\alpha} \in W_\alpha \subset \mathrm{End}(W_2)$ for $\alpha = 3, 4$ explicitly. We also set $K_{223} = K_{224} = 1$.

References

[1] Grabovsky Y, Milton G W and Sage D S 2000 Exact relations for effective tensors of polycrystals: necessary conditions and sufficient conditions *Commun. Pure Appl. Math.* **53** 300–53

[2] Grabovsky Y and Sage D S 1998 Exact relations for effective tensors of polycrystals II: applications to elasticity and piezoelectricity *Arch. Ration. Mech. Anal.* **143** 331–56
[3] Kwon N and Sage D S 2008 Subrepresentation semirings and an analog of 6*j*-symbols *J. Math. Phys.* **49** 063503 21
[4] Sage D S 2002 Group actions on central simple algebras *J. Algebra* **250** 18–43
[5] Sage D S 2005 Quantum Racah coefficients and subrepresentation semirings *J. Lie Theory* **15** 321–33
[6] Sage D S 2005 Racah coefficients, subrepresentation semirings, and composite materials *Adv. Appl. Math.* **34** 335–57

www.ingramcontent.com/pod-product-compliance
Ingram Content Group UK Ltd.
Pitfield, Milton Keynes, MK11 3LW, UK
UKHW051857100926
8556IPUK00005B/672